Natural Fiber Composites

Natural Fiber Composites

Processing, Characterization, Applications, and Advancements

Shishir Sinha and G. L. Devnani

CRC Press
Taylor & Francis Group
Boca Raton London New York

CRC Press is an imprint of the
Taylor & Francis Group, an **informa** business

First edition published 2023
by CRC Press
6000 Broken Sound Parkway NW, Suite 300, Boca Raton, FL 33487-2742

and by CRC Press
4 Park Square, Milton Park, Abingdon, Oxon, OX14 4RN

CRC Press is an imprint of Taylor & Francis Group, LLC

ISBN: 978-1-032-06306-5 (hbk)
ISBN: 978-1-032-06321-8 (pbk)
ISBN: 978-1-003-20172-4 (ebk)

DOI: 10.1201/9781003201724

Typeset in Times
by MPS Limited, Dehradun

Contents

Preface

Recent economic development and technological growth are inspiring academicians and researchers to look for newer materials which can compete with cutting-edge technology and at the same time should be sustainable and safe for environment also. Natural fibers reinforced in different polymer matrices are offering excellent mechanical and thermal properties and that is why they are getting more and more attention in this decade as an alternate of synthetic fibers as reinforcement in composites.

This book is intended to focus on the key areas and issues related to natural fibers and their reinforced polymer composites that can help academicians and researchers to have a holistic idea of this field. The first two chapters deal with introduction of composite materials and their classification and natural fibers and their different extraction methods. The third chapter elaborates different characterization techniques including atomic force microscopy and XPS. The major issue with these lignocellulosic fibers is the compatibility and adhesion with polymer matrix. Chapters four and five give the solution to improve adhesion between these natural fibers and different polymer matrix by applying different chemical, physical, and biological treatment methods. Fabrication of good-quality polymer composite meeting international quality norms and standards is very important for the commercialization of these novel materials. Chapters six and seven cover the various fabrication procedures and state-of-the-art characterization techniques for development and testing of these composites. Chapter eight, nine, and ten cover the different aspects like processing, development, and characterization of these natural fiber–based reinforced material based on different polymer matrices like thermosets, thermos, and biodegradables.

Application aspects are very important for successful commercialization of any new product. Chapters eleven and twelve explain the various application aspects like food packaging and structural and semi-structural applications of these composite materials.

An effort is made to cover all the recent developments in the major areas like processing, characterization, development and applications of natural fiber–reinforced polymer composites. We hope that this book will definitely help the students, academicians, researchers, and entrepreneurs who want to excel in this novel field.

Shishir Sinha
G. L. Devnani

Authors

Shishir Sinha, PhD, Director General Central Institute of Petrochemicals Engineering & Technology (CIPET) is a professor in the Department of Chemical Engineering, Indian Institute of Technology Roorkee. He is also adjunct faculty at the Centre for Disaster Mitigation and Management, IIT Roorkee. Prior to joining as director general CIPET, he held the position of director, Kamla Nehru Institute of Technology Sultanpur (an autonomous technical institute of Government of U.P.). Professor Sinha has more than 22 years of academic and industrial research experience with 85 reputed publications besides more than 100 in conference proceedings. Apart from this, he has written 10 books and several book chapters. He has successfully completed over 22 high-impact projects and consultancies valued at more than 152.79 million Indian rupees.

G. L. Devnani, PhD, is an assistant professor in the Department of Chemical Engineering, Harcourt Butler Technical University. He earned an MTech at IIT Kanpur and a PhD at IIT Roorkee. He has 17 years of teaching and research experience. His research area is development of polymer composites for diverse applications. He has published several research papers in journals of international repute and in conference proceedings related to polymer composites. He has authored several book chapters with reputed international publishers.

1 Introduction

Kajal Mishra and Shishir Sinha
Department of Chemical Engineering, Indian Institute of
Technology Roorkee, Roorkee, India

1.1 INTRODUCTION TO GREEN CHEMISTRY AND RENEWABLE BIO-BASED PRODUCTS

The resilient color green usually relates to life, productivity, health, and energy. Chlorophyll, a critical element of ecological systems, is green in appearance. In many nations, green is also aligned with the color of money. Becoming green has always been quite a rallying point for environmentalists and green political groups. Being green has recently become a popular marketing strategy for business companies. Green chemistry principles are crucial for chemists to apply to all aspects of the chemical sciences, including fundamental and applied research and practices, application mechanism, operational advancement, production, and education (Marteel et al., 2003). The requirement for food, fuel, energy, and commodities can cause major chemical development problems due to enormous quantities of dangerous chemicals and residues produced throughout the midst of a consistently growing economy with cutthroat trade rules. The whole pattern against "green computing" or "sustainable innovative technologies" necessarily requires a conceptual framework shift away from long-established theories of operational efficiency that target primarily on chemical yield, which accredit the economic benefit on reducing scrap at the source and evading the practice of noxious and dangerous materials (Saleh & Koller, 2018). Anastas and Warner from the US Environmental Protection Agency (EPA) created the phrase "green chemistry." The EPA formally selected its name to the US Green Chemistry Program in 1993, and it has since functioned as a prime focus for initiatives in the world. The terms "green chemistry," "green technology," and "sustainable development" are frequently interchanged to explain the idea of developing materials and systems that have a lower environmental effect and are (ideally) made from renewable resources. Moreover, a closer examination of such terms reveals significant philosophical divergence, which influences the viability of approaches and strategies in forming an environmentally responsible society (Anastas et al., 2002). These concepts primarily address how to conduct chemical reactions and produce chemical products and synthesize chemicals in an environmentally friendly way. Green chemistry emphasizes specific themes, such as the usage of mild and harmless intermediates like solvents for operations and extractions, decreasing the several stages, and the theory of atom economy, or combining all of the raw resources into the product (Anastas & Kirchhoff, 2002). The 12 standards were published more than two decades earlier and do not entirely represent current understanding. In the context of increased attention on product quality and renewable resources, other challenges such as toxicity and biodegradability are now playing an essential part in green chemistry (Anastas & Farris, 1994). Green chemistry indeed is reasoned as the scientific backbone of ecologically friendly manufacturing in many aspects. Corrêa et al. (Corrêa et al., 2013) investigated the progression of green chemistry, demonstrating that a significant amount of work was put toward a greener advancement in numerous fields of chemistry, including organic and inorganic synthesis, as well as analytical chemistry. According to the authors, different countries like India, Brazil, and South Africa have ideal circumstances for developing novel biomass modification mechanisms for biofuels and bio-based products. However, a few other current reviews (Krausmann et al., 2009) have revealed that these days academicians and researchers have a better understanding regarding environmental conservation, economic sustainability, as well as the aspect of green chemistry, resulting in improved analysis calculation and more consistent recognition of the purpose of green chemistry in the latest agribusiness approach. In recent years, there has been a switch to effective and renewable biomass utilization attributed to sustainable manufacturing processes to produce food and many other bio-based commodities with acceptable economic worth, low sources, strengthened ecological processes, waste reduction, and negligible environmental consequences and greenhouse emissions (Caldeira-Pires et al., 2013; Claudino & Talamini, 2013). Worldwide, 140 billion metric tons of agricultural biomass is produced each year. Also, the advantages of green schemes to yield better products could save approximately 50 billion metric tons of fossil fuels, which would significantly mitigate greenhouse gas (GHG) pollution and our reliance on non-renewable component. Minimal carbon emissions can also be achieved using appropriate tactics and high-density, rapid-growing crops like sugarcane and wheat straw. For example, palm oil farming for producing biodiesel in southern area of Brazil generates a CO_2 emissions ratio of roughly 208 kg CO_2 equivalent/1,000 kg crude palm oil annually (Höfer & Bigorra, 2008). Although an organic chemicals production sector dependent upon oil refining emerged in the early twentieth century, a similar organics industry that relies on biomass refining occurs in the twenty-first century. Energy is the driving force in both instances. The tremendous requirement for petroleum as an inexpensive, one-shot use fuel provided chemical production with a substantial amount and consistent source of hydrocarbons, allowing the petrochemical sector to grow. Also, chemical and engineering technology for splitting, segregating, re-configuring, polymerizing, and functionalizing enabled individual to use complex compounds of basic chemicals and convert them into a myriad of greater value compounds with an ostensibly infinite variety of applications, consisting of high quantity, relatively inexpensive plastics to limited amount yet high-cost drugs (Bozell, 2008). In the current world, novel, renewable resources are being pursued with tremendous enthusiasm; biomass, or renewable energy, is assured a role in the developing energy sector for the conceivable future. The expansion of the bioenergy (e.g., biomass incineration) and biofuels (e.g., biodiesel) operations will drive up the consumption of renewable carbon by the food sectors (Stevens & Verhe, 2004). Producing food is a wasteful procedure – from agricultural residues (e.g., wheat straw) to manufacturing (where significant losses exist) to market and export, an absurd amount of food gets wasted. Moreover, the discarded food from the industry can be supplied to the form chemicals, energy, and some other sectors. Wheat straw includes many functional wax components (fatty alcohols, alkanes, and so on), and the cellulosic part can be utilized to produce paper or ethanol. Also, rice straw from rice cultivation can produce energy for agricultural machinery, and the wastes are high in silica, which has a wide range of applications (Ragauskas et al., 2006). Utilized food oils can be extracted and converted into biodiesel via chemical transformation. A similar biodiesel production method, aided by tax subsidies and authorities' intent for biofuel consumption, yields glycerol as an intermediate product, which may be converted into a variety of higher-value goods with the appropriate chemistry actions (Deswarte et al., 2006). Food wastes materials cannot be buried in landfills because of health concerns. Thus they must be incinerated; however, they might be gasified and acclimated to operate a gas turbine to generate electricity. Therefore, the combination of heat and power divisions is an appealing prospect for the disposal of steep residues and pollutants. Controlled pyrolysis can be acclimated to manufacture microscopic organic compounds that could have worth in and of oneself, and even to make polymeric essentials, and as interface compounds for developing broader and more significant chemical products from the breakdown of food (and some other organic) wastes (Kalapathy et al., 2002). Through pre-treatment to cremation and committing several raw resources to bio-processing, refining the earth's regular harvest will give a gold mine of chemical potential. A biorefinery is a unit that acquires various biological raw materials and transforms them into various valuable outputs such as chemicals, power, and commodities (Figure 1.1).

Renewable resources are focused on the chemical implications of biomaterials, bioenergy, and biochemicals (Figure 1.2). Certain instances, such as a significant subset of naturally produced molecules (e.g., to be used in home healthcare goods

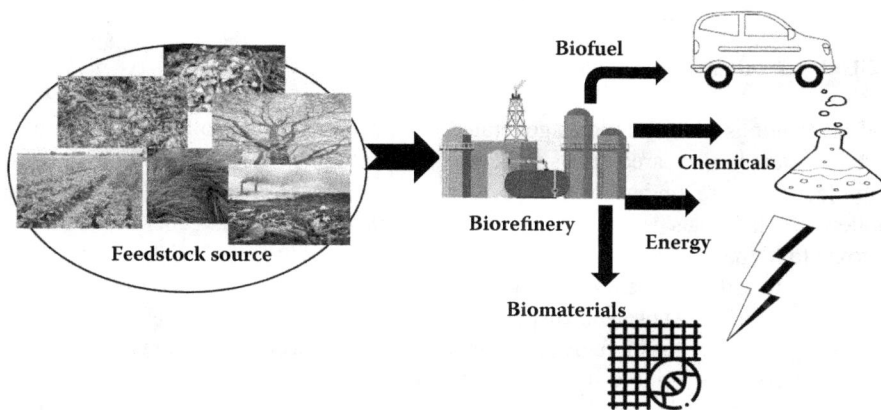

FIGURE 1.1 Biorefinery as a source of valuable products through merged feedstocks.

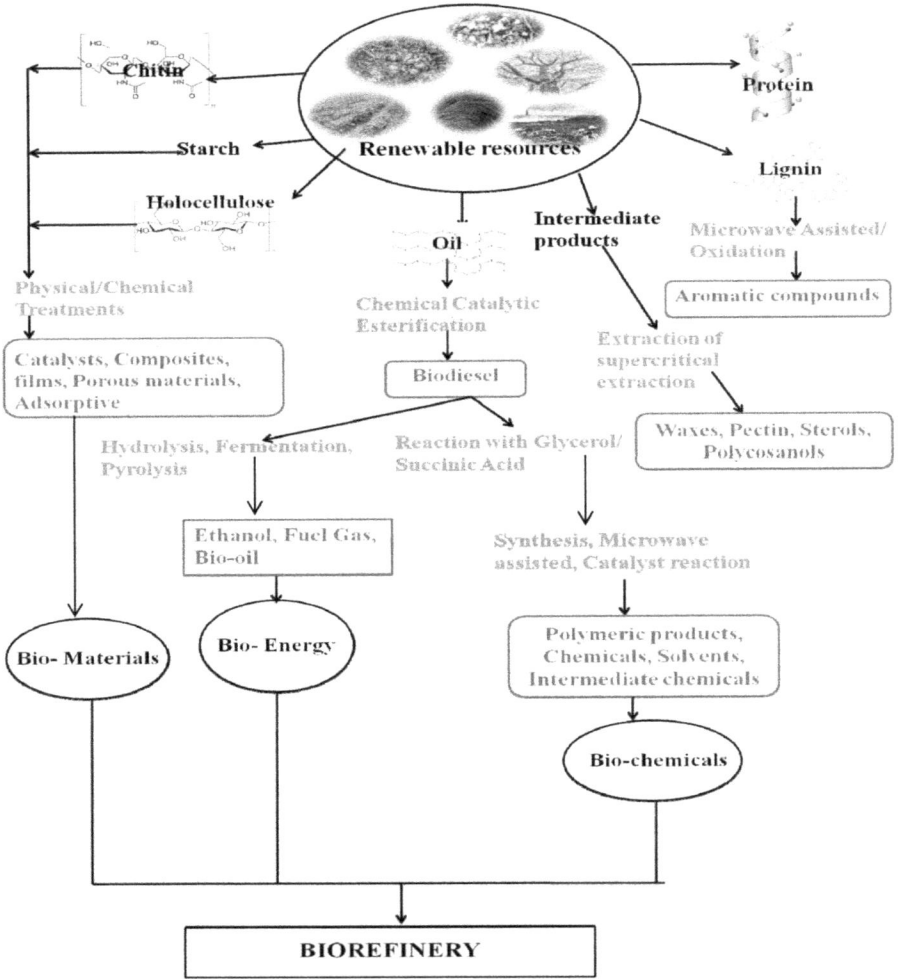

FIGURE 1.2 Renewable resources to the bio-based products.

and medications) and chemicals generated from coal, have controlled the sector as a carbon source (prepared to suppress trade obstacles presented in the era). Manufacturers are under massive pressure, particularly from users, to generate bioderived chemicals as substitutes for fossil fuels and materials generally regarded as toxic to humans or the surroundings (Audsley & Annetts, 2003). The shift to bio-based commodities is believed to provide a variety of benefits, such as the utilization of both renewable and dispensable sources, reduced emphasis on finite but increasingly costly fossil fuels, the capability for reducing greenhouse gas emissions, the ability of viable industrial activity, rural progress is aided by enhanced society health, rapid industrialization productivity through environmentally friendly products. Another very significant sub-division of the chemicals industry, according to the arbitrary advisory group for bio-based commodities, are active pharmaceutical ingredients (APIs), biopolymers, cosmetic products, coatings, and solvents. APIs, significantly, are predicted to be the biochemical category with the most excellent percentage revenue of biotechnologically manufactured goods, accounting for 33.7% of worldwide chemical purchases. The utilization of biomass as a fuel source will grow progressively essential as humanity step away from petroleum products.

There are many renewable resources present in our environment out of which one major source are the trees that act as a bio-based by-product factory. In several countries throughout the world, trees are a huge substitute energy resource. In the last decade, the overall volume (ground biomass) of wood generated in India has increased, and future projections indicate that this trend will continue over the next two decades (Dodson et al., 2012). The planted trees in between the years 1970s and 1980s will reach maturity in the coming years, but timber values have plummeted and are falling steadily. Increased stocks when costs are falling are about to aggravate existing issues, and several plantations are uncertain to be worth removing for fibers or pulp. Discovering opportunities to add value to such forest products will aid in the growth of rural businesses and preserve the rural community (Pfaltzgraff & Clark, 2014). This could involve using waste from the fiber and paper industries, such as oak, bark, and shrubs. Branches are another supply of wood currently underutilized, despite the fact that they can be widely dispersed, and their collecting and distribution will exhaust resources. Novel wood-using sectors, on the other hand, could conflict with current hardwood-based companies. Cellulose fibers for composites, film preparation for packaging industries, textile industries, and wood resins as adhesives are a few examples of innovative applications (Dodson et al., 2012). Another potential bio-based product that is unexplored is biopolymers. Polysaccharides, like cellulose, make up the majority of biopolymers. They can also be developed through microbial activities. Biopolymers are readily accessible renewable resources that are commonly used to make environmentally acceptable bioplastics (Azammi et al., 2020)]. They are mass-produced commercially for a variety of uses. Although biopolymers account for a minor portion of the polymer sector, it is expected that they will eventually overtake fossil fuel-supported polymers by 30%–90%. Biopolymers are used in certain fields depending on their affordability, accessibility, water absorption, heat resistance, fracture toughness, deterioration reliability, and biocompatibility (Christian, 2020). The biopolymer aspect of a biocomposite controls its chemical ingredients, molar mass, geometrical traits, mechanical strength, and fabrication methodology (George et al., 2020).

1.2 COMPOSITES AND REINFORCEMENTS

Researchers have become increasingly interested in developing recyclable polymers for over a quarter of a century, and many biomaterials have been produced and were used in numerous industries. Despite the fact that biomaterials have qualities equivalent to man-made materials, they can't replace man-made polymeric materials in all applications due to their exceptional value, short life span, and lack of processability. Reinforcing polymers on polymeric matrices improves their comprehensive characteristics, and these reinforcements of polymers are known as composites. Composites consist of a substrate called a matrix phase, a supporting component, and an interface that connects them all. Reinforcing components include natural and synthetic fibers, whiskers, nanoparticles, granule, and particles; on the other hand, matrix components can be polymeric materials, metallic and ceramic materials. Filament winding, hand-layup processes, resin transmission casting, compression, extrusion molding, and centrifuge are the most common ways to make composites (Dodson et al., 2012). In comparison with polymers, composites have superior physio-mechanical qualities. Non-biodegradable composites, semi-biodegradable composites, and biodegradable composites are the three categories of composites. Therefore, composite materials are composed of more than two chemically and physically different states segregated by an intermediate. Composites differ from mixtures in that they contain various stages that may be identified despite the reinforcement. The intermediate stage brings the matrix and reinforcing components together (Figure 1.3). However, they keep their unique features as these composites possess diverse characteristics like superior mechanical properties and light weight (La Mantia & Morreale, 2011). The matrix material has more excellent elasticity than the reinforcement material. They retain and distribute the reinforcing phase's load equally. The discontinuity of the reinforcing material makes it more complex than the matrix. Timber, for instance, is a wonderful description of a composite in which cellulosic components are reinforced in a lignin material (Jose & Joseph, 2012). Polymers are often utilized as a matrix phase, particularly epoxy, polyesters, and vinyl esters being the most prevalent. Reinforcing phases are designed to be durable and lend matrix stability and rigidity to the matrix material. Composite materials are characterized in various ways, such as a dispersion strengthened composites, particle strengthened composites, and fiber strengthened composites, all based on the reinforcing principle. Functional composites can be classified into active electrical composites, active thermal composites, active optical composites, and many more, depending on their function. The lamination process, convoluted structural process, pultrusion process and structured textile process, and many more are the preparation technique for fabricating composites. Throughout the mid-1970s, attempts were carried out to prepare metal matrix composites (MMC) using SiC flake reinforcements, primarily from aluminum alloys. The fundamental goal of these efforts seems to achieve better mechanical modulus and mechanical strength in aluminum alloys, which would significantly improve their properties. Particulate reinforcements were developed as a result of the exorbitant value of whiskers. The area of interest after completing the twentieth century was the need for economic reinforcements. The outstanding qualities of composite materials facilitate its employment in various sectors,

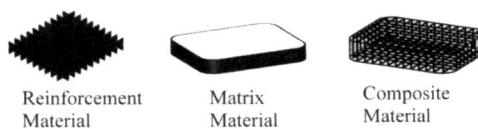

Reinforcement   Matrix   Composite
Material   Material   Material

FIGURE 1.3   Fabrication of composite material with a reinforcing phase and matrix phase.

particularly automobile, aviation, and packaging. Despite the fact that the reinforcement enhances the matrix's fundamental qualities, the matrix influences the reinforcement to obtain the required shape (Kandpal et al., 2015). At elevated temperatures, the composite constituents do not combine among themselves, causing the composite to collapse prematurely. Across humanity's civilization, the entire natural composite has been discovered in many types, such as bone tissues, wood, and wild crops. Specific composites are already in use, produced through creativity and productivity at the lowest possible cost, and are appropriate for ultra-demanding applications such as the space industry. The morphology of reinforcing, the form of the dispersion medium, the concentration arrangement, reinforcement assimilation, and size distribution are all elements that influence composite properties (Rimašauskas et al., 2019). The supply of a specific component is determined by the concentration, which determines composite features, whereas the composite symmetry is determined by orientation. Conventional monolithic composites have limitations in terms of conceivable combinations of characteristics. Galy et al. (El-Galy et al., 2017) investigated the production and structural characteristics of SiC-reinforced aluminum induced MMCs. The density will drop as the crystalline size lowers, while the durability will reduce as the crystalline size grows. Rao (Rao, 2017) investigated Al 7075 strengthened with SiC. If the particle size and SiC volume fraction are increased, the strength and toughness of the material will enhance. The friction and wear characteristics of SiC-reinforced aluminum alloy LM6 MMC were investigated by Pradhan et al. (Pradhan et al., 2017) in no moisture, mercerization, and aqueous media. Blend casting was used in the manufacturing process. No moist sliding had the least wear, followed by an aqueous solution and finally mercerization medium. Kandpal et al. (2017) investigated the characteristics of an Al 6061 composite reinforced with Al2O3. The finding demonstrates that when the Al2O3 concentration increased, the elasticity reduced.

The classification of composite material based on reinforcements and matrix is shown in Figure 1.4. The matrix phase can be classified as a polymer matrix, ceramic matrix, and metal matrix. The different matrix material adheres to the reinforcement material, defines mechanical performance, cohesiveness, transmits the stress of the reinforcing phase, and increases the composites' rigidity. Nature and situation determine the matrix used wherein the composite will be utilized. For commercialization, the polymer matrix is frequently employed. In elevated temperatures, such as engines, the other metallic, carbon, glass, and cement matrix materials are employed. Because the first investigation on metallic and ceramic matrices was oriented on seamless carbon or boron fiber, it was unable to produce suitable high-quality composites. However, constant advancements in the area of fibers have increased the interests of researchers. The matrix's purpose is to place the reinforcement's material in position while also supporting them. The reinforcement phase, generally, has an impact on physio-mechanical properties and any other customized properties that are enhanced by the matrix material (Baillie, 2004). A broad variety of properties can be obtained by mixing various potential reinforcement materials and matrix materials, allowing material qualities to be altered to fit particular criteria. The matrices keep the reinforcements in place to form the appropriate configuration, whereas the reinforcements improve the matrix's

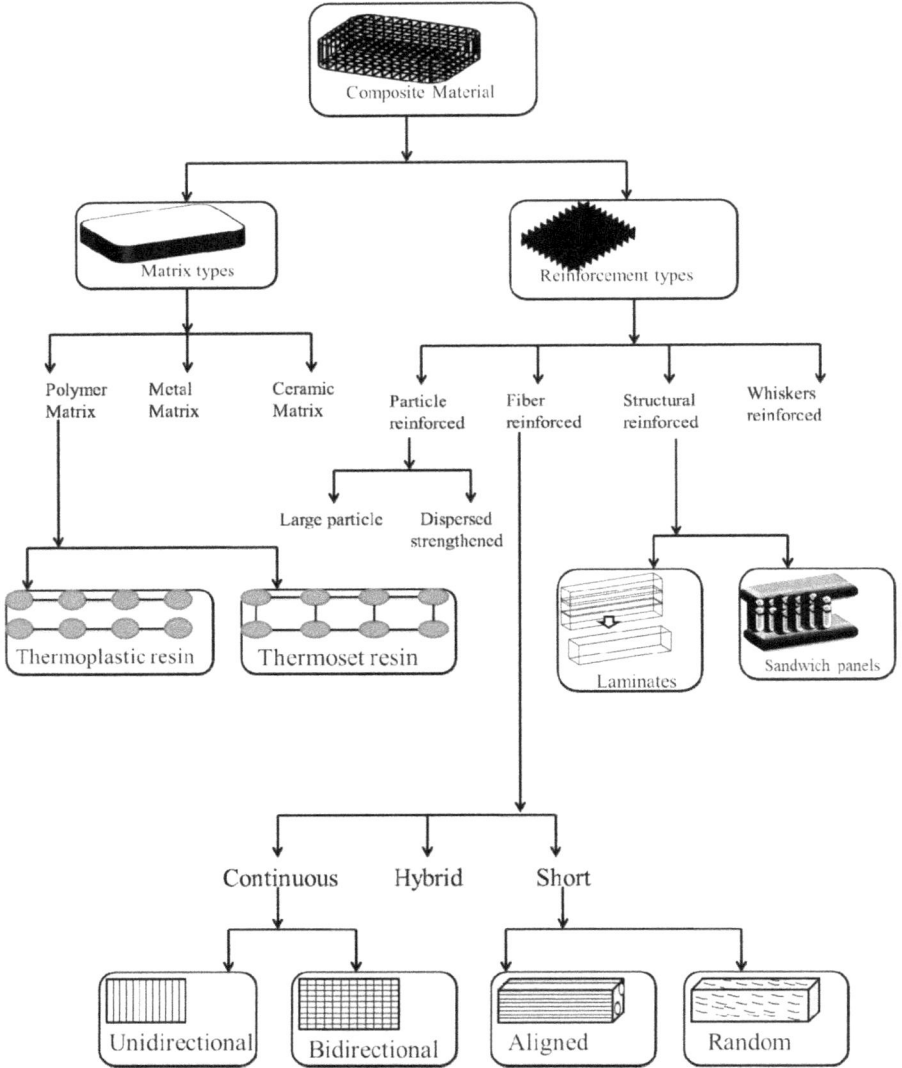

FIGURE 1.4 Classification of composite material.

overall mechano-chemical properties. The matrices are homogeneous substances in which reinforcements are usually incorporated and dispersed consistently across the whole substance. As matrices, materials like aluminum, Mg, Ni, Ti, and Co can be utilized. The matrix phase must be effectively bound to the reinforcement material used (Mallick, 2012).

The composite material consisting of a naturally polymerized matrix that binds several minor uniform threads simultaneously is classified as polymer matrix composites (PMC). PMCs are presently the most widely utilized composite of the existing composites. Ceramic fibers are commonly used to strengthen the substrate in PMCs because they have higher mechanical properties than the matrix phase. The various matrices, reinforcements, operating conditions, microstructure, configuration, and intermediate phases all influence the features of PMCs. PMCs are widely recognized for their low cost and simple manufacturing process. With a variety of manufacturing techniques, PMC producers may develop value-efficient goods. Every fabrication procedure contains characteristics that define the type of product that will be developed. Because of this knowledge, the producer is capable of providing the finest alternative for the buyer. PMCs are composed of a substrate comprising thermoplastic or thermosetting resin and more than one reinforcement phase like ceramics, metallic elements, or natural materials like fibers and whiskers (Singh et al., 2019). Polymers are effective materials because they are easy to process. Polymers are known for their low density. The qualities of polymer composite composites (PMCs) are diverse. Some of these are greater tensile and flexural strength, superior impact, compressive force, wear and strain qualities; value-efficient production and casting methods; remarkable chemical stability and corrosion resilience; and superior dimensional stability and mechanical features. Thermosets are extensively utilized polymeric resins that generate hydrophobic and hydrolyzable polymers when processed under thermal or chemical effects. Also, due to their outstanding electromechanical resilience and inexpensive value, unsaturated epoxy, acrylic resins are the utmost extensively used thermosets (Suaduang et al., 2019). Epoxy resins, cyanate compounds, polyphenols, polyurethanes, bismaleimides, and other thermosets are extensively utilized in innovative composite materials. Certain thermosets offer unique features and use that give excellent tenacity, endurance, chemical inertness, and heat resistance as epoxy polymeric materials. Cyanate polymers have high tenacity, excellent electrical characteristics, and durability while absorbing negligible humidity. Despite their increased price, polyphenol polymers are used to make flame-retardant aircraft interior surfaces and flame-renitent reduced smoke products (Weiss et al., 2015). The majority of thermoset resins are sourced from petrochemical outputs. Therefore, producing bioresins using natural derivations requires a significant number of investigations. Bioresins exhibiting characteristics equivalent to unsaturated polyester polymers have been created using polyols and alcohol derived from wheat, soybeans, rice, maize, and corn. Thermoplastics polymers can be modified and reformed due to their elasticity, making them re-accessible and re-consumable, giving them an advantage against thermoset polymers in terms of sustainable development. Thermoplastic polymers such as polystyrene, polyvinyl acetate (PVA), polybutylene terephthalate (PBT), polyamide, polybutadiene, polyarylsulfone, polyetheretherketone, polyphenylene sulfide, and polyethylene are utilized in industrial applications such as clothing, sportswear, orthopedic, artificial limbs, packaging, and automobiles require minimal production temperatures, higher water resistibility, and cost-effective (Bhargava, 2010). Table 1.1 gives a detailed work reported in literature on polymer matrix composites used with different reinforcements.

Metal matrices have higher durability, toughness, and elasticity. In comparison to polymeric materials, metal matrices are resilient at high temperatures in corrosive conditions. The metal matrix composite (MMC) durability can be increased by

TABLE 1.1

Polymer Matrix Composites Used with Different Reinforcements

S.No.	Matrix	Reinforcements	Applications	Findings	Reference
1.	Acrylonitrile-butadiene-styrene copolymers (ABS)	Thermoplastic starch	3D printing	Mechanical, thermal, workability, and organic pollutants are higher to those of industrial ABS strands	(Kuo et al., 2016)
2.	Polybutylene succinate (PBS)	Modified tapioca starch	Packaging industry	High crystallinity of composite with low mechanical properties	(Rafiqah et al., 2018)
3.	Polyvinyl alcohol (PVA)	Carbon fiber	Semi-structural	Mechanical, electrical properties were same as industrial carbon yarn	(Petkieva et al., 2020)
4.	Polytetrafluoroethylene (PTFE)	Carbon fiber	Wound dressing	Congelation of the blood at a quick rate	(Yadav et al., 2020)
5.	Epoxy matrix	Carbon fiber	Railcar application	Improved mechanical properties	(Daniyan et al., 2021)
6.	Polydimethylsiloxane (PDMS)	Carbon fiber	Aviation industry	Improved thermal conductivity	(Wei et al., 2020)
7.	Polyaniline (PANI)	Graphene nanofiber	Sensor, medical	High crystallinity and optical properties	(Mohammed et al., 2019)
8.	Polyurethane	ZnO	Coating application	Improvement in corrosion resistance	(Zhang et al., 2008)
9.	Polyvinyl acetate	ZnO +nanoparticle of metal oxide	Automobile industry	Improved dielectric and optical properties	(Iqbal et al., 2020)

using reinforcements resistant to high heat degrees. Metals such as aluminum, copper, magnesium, nickel, zinc, tungsten, and titanium have space engineering applications and can be used to reduce the matrix's higher load (Rohatgi, 2013). The toughness to load proportions of the metal matrix is greater than many of the alloys. The advantages of MMCs are: because of the exceptional strength of MMC, they are used extensively for structural purposes. The matrix has excellent mechanical properties and elasticity. They are extremely robust and impact-resistant. MMC is resistant to elevated temperature variations and thermal disturbances with greater temperature resilience. MMCs are resistant to superficial defects and have a long life span. At elevated temperatures, the strength of the matrix is stable and thermal deterioration occurs. The matrix has high environmental resilience. In metal matrix composites, the components are mostly distinct categories of alloys; in scarce circumstances, a specific metal is chosen (Brian et al., 2003). MMC is made up of a variety of elements that provides it with improved thermo-mechanical characteristics. Aluminum metals are extensively utilized in construction engineering applications, automotive and aircraft industries because of their good shaping and combining attributes, lightweight, fatigue resistance, and corrosion resilience. In order to improve the mechanical properties of aluminum and its alloys, particulates are integrated into composites. Titanium metals are extensively utilized in aircraft structures due to their exceptional durability and tenacity at elevated temperatures, and resilience to corrosion (Clyne & Withers, 1993). The metal, on the other hand, is costly. Magnesium metal is the cheapest non-ferrous alloy, and it's commonly utilized in electrical applications, blade cutter connectors, and transmission connectors for aeronautical purposes. Metal matrix composites have sparked a lot of interest in recent years due to their remarkable commercial operations (Sharma et al., 2018). Table 1.2 describes the MMCs potential uses in a variety of sectors.

Ceramic matrix composites (CMC) are mostly used for structural engineering functions that must withstand temperatures of up to 1600°C. CMCs have a large effective area and covalent interactions with reinforcement material, resulting in composite material with excellent mechanical properties, heat equilibrium conditions, barrier properties, corrosion resilience, and greater melting temperatures (Mallick, 1997). For elevated temperatures situations, ceramic matrices such as

TABLE 1.2
MMCs Potential Uses in Variety of Sectors

S.No.	Composite	Application	Fabrication Method	Reference
1.	Molybdenum-Ti alloy	Aerospace industry	Laser powdered, bed fusion	(Kaserer et al., 2020)
2.	Stainless steel-Ni alloy	Supersonic aircrafts engines	Electric discharge alloying	(Arun et al., 2021)
3.	Niobium carbide-Cu	Spot welding electrodes, superconductors	Powdered metallurgy	(Bian et al., 2021)
4.	Niobium-Nickel-Ti	Turbine blades, engine blocks	Accumulative pack roll bonding	(Ye et al., 2020)
5.	Carbon nanotube-Mg alloy	High-pressure vessels, turbine blades	Chemical vapor deposition, powder metallurgy	(Say et al., 2020)
6.	Aluminum oxide-graphite- aluminum alloy	Automotive parts	Powder metallurgy	(Simsek et al., 2019)

cement, glass are desired. They are employed in aerospace, automotive industries and in turbine applications for which elevated-temperature durability is desired. Ceramics are mostly crystalline but deficit the flexibility of materials like plastics, metallic elements, leading to deformation and fragility and insufficient strength in the ceramic matrix (Chawla, 2013). Only a minor fracture can break ceramic matrices. Reinforcing fiber improves the strength and tenacity of ceramics, concealing the low stiffness of ceramics in various ways. Discrete reinforcements such as whiskers, powder filler, and staple fibers, while other continuous reinforcing materials such as lengthy fibers can strengthen the ceramic composites. Assortment of reinforcing material relies on suitable characteristics ceramic matrices must acquire following their utilization. The advantages of ceramic composites over others are: Ceramic matrices are resistant to corrosive nature and abrasion in various environmental circumstances. These matrices are heat resistant and maintain toughness at high temperatures. These matrices have high energy to mass proportion, resulting in lightweight composites. In nature, ceramic matrices are the least sensitive and have excellent chemical resistance. Toughness is increased by strengthening ceramic matrices with good tensile and flexural strength fibers. Ceramic matrices are lightweight than metallic matrices, allowing them to be utilized in situations with low mass-to-toughness proportions required. The non-ruinous breakdown is observed in ceramic matrices (Rohatgi, 2013).The disadvantages of using ceramic matrices are: Manufacturing ceramic matrices filled composites at elevated temperatures necessitates extreme-temperature reinforcing materials. Although reinforcing phases enhance toughness, fragility is the major drawback of CMCs, as even minor damage can cause ceramic deterioration. Manufacturing CMCs necessitates extreme temperatures, which complicates the production operations and raises the operational value. Thermodynamic indices of growth of reinforcing materials and ceramic matrices differ, resulting in thermodynamic strains during quenching after fabrication (Cantor et al., 2003). The applications of ceramic matrices are ceramic matrices used in situations where elevated humidity, abrasion, and corrosive resilience are essential requirements. Fused silica reinforcement on aluminum alloy is utilized in edging instruments for fabricating hard elements. Ceramic matrices are often utilized in making segments for steam turbines, insulating devices, and aviation motors. They are also used for building heaters, fire collectors, and carrying segments with a significant risk of corrosion. The decelerating components such as brake pads and brake control devices applied in motor engines and aircraft that may be subjected to abruptly elevated temperatures and high vapor tubes are also made up of ceramic matrix composites (Rosso, 2006).

Reinforcements in composite materials can be classified into three categories according to their composition and characteristics. Fiber reinforcement, particulate reinforcements, and structural reinforcements are the three types. Particulate matter is granules that are embedded in different matrices. Particulate matter is mostly sizable fleck composites made by connecting these massive particles to a smooth matrix, preventing the matrix from moving. Also, dissipation-intensified composites are made using microparticles ranging in terms of size from 10 to 100 nm. Concrete is the most typical massive particulate composite, consisting of a cementitious matrix bonded to the particle of various intensity and magnitude such as crushed stone, shale. Ceramic to metal composites is another notable illustration of a large particulate, wherein rigid and fragile components are bound to a flexible and bendable metal matrix (Woodford, 2020). Ceramic to metal composites are accustomed to trimming tools to strengthen steel. In rubber tires, dissipation intensifies particle composite material such as strengthened rubber, which contains nanomaterials like CNT (carbon nanotubes) enhanced by natural rubber that accomplish more than regular tires.

Structural composites are a type of hybrid in which several components with varying qualities combine to generate a composite that balances for the specific constituent's deficiencies. Woods like hardwood and softwood are the utmost prevalent laminate component found naturally. Smooth lignin gets integrated with comparably sturdier cellulose, forming both the toughness and tenacity of the wood (Hinestroza & Netravali, 2014). Two sides, which are typically rigid and narrow with superior mechanical strengths, and a center section, which is broad, weak, and fragile, together compose structural composites. Adhesives are used to hold all of these components intact. Laminated composites are typically utilized for architectural and semi-structural purposes. Sandwich composite material refers to the framework in which multiple firm surfaces overlap a softer dense component. Sandwich composites are essential because the surfaces operate as a tension pair to prevent outward bending operation. At the same time, the internal structure opposes shear stresses and stabilizes the surfaces preventing creasing or swelling (Ratwani, 2010). The adhesive that holds the center and surfaces simultaneously is a constituent of laminate composites. In case the bonding within the surface and the center is insufficient, the exterior shear strain may be too much for them to handle. Structural composite materials are utilized in automobiles, off-road transportation, railroads, and other machines for heat barrier, mechanical pressure reduction, strength restoration, emissions, and reduced production expenses. Structural composites are also utilized in residences in flooring, paneling, and other items in addition to automotive. Laminate composites are one example of structural composites fabricated by integrating numerous levels of plates, one on top of the other to provide better mechanical properties, slight elastic rigidity, and enhanced heat barrier, among other properties (Mallick, 1997). Multiple sheeting comprises laminated composites, and sheet orientation is critical for improving mechanical properties and complete attributes. The total elasticity of the composites is enhanced by aligning the sheets in the same direction. Plywood illustrates this composite, wherein wooden boards are joined using natural fiber-enhanced polymers to fabricate hard-coated composites used in home decor. Roofs, ceilings, floor, home furniture, multiprotein ski structure, and other applications utilize laminated structural composites (Netravali & Chabba, 2003).

Synthetic fibers are man-induced fibers created through chemical processing and therefore are categorized as natural or artificial on the basis of their composition (Sathishkumar et al., 2014). Fiber components have substantially higher durability, toughness, and hardness than matrix components, creating these stress carrier components in the building frame (Jawaid et al., 2019; Rajak et al., 2019). Synthetic fibers such as glass fibers are among the most extensively utilized because they have exceptional toughness and structural integrity, heat resistance, wear resilience, barrier properties, dimensional stability, and fracture toughness. Moreover, when using traditional processing technologies, fabrication of glass fiber–enhanced polymers (GFEPs) is time-consuming, complex, and results in lower workpieces (Prakash, 2019). Glass fibers also have the drawback of being discharged after the end of their shelf life. Carbon fibers are used in place of glass fibers in certain situations where additional rigidity is desired. Other synthetic fibers, such as polyester, nylon, olefin, polyacrylonitrile fiber, polystyrene fiber, or polyurethane fibers, have certain benefits but are infrequently utilized thermoplastic discontinuous-fiber-enhanced polymers (DFEP). Carbon fibers have been employed in various applications wherein their suitable qualities are beneficial (Unterweger et al., 2013). Carbon fiber–enhanced composites have a wide range of utilization in aviation, automotive, sports, and various fields (Chung, 2017; Haim, 2017). When the mass fraction of carbon fibers was raised from 15% to 35%, the modulus of elasticity of aggregates and froth increased significantly by 79% and 123%. When propylene-reinforced carbon fiber was utilized to construct composite aerogels made by microporous injection molding technology, the porous configuration improved and increased 35% in modulus of elasticity. Graphite fibers are a novel development of highly efficient carbon-based fiber that exhibits excellent mechanical properties and improved thermal conduction compared to carbon fibers. Graphite fibers hold improvement in a wide range of applications, including ultralight conducting connectors, flammable supercapacitors, photovoltaic materials, and motors (Nobe et al., 2019). Modulus of elasticity, bulk modulus, and toughness of polymeric materials containing graphite strengthening increased by 152%, 26.6%, and 37%, accordingly, in a molecular dynamics simulation. Additionally, the drag and attrition frequency coefficient was reduced by 30% and 45%, respectively (Li et al., 2017). Basalt fiber surpasses fiberglass properties in terms of physicochemical and mechanical aspects. Furthermore, basalt fibers are much less expensive, unlike carbon fibers. The impact of reaction temperature on composites of polymer matrix enhanced by basalt fiber has been examined, on decrement in the temperature leads to increment in strength properties and durability at a specifically applied load (Zhao et al., 2018). Hybridizing twaron fibers with fiberglass or graphene fibers improves the thermal characteristics of twaron fiber-enhanced composites. At the same time, there are fewer investigations on the hybridization of twaron fibers and natural fibers. Twaron fiber–reinforced composites have high mechanical properties, such as tensile and impact strengths. However, they have weak compressibility relative to fiberglass and graphene fiber alternatives because of their asymmetrical structure (Singh & Samanta, 2015).

Natural fibers are a widely accessible and easy-to-find component in the environment. They exhibit biocompatibility, are inexpensive, and have high endurance and precise rigidity. Polymer composite material incorporating natural fiber reinforcements appears to provide many advantages over synthetic fibers, including lower weight, expense, toxic effects, and contamination. Given advanced applications in the modern world, natural fiber composites surpass human-made fiber composites in the capital, operating costs, and environmental impacts (Nair & Joseph, 2014). In addition, natural fibers possess comparable fabrication with varying constituents contingent on the variety. Smaller and longer natural fibers have been used in thermoset polymer matrices to generate highly efficient composites applications. The excellent tribological characteristics and sisal fiber polymer composites are often utilized for designing automotive internal spaces and equipment covering. An increment in tensile modulus with fiber content was perceived when sisal fiber was strengthened with polyester and polyethylene to form composites (Senthilkumar et al., 2018). In contrast with glass fiber–enhanced composite with a polypropylene matrix, hemp fiber–reinforced composite exhibited a 42% increment in overall strength properties (Shahzad, 2012). The mechanical strengths of a composite containing a mixture of 5 wt% maleic anhydride-grafted polypropylene (MAPP) and propylene matrices enhanced with 15 wt% mercerized hemp fiber increased by 37% for tensile strength and 68% of flexural strength (Sullins et al., 2017). Kenaf fiber–enhanced thermoplastic polylactic acid composite obtain mechanical strengths of 223 MPa of tensile and 254 MPa of flexural strength (Ochi, 2008). Moreover, eliminating excess moisture from the fibers prior to combining improves the mechanical characteristics of kenaf fiber composites. Earlier, polyester specimens with no enhancements had a flexural strength and modulus of 32.24 MPa and 4.61 GPa, respectively. In contrast, composites with a reinforcement of 11.1% mercerized natural raw kenaf fiber had an increment in flexural strength and modulus of 65% and 90%, respectively (Mlik et al., 2018). A sound transmission loss (STL) measurement was used to examine the acoustic behavior of flaxseed enhanced polypropylene composites. Because the composite has strong acoustic capabilities, the findings demonstrated improved elasticity, loss modulus, and mass-produced per unit surface (Goutianos 2006). The composites' tensile characteristics were improved by using micro flaxseed fiber. In addition, using 55 fiber assimilation, mechanical strength and moduli were enhanced by 20% and 45%, respectively (Mohamed et al., 2018). The natural frequency parameters of ramie fiber improved with polypropylene polymer matrix were studied. It was discovered that having more fiber in a matrix material causes a void among the fiber and matrix, enhancing the damping proportion throughout the flexural pulsation. This suggests that increasing the fiber load improves the damping characteristics of the ramie fiber–reinforced polymer composite. A protective covering grows near a rice grain throughout its development, termed *rice straw*, which is classified as crop residue. It is used as reinforcing material to explore mechanical and structural properties improvements. Reinforcing 5% rice straw in polyurethane froth provided superior sound attenuation efficiency for improving the composites' acoustic properties (Wang et al., 2018). After an impact characterization, a composite sheet containing 5% chicken plumage as reinforcing fibers in epoxy polymer matrix produced the best findings. Furthermore, these plumages combined with 1% carbon residue mixed with epoxy matrix to generate a hybrid composite showed significant improvements in mechanical properties such as tension, bending, and compressive strength (Verma et al., 2018). Mechanical properties of untreated jute stems reduce with height through the core to edge, with the core section-based composites having 45% and 25% increased tensile and bundle strength, respectively, compared with composites fabricated from the edge section of untreated jute stems (Das et al., 2019; Khan & Khan, 2015). Coir-polypropylene composite sheets with irregular arrangements have better damping capabilities over human-made fiber-reinforced composites. As a result of the increased resin load, greater damping attributes occur, whereas less fiber loading results in higher heat dissipation. The highest damping proportion of 0.4736 was achieved in a coir-reinforced polypropylene composite with 15% fiber loading. In contrast, an increment in fiber loading to 30% elevated the frequency response of the composite to 25.92 Hz (Verma et al., 2012). The fiber-reinforced matrix interphase interactions in palm fibers demonstrated remarkable qualities. Moreover, adding palm fibers with low-density polyethylene (LDPE) enhanced the modulus of elasticity relative to pure LDPE (Chollakup et al., 2013). Reinforcing abaca fiber is used to produce abrasion in composites, which has a remarkable wear resilience property with a wear rate of 3.124×10^{-7} cm^3/Nm for 5% fiber loading. In addition, when the amount of abaca fiber in the mixture increased, the density reduced (Liu et al., 2019). Reinforcing luffa fibers in composite material improved the barrier, interfacial, and physio-mechanical characteristics of the composites, such as tension, compression, flexural, fracture toughness, and moisture content parameters (Panneerdhass et al., 2014). Cotton fiber–enhanced epoxy resin was used to enhance a tube material's heat dissipation and load-bearing abilities (Laban & Mahdi, 2016). Table 1.3 shows the production processes and applications of various synthetic and novel natural fibers along with associated matrix materials.

In comparison with thermoset material, thermoplastic material enhanced with natural fibers demonstrates low mechanical efficiency. As a result, these natural fibers are often mixed with low quantities of human-made fibers to improve structural strength and regeneration capabilities, making these materials more acceptable for technological operations. Composite with fiber loading of 15 wt% loops of hemp and 5 wt% glass reveals the flexural strength and moduli as 211 MPa and 6.5 GPa, respectively. The composite's other properties, such as impact and tensile strength, and moisture content qualities were also considered to be enhanced (Panthapulakkal & Sain, 2007). Scanning electron microscopy (SEM) characterization of palm oil and kenaf cellulosic fibers enhanced with epoxy resin demonstrated strong interfacial adhesion among the fiber matrix, indicating that the composite's mechanical and barrier properties have improved. Hybridization of fiber is a potential method in which several kinds of fibers are blended in different matrices to reduce the disadvantages of one type of fiber while maintaining the advantages of the other fiber. The synergistic actions of hybrid fibers contribute to the composites' enhancement of characteristics that none of the parts possessed (Swolfs et al., 2014). The tensile strength of a hybrid composite comprised of an epoxy matrix with 25% banana pseudostem and 10% jute fibers was increased by 15% compared to pure epoxy composite. Composite with epoxy matrix hybridized with 22% cocoa husk and 16% jute fibers exhibited compressive strength of 34.7 MPa. The mechanical properties of composite enhanced as the amount of hybridized reinforcements increased (Abhemanyu et al., 2019).

After the disposition of composites and consequences to the environment, the effect of synthesis of matrices and reinforcement material on the ecosystem can be evaluated. Fabrication of matrices such as thermosets leads to an internal atmospheric impurity from contaminants such as volatile organic compounds (petroleum products, industrial lubricants, industrial chemicals, and dry cleaning agents all contain VOCs), and dumping compostable composites is potentially damaging to

TABLE 1.3

Production Process and Applications of Various Synthetic and Novel Natural Fibers along with Associated Matrix Materials

S. No.	Reinforcing Fiber	Matrix Materials	Fabrication Process	Applications	Reference
1.	Basalt	Epoxy, poly (butylene succinate, vinyl easters	Hand layup, compression molding	Lightweight automotive body parts, fuel cells, Insulation boards	(Carmisciano et al., 2011; Zhang et al., 2012; Zhao et al., 2018)
2.	Kevlar/graphene Kevlar/Glass	Epoxy, Polyethylene	Compression molding, Pultrusion, Injection molding	Tooling, bearings, bulletproof vests, automobile and aircrafts parts	(Elnaggar et al., 2020; Hallad et al., 2018; Singh & Samanta, 2015)
3.	Glass/CNT, Glass/egg shell	Vinyl resin, Epoxy resin	Vacuum assisted resin infusion (VARI), Wet hand layup technique	Automotive parts, Wind turbine, Gas tanks	(Shin et al., 2020; Wang et al., 2020)
4.	Sansevieria/carbon fiber	Epoxy	Hand layup method	Furniture, aerospace industry, structural applications such as window and door frames	(Anjum et al., 2020)
5.	Arenga pinnata	Epoxy, Polyurethane, Polylactic acid (PLA)	Hot mold press, Hand layup, Injection molding	Trim/door panels, headliner panels, heat padding	(Ali et al., 2010; Atiqah et al., 2018; Sanyang et al., 2016)
6.	Prosopis Juliflora	Epoxy, PLA	Injection molding, Filament winding, Resin transfer molding	Lounge furniture, helmet shell, aircraft structural parts	(Benin et al., 2020; Raj et al., 2020)
7.	Acacia Arabica/Pencil Cactus	Polyester resin	Hand layup technique	Thermal insulation material, acoustic panels	(Kulandaisamy & Govindasamy, 2020)
8.	Muntingia Calabura	Epoxy	Compression molding	Structural and textile applications	(Vinod et al., 2021)

(Continued)

TABLE 1.3 (Continued)
Production Process and Applications of Various Synthetic and Novel Natural Fibers along with Associated Matrix Materials

S. No.	Reinforcing Fiber	Matrix Materials	Fabrication Process	Applications	Reference
9.	Moringa Oleifera seed	Epoxy	Hand layup, Hot mold press, Compression molding	Automobile parts such as headliner, door panels, semi-structural applications like window frames	(Mishra & Sinha, 2020)
10.	Moringa Oleifera leaf	Polyvinyl alcohol (PVA), PLA	Solution casting, Spin coating	Food packaging, wound healing	(Islam et al., 2017; Mishra & Sinha, 2021)
11.	Pennisetum Purpureum	Polystyrene, Polyesters	Manual mixing, Hand layup, Hydraulic compression molding	Insulation boards, socket prosthesis, civil works, automotive parts	(Adeniyi et al., 2020; Ridzuan et al., 2016)
12.	Cortaderia Selloana	Polyethylene	Twin screw extruder, Injection molding	Ropes, textile industry, food packaging, wound healing	(Jordá-Vilaplana et al., 2017)
13.	Cynara Cardunculus	PLA, Polyurethane	Film-stacking, Compression molding	Automobile structural parts such as interior door paneling, mudguards, engine insulation	(Botta et al., 2015; Olcay & Kocak, 2020)
14.	Manicaria Saccifera	PLA, Polybenzoxazine	Hand layup, single screw extruder, Injection molding	Window and door frames, heavy equipment packaging, semi-structural	(Oliveira et al., 2020; Porras et al., 2016)
15.	Ricinus Communis	Polypropylene	Compression molding	Civil works such as, building boards, insulation boards, packaging indsutry	(Vinayaka et al., 2017)
16.	Luffa Aegyptiaca	Concrete	Hand layup	Ceiling, aerospace	(Quadri & Alabi, 2020)

marine and other habitats. Issues regarding composites include the fact that when examining composites' degradation or removal, both the matrix and fiber qualities must be taken into account. Strengthening components such as natural fabrics enhanced on polyvinyl chloride or polyurethane are utilized to manufacture sustainable composites compounds, although this results in partly biodegradable materials with dumping concerns. The reinforcements and matrices in green composites are naturally compostable under ordinary external conditions. Agropolymers derived from sustainable resources mainly synthesized compounds derived from agro-based compounds and compounds produced by bacteria and fungi such as polyhydroxy butyrate. Significant advancements in building entirely biodegradable green composites include starches and wood fiber–enhanced composites and soybean protein composites with natural fibers. Therefore, going green with natural fiber–based composites is of new interest to researchers (Lodha & Netravali, 2002; Netravali & Chabba, 2003).

1.3   INTRODUCTION TO NATURAL FIBERS

Since the dawn of civilization, natural fibers have been used to construct strands, cordage, and textile materials and have contributed significantly to society. In addition, natural fibers are environmentally friendly since they are renewable, carbon-free, and compostable (Azammi et al., 2020). Natural fibers are a viable substitute for human-made fibers in the composite market because of their adaptability and easy accessibility (Getme & Patel, 2020). Natural fibers, contrary to human-made fibers, are disposable and environmentally benign. They are sustainable sources with unique qualities like simple structure, minimal density, ease of manufacturing and processing, and inexpensive (Hosseini, 2020; Vinod et al., 2020). Natural fibers are presently prioritized above human-made fibers because they cause reduced corrosion on machines and their component throughout fabrication and also don't pollute the environment or aquifers (Vinod et al., 2020). Natural fibers are commonly utilized as reinforcing elements in polymeric materials, in which resins are used as a support component to grip fibers in place and offer structural integrity (Asim et al., 2018). Natural fibers are a good and appealing alternative as a reinforcement component because of their remarkable strength per unit mass (Bari et al., 2018). Natural fibers are generated in enormous amounts year after year. They are utilized as a desirable feedstock in clothing and accessories, cardboard, packaging, sporting goods, automotive, and construction products (Jawaid & Khalil, 2011). All fibers became prominent as they were utilized in several distinct ways. A few of these were utilized to construct paddle or twine for trailing, angling, fencing, scaling, transporting, and specific different purposes, as well as fabric production. The primary considerations for excising fibers in prehistoric days may be in the development of fabric. Humanity encountered cooler weather in their unfamiliar environments in prehistoric times due to their migration from one continent to another. Therefore, it was necessary to shield their bodies against gusty winds and cold weather.

Consequently, people began to experiment by using natural fibers such as woven fabric and mammal fur skin and wools to handle the chill. Natural fibers were extracted from the wild, mainly in the early stages. Certain fibril stems or shrubs were utilized. However, the invention of plant-based fibers such as cotton and kenaf transformed the globe's future. Then, due to massive industrialization, fibers from grown plants rendered a couple of changes for easier production. Thereby, natural fibers became an integral part of human life in a relatively short time span. Natural fibers were used by humankind long before ancient times, according to archaeological data. Many experts consider hemp to be the earliest plant fiber. It was first used in 4500 BC, as per archaeological evidence from southern Asia (Hosseini, 2020). Around 3400 BC, Egypt had developed sewing and twirling with cotton (Asim et al., 2018). Food crops have been used from the dawn of time. People were accustomed to using species of plants as feed ingredients, but then, they began to investigate the distinctive characteristics of crops. Flora's therapeutic properties provided them a long service life after they were fed, while other applications of crops improved living. As a result, our ancestors experimented with novel plant characteristics daily. Around 10 millennia BC, humanity began to cultivate organically, and crop production spread to many various places of the globe. Initially, Mesopotamian culture served as a beginning place for cultivation. Later, it was joined by China and India. Mexicans and the mountainous areas in the west were essential in the diversity of plants.

Given these actions, agriculture evolved spontaneously and gradually in various locations (Bari et al., 2018). Most research believes that numerous fiber crops were cultured widely in multiple countries, but cotton, kenaf, ramie, hemp, nettle, and jute were the most common sources of fibers. Cotton was initially developed around 8,000 years ago in the Kachi Valley of Karachi, Pakistan. The primeval proof was identified in Mehrgarh's Neolithic civilization. Production of the first cotton plant (*Gossypiumbarbadense*) originated in southern India and Pakistan's Indus Valley and later expanded throughout the southwest part of Africa and Asia. One other cotton variety (*Gossypium herbaceous*) was domesticated in southern Arabia and Syria. Cotton varieties are genetically distinct from one another. Jordan and the Northeast Caucasus were also prominent cotton cultivators in the seventh millennium BC. Cotten-based textiles were discovered in Iran, Armenia, and Ireland, subsequently in the first millennium BC. Cotton was cultivated in Mosul, as per Syriac chronicles of Sennacherib, although the yield was not significant because of the severe winter. Cotton was produced in the Iranian Gulf, Lebanon, Tajikistan, and China over the millennium. In the middle of the tenth and eleventh centuries after Christ, Saudis introduced cotton in South America and the Middle East (Mhatre et al., 2019). Cotton was initially used to build angling nets by Native Americans in the coastal areas of America since the economic system was centered on oceanic culture. Flax fibers are cellulosic crops that are also crystalline in appearance. These fibers can reach a length of 80 cm and a radius of 6–8 m. The primary producers are Holland, Germany, and France. It is mostly utilized in household products, textiles, bedclothes, linen, architectural decor components, etc. (Van de Weyenberg et al., 2006). For humanity, hemp is amongst the most significant plants. It is unique since it may be used in a variety of ways. It is a considerable fiber plant and adequate nutrition, therapeutic, and narcotic crop. Hemp is among the earliest food crops in recorded existence because of these characteristics. Also, these are primarily farmed in Eurasia. It can reach a height of 2.2–5.5 m and a radius of 2 cm. The center surrounds the internal width, and the surface sheet is the bast fiber, linked to the inner lining by an adhesive material called pectin. Jute is a popular natural fiber due to its low cost. However, it is a relatively recent variety as a fiber crop. There were no jute-related artifacts discovered in prehistory. Comparing it to other traditional natural fiber, it is a novel fiber. Jute is farmed throughout Asia, particularly India, Pakistan, China, Indonesia, and Bangladesh (Khan & Khan, 2015; Shahinur & Hasan, 2019). The crop develops to a height of 20–25 cm in 5 months, and the fibers are removed 5 months following cultivation. Kenaf fibers are another most prominent fiber in the natural fiber category, and they are mostly utilized in manufacturing paper and rags (Omar et al., 2019). These are plants with a fibrous texture. They are rigid, sturdy, and robust, with strong pesticide resilience. These plants have been grown in southern Africa, India, China, the United States, and countries in Europe for about four centuries. Kenaf fibers are easily recyclable, making them eco-friendly. These fibers were once used to make fabrics, strings, ropes, and firm bags, and Romans used them to make watercraft. These fibers are currently employed in automobiles, architecture, packaging, housing, fabrics, carpets, paper pulp, and other industries as composites with different components. Table 1.4 lists some of the most significant fiber sources with their origin.

In 2018, global fiber quantity demanded around 110 million metric tons, with 31 million tons of natural fibers and 75 million tons of cellulosic and non-cellulosic fibers. According to Terry, chairperson of the Discover Natural Fiber Initiative on March 2019, reported to the annual conference in Frankfurt, Germany, the human-discovered fiber accounted for 60% of overall fiber supply in 2018 (Townsend, 2020). The two pie charts represent the overall production of natural fiber and synthetic fiber in the world (Figures 1.5 and 1.6).

Natural fibers have a strong and historical background in the global market, and given the rise of artificial substitutes, they maintain thriving enterprises with hundreds of millions of workers. Almost everyone on the planet comes into contact with natural fibers on a daily basis. Natural materials, on the contrary, are being endangered by the development of human-made fiber substitutes. Everyone working in the natural fiber industry is critical in how these industries will compete with polyester, polyamide, acrylic, polyethylene, and certain other human-made fiber materials. Natural fiber output increased from 12 million tons in the mid-1960s to greater than 25 million tons by 2004, and it remained at a similar position in 2018. Human-made fiber supply, on the contrary, expanded from some million tons in the late 1970s to more than 80 million tons in 2018 and is still growing rapidly (Townsend, 2020). The scientific advancement that began with orientated re-production and proceeded with industrialization, pesticides, fertilizers, and plant shield chemicals, including the use of genetically engineered equipment to enhance plant properties, must move for natural fiber industrial sectors to compete effectively (Khan & Khan, 2015).

Natural fibers' chemical constitution changes depending on their size (Maepa et al., 2015). Natural fiber physicochemical properties vary by variety, lifetime, and maturity among various cells of the similar plant and soil variety, environmental factors, and fiber harvesting and degrading procedure (Anastas et al., 2002). Different fibers like animal fiber are made up of proteins like epidermis, fibronectin,

TABLE 1.4

Significant Fiber Source and Origin

Fiber Source	Species Names	Common Names	Origin	Places
Arenga Pinnata	A. pinnata	Sugar palm	Fruit, leaves	Tropical Asia (India, Malaysia, Indonesia)
Acacia Arabica	Vachellia nilotica	Babool, Gum arabic	Flower, fruit	Africa, Middle East, Indian Subcontinent
Artichoke	Cynara cardunculus	French artichoke	Fruit	Saudi Arab
Arundo Donax	A. donax	Giant cane	Stem, leaves, Flowers	Southern America
Abaca	Musa textilis	Manila hemp	Stem, leaves	Philippines
Broom root	Ruscus aculeatus	Cladodes	Root	Europe and Asia
Banana	Musa indica	–	Leaves	Europe and Asia
Cortaderia selloana	C. selloana	Pampas grass	Grass	Southern America
Cantala	Agave cantala	–	Leaves	Southern America
Coir	Cocos nucifera	Coconut Fiber	Fruit	India, Arab, China
Dracena trisfasciata	D. trisfasciata	Snake plant	Flower, leaves	West Africa, East Nigeria
Euphorbia tirucalli	E. tirucalli	Pencil Cactus	Roots	Africa, Sri Lanka, India
Furcraea foetida	F. foetida	Mauritius hemp	Stem	Central America, Columbia, Guianas
Henequen	Agave fourcroydes	–	Leaves	Mexico, Italy, Costa Rica, Cuba
Isora	HelicteresIsora	Marorphali	Stem, Flower	Southern Asia, Northern Oceania
Juncus effusus	J. effusus	Common rush	Leaves	Africa, Europe, Asia
Kapok	Ceiba pentandra	Silk cotton	Fruit	Mexico, United States, West Africa
Kans Grass	Saccharum spontaneum	Wild sugarcane	Grass	India, Bangladesh, Nepal
Kudzu	Pueraria thunbergiana	Arrowoot	Stem, leaves	Asia, Pacific Islands
Luffa cylindrica	L. aegypticaa	Dishrag gourd	Fruit, Leaves	Southern Asia
Muntingia	M. calabura	Jamaica cherry	Leaves	Mexico, Bolivia
Moringa	M. oleifera and other >13 species	Sahajana, Drumstick	Seed, flower, Leaves	Asia and Western Africa
Manicaria	M. martiana and M. saccifera	Palma Real	Leaves, fruit	Trinidad, Southern America, Brazil
Nettle	Urtica dioica	Sringing nettle	Stem	Europe, Asia, West Africa

TABLE 1.4 (Continued)
Significant Fiber Source and Origin

Fiber Source	Species Names	Common Names	Origin	Places
Prosopis juliflora	*P. juliflora*	Mesquite	Leaves	Mexico, South America, Caribbean
Pennisetum purpureus	*Cenchrus purpureus*	Napier Grass	Grass	Africa
Piassava	*Leopoldinia piassava*	–	Fruit, leaves	Europe, Southern America
Ricinus	*R. communis*	Castor oil plant	Flower	Mediterranean Basin, Africa, India
Urena	*U. sinuata*	–	Flower, leaves	Asia

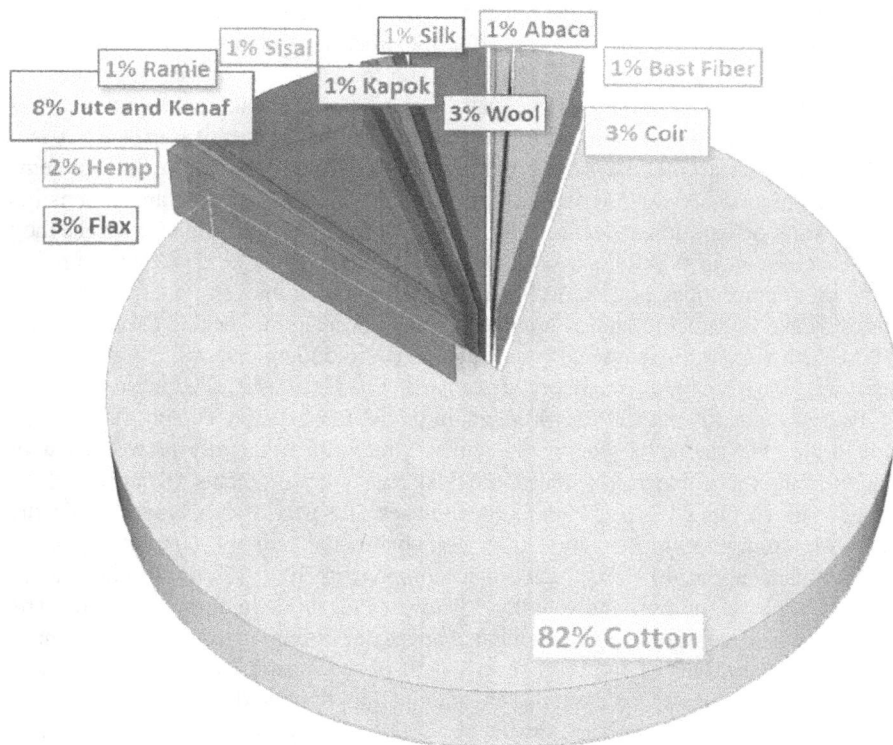

FIGURE 1.5 Overall world natural fiber utilization.

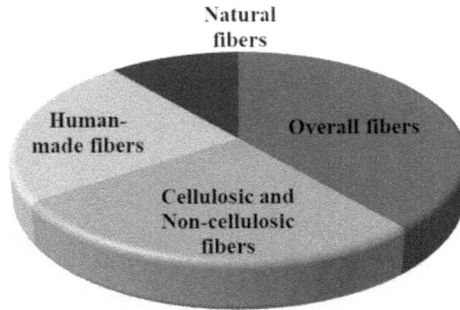

FIGURE 1.6   World natural and synthetic fiber production.

gelatin, and chitosan, determining the fiber's toughness and durability. The composition of proteins differs with various species (Maheshwaran et al., 2018). Natural fibers are mostly made up of holocellelulose, ash, pectin, moisture, lignin, and waxy components linked together by hydrogen bonds to produce rigidity and tenacity (Malkapuram et al., 2009). The primary skeletal constituent of natural fibers is cellulose. Cellulose is a hydrogen-bonded linear polycarbohydrate chain made up of 1,4-coupled glucose molecules that offer the fiber its toughness, flexibility, and corrosion resistance (Ashok et al., 2015). When the cellulosic matter is bundled simultaneously by pectin, the cellular wall becomes more robust, contributing to resistance and durability to rupture in the hydration process (Maheshwaran et al., 2018). Cellulosic polycarbohydrates have high thermal stability with a breakdown temperature variation of 310°C to 450°C. Hemicelluloses, on the contrary, are responsible for the water penetration and thermo-biological destruction of natural fibers since they are less resistant than cellulosic components (Malkapuram et al., 2009). Lignin is a networked, amorphous polymeric linkage made up of an uneven pattern of bonded hydroxylated and methoxylated alternated phenylpropane molecules (Maepa et al., 2015). Although lignin is crucial for ultraviolet destruction in natural fibers, it is heat resistant, with a breakdown temperature range of 170°C to 950°C. Pectin is an integral acidic heterogenous polycarbohydrate chain made up of rhamnose and transformed glucuronic acid remnants. Pectin molecules, which are frequently networked with calcium-binding, enhance the morphological constancy of plants (Khan & Khan, 2015). The combined activity of lignocelluloses and pectin as a binder holds the cellulosic configuration in natural fiber composites collectively. Natural fibers are usually made up of 40%–70% cellulosic component, 3%–15% lignin constituent, 20%–30% hemicellulose, and roughly 15% water (Maheshwaran et al., 2018). The chemical constituents of each component changes depending on the variety of plant or crop and the portions of the crop from which fiber is taken (Maepa et al., 2015). Chemical components are essential in anticipating fiber and composites thermo-mechanical operation (Malkapuram et al., 2009). Diverse approaches are used to examine the chemical components of fibers. Chemical technique is conducted to identify the mass fraction of different components present in the natural fiber (Maheshwaran et al., 2018). The Technical Association of the Pulp and Paper

Industry (TAPPI) specifications like T264 cm-973 (moisture content) and TAPPI UM250 (acid soluble lignin) to assess distinct chemical constituents (Schoning, 1965). The thermal technique equipment is commonly used to specify the overall cellulosic component. The chemical constituents can be analyzed using several American Society for Testing and Materials (ASTM) professional regulations. The ASTM E1721-01 (2020) and ASTM E1755-01 (2020) standards are used to determine lignin and ash content, respectively. Lignin removal and alkylation as surface modification procedures are used to evaluate hemicellulose content according to ASTM D1166-84 (2001). An electrical humidity tester can be used to examine the water content of fibers. The residual weight percentage calculated following dry oxidation represents the ash constituent of fibers (Malkapuram et al., 2009). Conrad technique with Soxhlet apparatus for extraction is used to measure the distribution of waxy component and other fatty compounds (Ashok et al., 2015). Table 1.5 gives the physicochemical properties of the new and traditional fibers.

1.4   CLASSIFICATION OF NATURAL FIBER

Natural fibers are divided into three different forms animal, mineral, and plant fibers based on their organic source. Animal fibers are mostly made up of long chains of biomolecules or macromolecules of the amino acid group, whereas plant fibers are mostly made up of lignocellulosic compounds. Plant fibers have attracted interest as a reinforcement phase for biocompatible composites due to their accessibility and low price. Animal fibers have considerable attention due to their elasticity, high durability, the enhanced surface-to-volume proportion, and reduced hydrophilicity (Mhatre et al., 2019). Plant fibers such as cannabis, bamboo, wheat, jute, agave, and animal fibers such as camel hair, spider silk, and wool are commonly utilized in the clothing industry (Vinod et al., 2020). The classification of natural fibers is shown in Figure 1.7.

According to different studies, natural animal fibers are primarily hair or skin of the animals like camels, spiders, bird plumage, and silk cocoons. Animal fibers are a promising strengthening phase in composite materials due to their remarkable physiochemical and structural characteristics. Sheep, camel, bovine, Ankara bunny, Changthangi goats, and other furry creatures produce wool. It's a prevalent fabric fiber, and Chinese, New Zealanders, and Australians are the dominant suppliers (Ramamoorthy et al., 2015). Wool has different physical and chemical qualities depending on where it derives. Ankara, Vicugna pacos, muskox, and pashmina wool fibers, have diameters of 12–16 μm, 10–30 μm, 15–25 μm, and <19 μm, respectively (Ramamoorthy et al., 2015). Wool fibers are water absorbent and soak up to the one-third the portion of the overall mass in water. The amount of flame expansion, thermal efficiency, and calorific value are low in wool animals.

Sheep, camel, and yak furs and hairs are indeed extensively utilized animal fibers. Bird fibers include plumage and wing fibers from chickens and other mammals. Chicken plumes, primarily made up of protein, are waste material from factory farms (Vinod et al., 2020). The fiber in chicken plumes is made up of 91% protein, i.e., keratin, 16% serine, 8% water, 6% moisture, 1.5% fat, 1% lipids, and many other amino acid groups (Ramamoorthy et al., 2015). Due to the extreme

TABLE 1.5

Physicochemical Properties of New and Traditional Fiber

NF	Chemical Composition						Physical Properties		Reference
	C (%)	HC (%)	L (%)	MC (%)	AS (%)	O (%)	Density (g/cm³)	Diameter (µm)	
Ferula communis	53.3	8.5	1.4	24.8	7	36.8	1.24	90–300	(Seki et al., 2013)
Hierochloe odorata	70.4	21.5	8.1	–	–	–	1.16	136.71	(Dalmis et al., 2020)
Corypha taliera	55.1	21.78	17.6	7.1	11.25	–	0.86	45.548	(Khondker et al., 2010; Tamanna et al., 2021)
Calamus manan	42	20	27	64.3	0.8	–	0.45	400	(Ali et al., 1993; Ding et al., 2021)
Castor fiber	65.5	–	5.56	10.17	4.89	0.87	1.18	120	(Li et al., 2014; Nijandhan et al., 2018)
Artichoke	75.3	11.6	4.3	–	2.2	–	1.579	260	(Fiore et al., 2014; Gominho et al., 2018)
Arundo donax	43.2	26.4	17.2	6.4	1.9	6.1	1.168	–	(Fiore et al., 2014; Giudicianni et al., 2014)
Sansevieria ehrenbergii	80	11.25	7.80	10.55	0.60	0.45	0.887	90–250	(Loganathan et al., 2020; Sathishkumar et al., 2013)
Arenga pinnata	43.88	7.24	33.24	8.36	1.01	2.73	1.50	212.01	(Ilyas et al., 2019)
Acacia arabia	68.1	9.36	16.86	–	–	0.49	1.028	–	(Manimaran et al., 2016)
Abaca	56–63	20–25	7–9	14	0.19	3	1.5	10.30	(Gurunathan et al., 2015; Pereira et al., 2015)
Roselle	64.50	20.23	6.21	5.8	1.25	–	1.332	40–100	(Razali et al., 2015)
Phormium tenax	45.10	30.10	11.20	10	–	0.70	1.30	159	(Komuraiah et al., 2014)
Borassus flabellifer	68.94	14.03	5.37	6.83	–	0.64	1.256	287.23	(Singh et al., 2021)
Banana pseudostem	38.48	25.36	5.77	7.98	9.36	–	0.22	293	(Kabenge et al., 2018; Silva et al., 2020)
Cortaderia Selloana	53.7	14.43	10.32	7.6	4.2	3.1	1.261	–	(Khan et al., 2020)
Coir	43.44	0.25	45.84	10.50	2.22	3–4	1.40	0.1–1.5	(Madueke et al., 2021)
Pennisetum purpureum	34.67	20.23	19.33	–	–	–	–	–	(Haldar & Purkait, 2020)
Dracaena reflexa	70.32	11.02	11.32	5.19	6.23	0.23	0.790	176.20	(Manimaran et al., 2019)

	C	HC	L	MC	AS	O			
Sida cordifolia	69.52	17.63	18.02	8.51	2.62	0.42	1.330	–	(Manimaran et al., 2019)
Grewia tiliifolia	62.8	21.2	14.9	2.3	–	–	–	–	(Jayaramudu et al., 2010)
Heteropogon contortus	64.87	19.34	13.56	7.4	–	0.22	0.660	–	(Hyness et al., 2018)
Pennisetum orientale	60.3	16	12.45	8.5	–	1.9	1.045	213.10	(Vijay et al., 2021)
Kudzu	33	11.60	14	–	–	–	–	–	(Uludag et al., 1996)
Piassava	28.60	45	25.80	–	–	–	–	–	(Rowell & Stout, 2007)
Aristida adscensionis	70.7	10.5	8.91	7.49	2.23	0.18	0.79	–	(Manimaran et al., 2020)
Azadirachta indica	68.42	13.72	13.58	–	–	0.43	–	91.15	(Manimaran et al., 2018)
Cissus quadrangularis root	77.17	11.02	10.45	7.3	–	0.14	1.510	610–725	(Indran et al., 2014)

*C = Cellulose; HC = Hemicellulose; L = Lignin; MC = Moisture Content; AS = Ash; O = Others; NF = Natural fiber

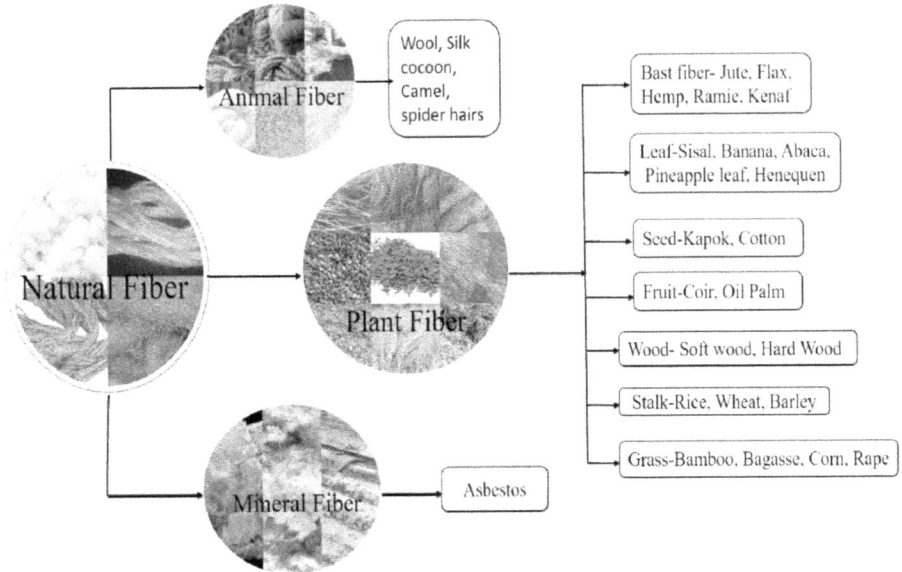

FIGURE 1.7   Classification of natural fibers.

absorption characteristic of keratin, chicken plume fibers are employed in micro-biological corrosion resilience applications and the fabrication of permeable scrubbers and composite scrubbers for washing glasses and perfectly reflective materials (Mohamed et al., 2018). Also, due to the chemical constituents, higher permeability, and interfacial characteristics, chicken plume fiber-strengthened composites have been shown to have improved tensile and flexural strengths and sonic and electrochemical properties (Ramamoorthy et al., 2015). Approximately 14,000 varieties of butterflies, caterpillars, and about 4,000 kinds of spiders are used to make silk (Vinod et al., 2020). Moreover, throughout the formation of a swaddle, glands around the throat of insects release silk. Silk contains chitin, an extremely organized protein that provides chemical stability, resilience, and strength. The two constituents of the utmost used silkworm or silk moth are 70%–80% fibroin and 20%–30% sericin. There are 5,263 long networked chains and 262 small chains of amino acids linked with disulfide bonds in fibroin. The tenacity of fibroin is approximately 600 MPa, which is higher than widely used plant fibers. Silk fiber–strengthened polymeric materials are employed in the biogenic field, such as regenerative medicine and the fabrication of therapeutic scaffolds (Ramamoorthy et al., 2015). Gigantic forest spiders or phoneutria spiders generate major-ampullate spider silk. The major-ampullate silk fiber is made up of spidroin I and II; both are highly recurring proteins. The fiber is relatively noncrystalline and exhibits a less crystalline nature than bombyx Mori (Ramamoorthy et al., 2015). Major-ampullate (dragline) silk is among the finest silk fibers across numerous species, with a tenacity of 1.1 GPa, comparable to high strength metal such as steel with a tenacity of 1.5 GPa (Mohamed et al., 2018).

Plant fibers have long been utilized in making garments, automotive furnishings, division panels, food packaging, and many other applications. Therefore, it functioned as a reliable source material for a variety of purposes (Loureiro & Esteves, 2019). Plant fibers are largely made up of lignocellulosic components such as holocellulose, lignin, waxy components, pectin, ash, and moisture that vary in quantity between different varieties of plant fibers and also between other portions of the same fiber (Asim et al., 2018). The physiochemical characteristics of plant fibers are influenced by geographical locations and growing conditions, as well as the variety of species, its maturity and development, the section of the plant from which fiber is extracted, and the method of fiber extraction (Loureiro & Esteves, 2019). Plant fibers are in high competition as a replacement for human-made fibers and as a sustainable strengthening material in bio-based composites. Plant fibers such as jute, hemp, cotton, and ramie are frequently substituted for human-made fibers such as glass, polyester fibers in the production of composite materials (Sanjay & Yogesha, 2017). Plant fibers are classed as major or subsidiary fibers based on their function. Ramie, hemp, agave, sugarcane, cotton, and other fundamental fibers occur naturally that are consumed directly as fiber. Auxiliary fibers are rice husk, coconuts, corn, and pineapple fibers that are derived as side derivatives from their selected sources (Vinod et al., 2020). From the commercialization point of view, plant fibers are divided into different categories based on their geographical basis, as shown in Figure 1.7.

Bast fibers are those derived from a plant's shank such as roselle, hemp, sunn, urena, jute, flax, kenaf, and many more; leaves such as henequen, maguey, abaca, phormium, cantala, pineapple, banana leaf fibers, and many more; seeds; hairs such as coir, corn, rice husk, and many more; fruits such as sponge gourd, coir, macaw-fat, and many more; stalks such as wheat, rape, rice, and many more; grasses such as pampas, kans, bagasse, Napier, and many more; and lastly fibers from hardwood and softwoods such as eucalyptus beech, birch, maple, and so on (Hosseini, 2020). Several plants, on the other hand, have more than one variety of fiber. For example, bast and fibrous cores may be found in roselle, hemp, sunn, urena, and jute, whereas shank and fruit fibers can be found in coir, macaw-fat, and agave. Also, shell and bast fibers can be found in wheat rice and rape (Mohamed et al., 2018). The bast fibers have a higher tensile and flexural modulus; meanwhile, the fibrous leaves have a high impact resistance (Sanjay & Yogesha, 2017). Kenaf (*Furcaria cannabina* Ulbr.) is grown mostly for its fiber and seed extract in warm climates. It is predominantly a central African and southeast Asian crop, and it is currently majorly cultivated in India and Bangladesh. It is a relatively novel crop in America, but it's already looking promising in the fabrication of bio-based composites. Kenaf has been produced from roughly 4000 BC, as per history. The crop can mature to 10 cm each day in natural circumstances and culminates in 4 months (Nishino, 2004). Because it is a rapidly growing plant, this could produce up to 4,046.856 kg/acre per year, while the current kinds can produce up to 12,140.569 kg/acre per year. The typical crop reaches a height of 10 ft and has a bushy bottom with a 1.5–2.5 cm radius. The bark makes up around 50% of the crop, from which the fibers are removed, with the remainder inner wood (Lee & Eiteman, 2001). The internal wood has a noncrystalline pattern, while the shell has a solid crystalline structure. The shank is usually 6 cm in radius. It offers ecological and commercial benefits; producing 1 kg of kenaf consumes 4.2 kWh of energy while it takes 15 kWh for glass fibers, and the fibers are significantly cheaper (Nishino, 2004). Kenaf fiber is a prominent bast fiber with nearly 75.7% holocellulose content, 12% lignin, and 10% other components (Mohamed et al., 2018). The fiber has high fracture toughness and mechanical strength, it can be used to replace glass fiber and other human-made fibers (Sanjay & Yogesha, 2017). Fabrics, carpeting stuffing, automotive panels, perforated media, and other extruded or injected formed polymer products can benefit from kenaf fiber (Mohamed et al., 2018). Flax (*Linum humile* Mill.), a member of the Linaceae family, is among the earliest agricultural residues in tropical environments (Faruk et al., 2012).

Flax fiber is made up of 85% holocellulose, 8% lignin, and 5% of other waxy components. The mechanical characteristics are comparable to glass and carbon fibers (Mohamed et al., 2018). Flax fiber is used as a substitute for human-made fibers in a wide range of applications, including cement and polymeric strengthening, asbestos substitution, packing products, insulating materials, circuit breakers, and the production of lightweight components in the automobile sector (Mohamed et al., 2018, Terzopoulou et al., 2015). Hemp (*Cannabis sativa L.*) is a periodic herbal bast fiber crop mostly grown for fiber, extracts, and seeds in Asia and France. Cannabaceae crops are native to Eurasia, and they are assumed to have arrived in Europe around the middle ages (Barber, 1991). It is now extensively distributed in nations with temperate climates, like Chilean, Korean, Indian, Japanese, and many European areas. The Eurozone is considering subsidizing hemp growth in member nations and is anticipating additional advancements. Cannabis is now utilized for various purposes, including fabrics, tissue, fiber-reinforced to composites, seed grain, fuel, waxy components, polymer, pulp, bioenergy, and more. Also, it is mostly determined by the level of the industrial hemp crop (Shahzad, 2012). The cannabis crop produces trace levels of tetrahydrocannabinol (THC), the psychoactive ingredient in opium. Hemp grows to a level of 3 m and has a relatively limited breadth due to the proximity of the crop. Numerous researchers have studied the hierarchical arrangement of industrial hemp fiber and fiber bundle thickness. The mean fiber bundle radius is 12 mm, and the mean fiber bundle height is 0.025 m (Olesen & Plackett, 1999). Hemp fiber holocellulose constituent has been found in between 65%–70%. Cannabis fiber is a reinforcing material in biopolymers composites due to its enhanced strength and flexibility (Kandachar & Brouwer, 2001). The chemical constituents of hemp fibers vary for numerous reasons, resulting in a wide range of mechanical qualities. Researchers found that diverse plant groups grown under various environmental circumstances resulted in varied hemp fiber physio-morphological qualities (Svennerstedt & Sevenson, 2006).

Jute fiber is a multifunctional, tough, high cellulosic bast fiber cultivated in rainy climates in Asian countries like India, Pakistan, Indonesia, Bangladesh, Japan, and China (Faruk et al., 2012). It comes from the periodic crop *Corchorus olitorius*, which belongs to the Malvaceae family. During the retting procedure, the crop is chopped and immersed in water at all times of the year. Reduced volumetric thickness, reduced abraded spot, reduced heat conduction, medium moisture resilience, and strong mechanical characteristics describe jute fibers. It also has good electrostatic and isolation attributes (Mohamed et al., 2018). Backpacks, rags, and polymeric reinforcing material all contain jute fibers. Ramie (*Boehmeria nivea L.*) is a blooming crop in the *Urticaceae* genus that produces the most readily available bast fibers (Varghese & Mittal, 2018). Ramie fibers are extracted from the shank beneath a fine overlay of bark. It has around 85% holocellulose content, 2% lignin, and 1% other waxy components. Ramie fibres are thinner and tougher in dry areas, and they grow much stiffer when moist (Mohamed et al., 2018). In contrast with other plant fibers, ramie contains more cellulose and has a finite lifetime (Varghese & Mittal, 2018). Furthermore, the qualities of ramie fibers are similar to those of other bast fibers such as kenaf and hemp. Ramie, kenaf, and flaxes are the most bug and pest resilient of all the plant fibers, and they don't need the specific growing environment to flourish (Mohamed et al., 2018).

Sisal (*Agave amaniensis* Trel and Nowell), a member of the *Agave* genus, produces a tough fiber when decorticated, which is historically utilized to make thread and cord (Sanjay & Yogesha, 2017; Varghese & Mittal, 2018). Asian communities like India, Indonesia, Brazil, Haiti, and East Africa are among the people that cultivate it economically. Sisal is a robust, abrasive, and tough fabric. It supplies 50% of all textile fibers produced each year. Sisal plants are very smooth to cultivate because they can survive extreme heat and have the ability to flourish in desert environments. It can thrive in every growing condition except mud and is resilient towards bug or pest infection (Naveen et al., 2019). Sisal fiber constitutes around 75% of holocelluloses, 10% lignin, and 1% of other waxy and pectin components (Varghese & Mittal, 2018). Sisal fiber have comparable mechanical strengths and modulus as aramid fibers. Interior motor covers, harness straps, side view mirrors, package compartments, cap hooks, door handles, and external or even underground paneling are examples of where sisal fibers are employed in polymeric materials in automobile applications. It is also utilized in the aviation industry for interior paneling (Naveen et al., 2019). The abaca fiber extracted from the *Musa* genus of the banana plant (*Musa amboinensis* Miquel) is strong and resilient to saltwater. Ecuadorians and Philippians cultivate the toughest commercially accessible cellulose fibers (Faruk et al., 2012). Abaca fiber manufacture ropes can be utilized as a substitute for glass fibers in automotive (Vinod et al., 2020). *Ananas bracteatus*, a perennial herb from the *Ananas* genus belonging to Brazil, yields pineapple leaf fiber (PALF), precisely high in cellulosic components. PALF is a widely accessible by-product derived from pineapple production that is now employed in a variety of polymer reinforcing operations (Faruk et al., 2012).

Coconut fiber, often known as coir, is a cellulosic fiber found between the external layer and the hull of the cocoa tree (*Cocos l.*), which belongs to the Arecoideae sub-family (Sanjay & Yogesha, 2017). It's a prolifically accessible side-derivative of cocoa processing that's primarily utilized in fabrics, strands, handbags, buckets, and carpets, among other things. Coir is an excellent reinforcing phase in polymers because of its endurance and hardness (Varghese & Mittal, 2018). Mattresses for automobile and bicycle seats are made from coir fiber strengthened polymer composites (Vinod et al., 2020). Oil palm fiber is a cellulosic fiber derived from the stem, fronds, flower pericarp, and fruit peels of the *Elaeis* genus and *Elaeis guineensis* species. Because of its unique qualities, oil palm fiber is increasingly being used as a reinforcing material in polymer matrices (Varghese & Mittal, 2018). Bamboo (*Bambusoideae luerss.*) is an evergreen periodic blooming grass of the *Poaceae* genus that can reach a length of 132 ft. in a tropical environment (Faruk et al., 2012). Bamboo fibers are extracted primarily using steam explosion followed by mechanical modification as crushing, grinding, and retting from the rigid stem, limbs, and leaves (Varghese & Mittal, 2018). Bamboo fibers are ultraviolet radiation absorbent and are used in custom-built paper, carpentering, and structural applications (Frauk et al., 2012; Mohamed et al., 2018). Packs of bamboo fibers displayed significant tensile strengths equivalent to typical glass and aramid fibers (Faruk et al., 2012). Bagasse biomass fiber is a waste left over from the grinding of cane sugar. *Saccharum* genus and Andropogoneae family grinds between the toothed blades in the grinder mill post cutting. About 60% of bagasse is made up of holocellulose constituents (De Resende & Da Costa, 2020). Many academicians are beginning to believe that bagasse fibers can be employed as a reinforcing phase in composite fabrication. Wood fiber is commonly utilized as a natural filler in a multitude of scenarios (Getme & Patel, 2020). After a basic sifting procedure, wood fibers, usually produced from wood chips from various areas such as paper production, woodwork, engineering and structural, and so on, are used to reinforce functions after sieving (Hosseini, 2020). Several new and traditional plant fibers have recently been discovered; Table 1.6 gives a brief idea of these fibers' mechanical and morphological properties.

Mineral fibers, primarily occurring or moderately altered fibers extracted from mineral deposits, are classified into the succeeding categories. Asbestos is a collection of mineral compounds that assuredly transpire as bunches of fibers in the ecological system. These are heat resilient, flame resilient, and poor conductors of electrical energy. Asbestos is a silicate mineral with silicon dioxide in its chemical pattern. There are various forms of asbestos: chrysotile and amphibole. Amphibole is composed primarily of five fibroid minerals. Grunerite, fibrous riebeckite, fibrous tremolite, fibrous actinolite, and fibrous anthophyllite are the five types of asbestos. The second significant variety of asbestos discovered in construction products is fibrous amosite. It is occasionally known as "brown" asbestos. The native occurrence of magnesium and iron gives it a specific color. It was commonly utilized in insulating materials as a flame decelerant (Kelse & Thompson, 1989). Crocidolite, commonly referred to as bluish asbestos, is the type of asbestos less commonly employed for commercialization (Gualtieri & Tartaglia, 2000). Tremolite is formed through the metamorphic rocks of dolostone and quartz-rich deposits. The color of magnesium silicate tremolite is milky white, but when the ferrous component increases, the color changes to deep green. It is poisonous at elevated temperatures and converts to diopside (Orden, 1964). Actinolite, for example, is originated from the gibberish word *aktis*, which means "ray" or "gleam." Actinolite is mainly encountered in metamorphisms such as aureoles and hard intruding igneous rocks. Metamorphic rocks are magnesium affluent, and dolostone shales produce anthophyllite. White asbestos is a kind of serpentine asbestos. It's a sheet silicate mineral that's incredibly soft and rubbery. It is one of the most often utilized types of chrysotile. This fiber is made up of long cylindrical tubes that are extremely strong (Orden, 1964).

TABLE 1.6

Mechanical and Morphological Properties of Natural Fibers

Natural Fiber	Mechanical Properties			Crystallinity Properties		Reference
	Tensile Strength (MPa)	Tensile Modulus (GPa)	Elongation at Break (%)	Crystallinity Index (%)	Crystallinity Size (nm)	
Ferula communis	475.6 ± 15.7	52.7 ± 3.7	4.2 ± 0.2	48	1.6	(Seki et al., 2013)
Hierochloe odorata	105.73 ± 35.42	2.56 ± 0.98	2.37 ± 0.95	63.8	–	(Dalmis et al., 2020)
Corypha taliera	53.55	0.451	–	62.5	1.45	(Tamanna et al., 2021)
Calamus manan	273.28 ± 52.88	7.80 ± 1.70	9.40 ± 3.67	48.28	1.9	(Ding et al., 2021)
Castor fiber	356 ± 23.87	34.931 ± 0.45	1.02 ± 0.053	48.88	4.8	
Artichoke	201	11.6	–	56	–	(Cabral et al., 2018; Scalici et al., 2016)
Arundo donax	193.47	10.4	1.53	66.7	–	(Abhemanyu 2019; Scalici et al., 2016)
Sansevieria ehrenbergii	278.82	9.71	2.81	52.27	–	(Sathishkumar et al., 2013)
Arenga pinnata	190.29 ± 46.77	3.69 ± 0.54	19.6 ± 67	57.1	–	(Bachtiar et al., 2010; Saputro et al., 2017)
Acacia arabia	71.63	4.21	1.3	51.72	15	(Dawit et al., 2020; Manimaran et al., 2016)
Abaca	400	12	3–10	68.7	–	(Azwa et al., 2013; John & Anandjiwala, 2008)
Roselle	147–189	2.76	5–8	76.2	–	(Athijayamani et al., 2009; Kian et al., 2018)
Phormium tenax	770	23.89	5.04	–	–	

(Continued)

TABLE 1.6 (Continued)
Mechanical and Morphological Properties of Natural Fibers

Natural Fiber	Mechanical Properties			Crystallinity Properties		Reference
	Tensile Strength (MPa)	Tensile Modulus (GPa)	Elongation at Break (%)	Crystallinity Index (%)	Crystallinity Size (nm)	
Borassus flabellifer	50.9	1.21	41.2	27.46	–	(Reddy et al., 2009; Singh et al., 2021)
Cortaderia selloana	20 ± 1.0	8.88	–	22	–	(Khan et al., 2020)
Coir	95–230	3–6	15–51	50.8	3.98	(Dittenber & GangaRao, 2012; Manjula et al., 2018)
Pennisetum purpureum	73	5–7	1.4	76.2	–	(Ridzuan et al., 2016; Sucinda et al., 2020)
Dracaena reflexa	829.6	46.37	2.95	57.32	19.01	(Manimaran et al., 2019)
Sida cordifolia	703.95 ± 25.73	42.84 ± 2.1	2.89 ± 0.24	56.92	18	(Manimaran et al., 2019)
Grewia tiliifolia	65.2	4.56	1.6	8.8	–	(Jayaramudu et al., 2010)
Heteropogon contortus	476 ± 11.6	48 ± 2.8	–	54.1	–	(Hyness et al., 2018)
Aristida adscensionis	–	–	–	58.9	11.5	(Manimaran et al., 2020)
Cyperus pangorei	196 ± 56	11.6 ± 2.6	1.69	41	–	(Manimaran et al., 2019)
Hibiscus sabdariffa	–	–	–	51.72	–	(Manimaran et al., 2019)
pongamia pinnata L	322	9.67	2.09	45.31	5.43	(Umashankaran & Gopalakrishnan, 2021)
Lagenaria siceraria	257–717	7–42	1.38–4.67	92.4	7.2	(Nagappan et al., 2016)
Lygeum spartum L.	64.63–280.03	4.47–13.27	1.49–3.72	46.19	–	(Belouadah et al., 2015)
Conium maculatum	327.89 ± 67.41	15.77 ± 0.13	2.67 ± 0.53	55.7	8	(Kılınç et al., 2018)

1.5   EXTRACTION OF NATURAL FIBERS

The fiber extraction procedure is a crucial stage in determining the quality and quantity of fiber. The fiber extraction process is determined by the characteristics of the fiber and its applicability. Depending on the root and source of the natural fibers, different extraction procedures are used. Wool, for example, is removed using physical labor and then rinsed to eliminate contaminants. The bamboo fiber is first extracted through the steam explosion method, followed by crushing grinding and water retting. Silk from a cocoon is removed by gently heating the cocoon in a moderate soapy mixture. Tranquilizing the spider and twisting it inside out to reveal the spinnerets yields silk. The wool and silk from animals are then extracted and purified using a sponge and a magnifier (Vinod et al., 2020). Additional strangling methods for silk extraction from the cocoon described by the investigators are water vapor stimulation, sunlight revelation, and dry thermal exhibition (Aznar-Cervantes et al., 2019). Plant fibers can be extracted in a variety of methods, similar to animal fibers by detaching, disintegrating, and deteriorating impurities, waxy components, pectin, and other sticky compounds. Retting separates and removes fibers from non-fibrous cells of crops and plants (Mohamed et al., 2018). The productivity, purity, morphology, chemical constituents, and characteristics of the fibers are determined by the retting process. Two important considerations for retting are to totally distinguish fibers from pectin and impurities. Also, it stops the retting operation at the appropriate moment to avoid excessive retting (Mohamed et al., 2018). As a result, ensuring the purity of fibers requires a constant scanning retting method. Microbial retting, physically retting, mechanized retting, chemically retting, protein retting, and enzymatic retting are all forms of retting. Hydro retting and field retting techniques were the most often used retting procedures in the past. The crop portion requires 2–3 weeks to break down into separate fiber layers using these procedures (Vinod et al., 2020). Alternative procedures such as physical, biological, protein, and chemical processes have been promoted because they are labor-intensive. The microbial population and wetness of the plants during the retting method enables a vast portion of the cell tissues and binding compounds that cover the fibres to be disintegrated, allowing specific fibers to be separated from the plant (Hyness et al., 2018). When utilizing field or hydro retting, the response time must be thoroughly assessed since inadequate retting might make it hard to separate specific fibers or reduce fiber strength (Tahir et al., 2011). Researchers suggested two easy and ecologically friendly extraction procedures for agave fiber. The fiber leaves were immersed for 90 days at a level of 0.3–0.4 m in the ground during the first procedure. The fibers were soaked in a vessel with water for 1–2 weeks in the second technique (Figure 1.8).

1.6   VARIOUS EXTRACTION METHODOLOGIES

The fiber bunches are extracted from the cultivated shank during the retting procedure, commonly termed *degumming*. Numerous retting procedures have been employed until present; other very conventional and quite extensively utilized approaches, including water and dew retting (Tamburini et al., 2004), are centered on

FIGURE 1.8 Water retting process.

bacteriological retting. Retting can also be done mechanically, physically, chemically, or enzymatically. The latter is quite intriguing, but it has yet to be implemented on a large scale. Figure 1.9 depicts a systematic structure of all retting techniques. Bacteriological or biological retting is a well-known and widely used retting technique. Dew and water decanting are the two most common varieties of biological retting. Pectin enzymes released by native microorganisms perform both functions. Cultivated plants are finely disseminated out for 20–70 days in areas during dew retting, also known as field decanting. Microbes, primarily cylindrical fungi or aerobic pathogens found in the ground and on plants, target non-cellulosic cellular processes at this time, eliminating enzymes and hemicelluloses from vascular tissue and the intermediate lamellae but not cellulosic components. The colonizing fungus

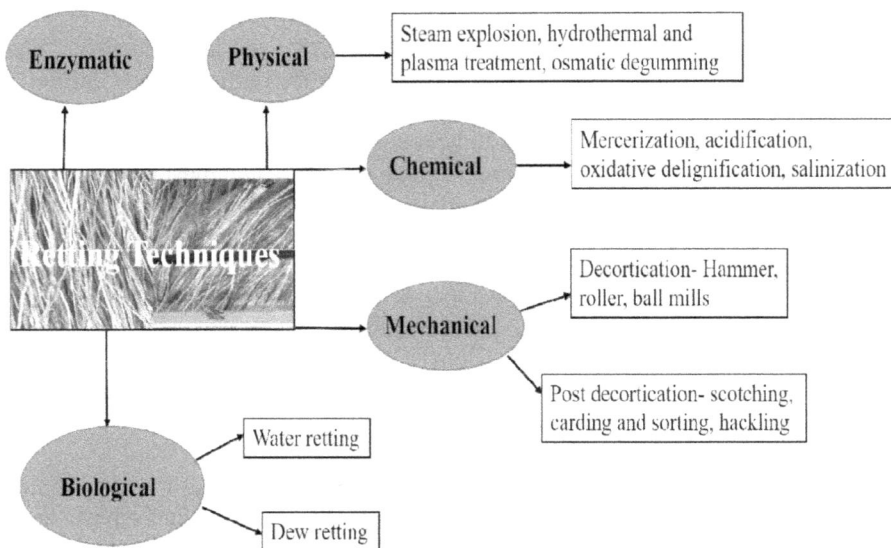

FIGURE 1.9 Retting methodologies.

has a greater degree of pectinase operation and the ability to permeate the cuticular layer of the shank during this phase, resulting in fiber bunches being split into shorter bunches and isolated fibers. *C. cladosporioides, C. herbarum, Mucor sp., Aspergillus, Chrysonilia sp., Verticillium sp., Rhodotorula sp., Geotrichum sp., and Penicillium sp.* were among the microorganism groups recovered from dew-retted plants (Ahmed & Akhter, 2001; Ribeiro et al., 2015). Various fungi like *Cladosporium herbarum, Fusarium lateritium, Epicoccum nigrum, Fusarium equiseti,* SC Orange (yeast) *Aureobasidium pullulans, Trichoderma virens, Rhizomucor pusillus, Fusarium oxysporum,* and *Rhizopus oryzae* were isolated during flax dew retting (Booth et al., 2004; Xiao et al., 2008). Due to the reduced price, field decantation is presently the most widely used procedure for the commercialization of bast fibers, primarily flax and hemp. Consequently, the design is only applicable to areas where the environment is conducive to fungal growth. Furthermore, like water decantation, the fiber characteristics are frequently weak and irregular in contrast to other procedures. Inadequate retting and excess retting have dangers: they can make extraction difficult or degrade the fiber (Jankauskiene et al., 2015). Cellulosic enzymes released by the bacteria or fungi, such as cellobiose, can destroy the fibers if exposed for a longer time.

To enhance the integrity of the fibers, the retting operation must be closely monitored. Other disadvantages of this procedure include the utilization of fields over the number of days during retting and the existence of a result polluted with bacteria and fungi. Researchers have been looking into using fungi in a better-governed setting to separate natural fibers over the past number of years in artificial field decantation. The variety of microorganisms, temperature, and modification length are all regulated criteria to deliver retting treatments that are less expensive, more effective, and ecologically sound. Fungi like *Hanerochaete chrysosporium* can degrade lignin chemicals from natural fibers, increasing the mechanical characteristics of the subsequent natural fiber strengthened polymer composites (Pickering et al., 2007). One new survey showed that using bespoke enzymatic mixes (Texazym® SER) improved the dew retting mechanism by enzymating it. INOTEX catalysts are applied on the field before the withdrawal or during the initial 3 days of dew retting. This approach has been discovered to boost flax fiber yields beyond 50%. In this situation, microorganisms combined with gentle mechanical operations can substitute severe and energy-consuming processes such as Laroche cottonization (Antonov et al., 2007).

Stalks are immersed in aquifers, usually in enormous tanks, rather than lakes or pools, in water retting, which was popular half a century ago. The water seeps into the center side of the shank, rupturing the cultivated plants' external coverings. Vast cell structures, pectin, and different waxy components around the fibers that are soaked in rivers, pools, tiny ponds, or man-made water reservoirs or tubs are broken down by bacteria existing within the plant's shank (Figure 1.10) (Vinod et al., 2020). Water gets inside the center stalk section of the fiber throughout the treatment, which lasts 1–2 weeks, by shattering the outer surface, causing enhanced hydrophilic nature and establishing a cellulolytic microbial population. The modification time is determined by the water variety, temperature, and microbial inoculation. The initial stage of the procedure involves the development of aerobic

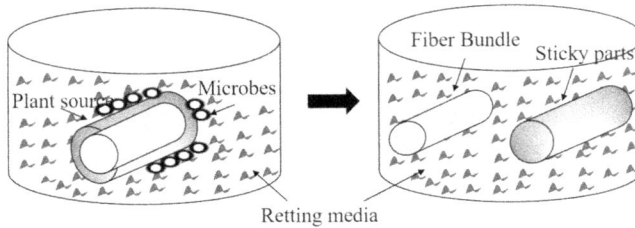

FIGURE 1.10 Microbiological retting process.

bacteria that absorb the majority of the oxygen levels, resulting in a suitable situation for the development of anaerobes (Ramamoorthy et al., 2015). Grampositive bacteria (Bacillus) from the family of Bacillaceae, its species like *B. stratosphericus*, *B. polygoni*, *B. subtilis*, *B. mesonae*, *B. macquariensis,* and *B. mesentericus* have all been discovered to be effective in the aerobic propagation fermentation (Tamburini et al., 2004). Various *Fusobacterium* species, including *F. canifelinum, F. gonidiaformans, F. naviforme*, and *F. perfoetens,* have been recovered from water decantation during the anaerobic propogation (Munshi & Chattoo, 2008; Tamburini, 2004). Although water retting generates better fibers than field retting (Amaducci & Gusovious, 2010), the water decantation procedure has ecological consequences due to the intake and impurity of excess water and energy (Van der Werf & Turunen, 2008). Since water sources are getting extremely rare, finding a new or improved method of water retting will be essential in driving water shortages and contamination minimization. Enzyme retting, for instance, has been investigated as a potential substitute for present retting procedures. Furthermore, Zhang et al. (2008) investigated the feasibility of replacing water with saltwater from the sea, which is an infinite and plentiful commodity. They showed that saltwater decantation modification provides acceptable retting effects and efficient ligninolytic isolates like *P. aeruginosa* and *O. antrophi* species. Artificial water decantation has also been utilized to generate homogenous and pure elevated characteristics fibers within 1 week using hot water and microbial inoculation (Sisti et al., 2016).

   Enzymatic modification, commonly known as bioscouring, is a variation of water in which decaying enzymes are administered straight to container water or in a bioreactor. In the context of time efficiency, environmental friendliness, and convenience, this technology is a potential substitute for standard retting procedures. Enzymatic retting might last anywhere between 10 to 48 hours. The high-intensity intake and non-recyclability of enzymes are the key factors that impact the process's economic rationality (Tahir et al., 2011). Pectinases are the most common enzymes used in decantation, which separates fibers from diverse tissues. Pectic enzymes have become increasingly important as bioengineering methods have developed because of their wide range of applications, including fruit juice separation and purification, cotton bioscouring, dewaxing of natural fibers, wastewater modification, vegetable oil exploration, tea and coffee pasteurization, paper discoloration, and poultry dietary supplements and in sectors of alcoholic appetizers and food. Pectic enzymes are a diverse set of catalysts that hydrolyze pectic compounds, which are typically found in plants. Because they aid in cell membrane assembly and weakening specific plant tissues throughout development and preservation, they are extensively disseminated in upper plants and microbes (Mohnen, 2008). They also contribute to environmental equilibrium by allowing unwanted plant materials to decompose and be recycled. Other prominent symptoms of pectic enzymes include soil pathogenicity and the withering of plant-based foods. Researchers reported that pectin depolymerase function is significant throughout the retting procedure (Evans et al., 2002) while polygalacturonic transeliminase has also been valuable decanting bast stems (Akin et al., 2007). Polygalacturonic transeliminase initiates a non-hydrolytic decomposition of pectates and pectolyase by a trans-elimination splitting of the pectin polymer. In contrast, pectin depolymerase facilitates the pectic acid chain. Bacterial cultures like *Bacillus sp.* and *Pseudomonas sp.* (Kapoor et al., 2001), yeast and fungal cultures like *Aspergillus niger, Rhizomucor sp., Saccharomyces sp., Schizosaccharomyces sp.*, and *Penicillium chrysogenum* (Blanco et al., 1999) can all produce pectinases. The majority of marketable pectinase formulations come from fungi. Enzymatic retting using pectinases can produce greater strength recyclable fibers with varying aspect values, which can be used in new polymers. The following are some of the most widely used industrial enzymes (Jayani et al., 2005). Novozymes' Viscozyme® L is a multi-dimensional enzyme mixture that contains cellulose, polygalacturonase, β-glucanase, polyose, and endoxylanase, among other carbohydrases. It has been put to trial in the separation of flax fiber by many researchers (Akin et al., 2007; Bacci et al., 2010). Scourzyme® L is an alkali polygalacturonate lyase produced by Novozymes through bioscouring technology. It dissolves the pectic polysaccharides from the fundamental cell wall of fibers by not destroying the fiber structure (Ouajai & Shanks, 2005). Flaxzyme®, an industrial enzymatic solution produced by Novo Nordisk, was established particularly for enzymatic retting and contained polygalacturonase, polyoses, and cellulases (Akin et al., 2007). Inotex produced Texazyme® BFE for the elementary process of bast fibers by degrading pectic polysaccharides layers. It's a multi-dimensional constituent product that doesn't have any cellulase action (Foulk et al., 2008). Because the action of individual enzymes varies drastically depending on acidic or basic nature, temperature, and enzymatic concentration. In addition, chelating agents and surfactants are frequently used in compositions to increase action. The importance of calcium chelating agent, such as ethylenediaminetetracetic acid (EDTA), in improving decantation, notably in eliminating cuticle layer covering the epidermis part of fiber and fiber bunches, is well characterized (Akin et al., 2007). In particular, EDTA disrupts cell membranes by weakening junctions connecting calcium and pectin lyases. According to Foulk et al. (2008), enzymatic retting can be utilized to design fibers bundles with certain attributes, like toughness and smoothness, and for specific purposes by knowing the constituents of the enzyme solution. Decantation with highly potent pectinase, for example, pectin depolymerase, protects toughness, which is crucial for many purposes. Moreover, for purposes in which the fibers will be shorter, like injection molding, a combined enzyme formulation comprising cellulase could be employed. Additionally, it has been discovered that enzyme preparations including cellulase, such as Pectinex®, can damage the bast fibers, because the junctions of the fibers are highly susceptible to the enzyme's attack (Foulk et al., 2008). As a result, the decantation formulation is determined by the end purpose.

The **mechanical retting** procedure is mechanized in comparison with the traditional handling (Figure 1.11). The stems are initially separated by cutting, which involves passing them through grooved rollers that shatter and crack the woody center into small segments (hurds). Scutching (done historically with panels and hammers) separates the hurds and cracked fine fibers from the residual continuous fibers by gripping the fiber bunches among rubber straps and carrying them former rotating drums with protruding bars that strike the fiber bunches. Lastly, dense fibers are split by pushing the fibers via a succession of crushers of increasing refinement to cleanse and straighten the continuous fibers and segregate the leftover tow and waste in the hackling (this was previously accomplished by dragging the fibers via a pair of connectors). The quality of the fibers is retained in current mills by untangling and straightening the fibers with no length reduction. Extraction, which can be done using a hammer or roller mills, is an alternative technique for mechanically separating the fibers (Deyholos & Potter, 2014). One or more contemporaneous drums revolving with hammers protruding crosswise from the drum edge strike the stems (Figure 1.12) until the segregated hurd and fiber pieces can easily transit through specific screens put within the machine. Cylindrical-shaped perforated rollers are arranged in such a way that they break the shank stalks with causing the least harm to the fiber (Figure 1.13). The advantages and disadvantages of the two techniques vary: although hammer crushing has a better performance capacity, roller crushing has considerably better length precision, generating uniform fibers and will protect the fiber quality, avoiding fractures or interferences. The variety of fiber, its eventual utilization, and subsequent treatments all influence

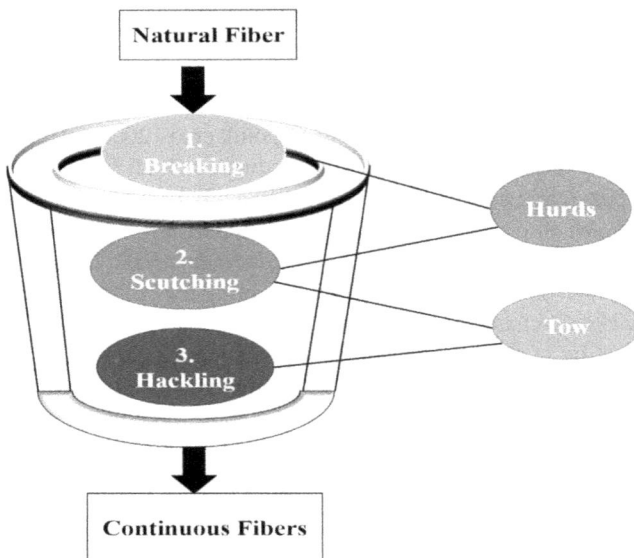

FIGURE 1.11   Mechanical retting process.

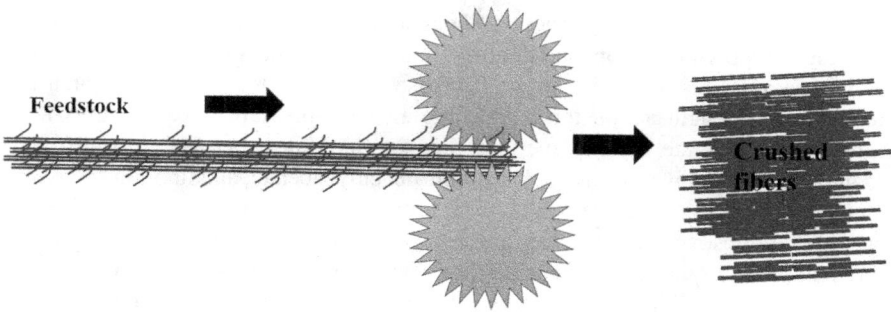

FIGURE 1.12 Fiber crushing through hammer mill.

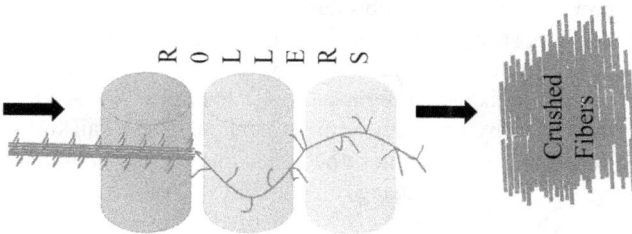

FIGURE 1.13 Fiber crushing through roller mill.

which mechanical decantation method is best. As a result, it is constantly being investigated with these numerous elements in consideration. Mechanical retting, on the other hand, is recognized to destabilize fiber cell membrane constructions, resulting in misalignments, bondage bands, all of which have a detrimental effect on tensile, flexural, impact properties, and may even hinder composite quality (Hänninen et al., 2012; Jankauskiene et al., 2015). Nevertheless, it has been identified that the magnitude of mechanical retting flaw is closely reconnected to preceding decantation modification (Van der Werf & Turunen, 2008). Several other authors have discovered that fibers can be extracted, avoiding significant degradation, and developed with greater quality even though no treatment methods were performed (Bacci et al., 2010; Kengkhetkit & Amornsakchai, 2012). These results are significant and remarkable: undoubtedly, eliminating pre-separation enzymatic or bacteriological modifications should minimize fiber performance, variability, productivity concerns, and the time required for the area to be ready for the subsequent crop, all of which should lower marketing prices. Furthermore, novel extraction technologies can be established to get the greater performance of fibers while lowering manufacturing costs by shortening the rates and enhancing the quantity of treated natural fiber.

Techniques involving electromagnetic radiation with elevated pressure and temperatures might be classified within the **physical retting** of fibers. Steam explosion works on automated hydrolysis procedures involving the utilization of dry steam at elevated pressure resulting in rapid decompression, resulting in significant lignocellulosic structural destruction, hemicellulose portion degradation, lignin constituent depolymerization, and defibrillation. Fiber fineness improves with higher decompression rates, but fiber length decreases. Elevated temperatures lighten the product, and strength action throughout the elevated-pressure release causes fiber segregation: moderately depolymerized lignin turns almost permeable in different organic solvents, such as acetic acid, acetonitrile, ethyl acetate, alkaline solutions, and chloroform, whereas cellulose, which is far more resilient to hydrolysis unlike pectic and hemicellulose polysaccharides, maintains its integrity. The steam explosion modification is a quick and adequately supervised procedure with cheap costs and quite adjustable modification parameters that are ideally suited for the preparation of various fibers, such as those that have not been retted before (Zhang et al., 2008). It's been used on banana pseudostem fibers (Sheng et al., 2014), pre-retted hemp fibers (Thomsen et al., 2006), and coir fibers with tremendous results. The process can be regarded as a preparation for the subsequent decantation operations (Zhang et al., 2008). It can be done post-alkali washes, bleach, and occasionally acidic hydrolysis to decompose the hemicelluloses, impurities, waxy, and lignin components (Sheng et al., 2014). To favor the breakage of lignocellulose linkages, the steam explosion is commonly accompanied by an alkali semi-immersion. The alkali-lignin solvent becomes more soluble, and hemicellulose becomes more water-soluble due to the interaction (Sheng et al., 2014; Zhang et al., 2008). Another intriguing physical technique for extracting fibers is the hydrothermal approach, which involves degrading lignocellulosic components such as hemicellulose, wax, pectin, and lignin with freshwater at elevated temperatures and pressure. This method was used on hemp fibers (Zhang et al., 2008) and flax fibers (Stamboulis et al., 2000) and observed that the curing procedure enables fibers to be easily separated from the stalk by an easy cracking and scutching operating condition. Fiber bunches, instead of individual fibers, are created using such techniques, and they have superior barrier properties and a slightly greater consistent strength. Another recently discovered retting method is osmotic dewaxing (Konczewicz & Wojtysiak, 2015). The dewaxing mechanism relies on water diffusing into the stalk, wherein lengthy bunches of cellulosic fibers are grouped in chunks with carbohydrates, primarily pectins. The extremely absorptive pectic polysaccharides expand their capacity multiple times, resulting in a significant enhancement in hydraulic pressure within the stalk and pushing the epidermal layer. Because circumferential strain is greater than longitudinal strain, epidermal fractures develop vertically rather than fracturing and reducing the fibers. The industrial liquid is treated to adequate purification, which further contributes to retrieving pectic polysaccharides for more utilization in the cosmetic sectors because the pectic polysaccharides are diluted and dissolved (along with other bast components) in freshwater. Fibers with high tensile strength, dimensionality, and smooth texture are produced using this osmotic technique. Physical treatments are undoubtedly in their infancy, with only some studies describing their application. They are, nevertheless, distinguished by high speed, adaptability, and process suppleness, making this type of retting deserving of future exploration (Jayani et al., 2005; Thomsen et al., 2006).

Chemical procedures are often preferred to water decantation because they generate fibers with better-persistent quality independent of climatic conditions and generally in less duration. Based on the fiber orientation, the retting technique to be used, and the ultimate purposes, various chemical retting modifications can be performed to the fibers. Mercerization is the most often utilized chemical technique for eliminating hemicelluloses. It is normally done using sodium hydroxide (NaOH), which is supplied as an aqueous mixture at a changeable concentration of 1–25 wt%. When it comes to coir fibers, the sodium hydroxide impact is often confusing, and the published results are mixed: although the alkaline pretreatment appears to improve elongation at breakage and surface quality (Silva et al., 2000), whereas raising the sodium hydroxide concentration improves the mechanical strength, early modulus, electrical characteristics, and heat resistance; overuse of alkaline solution can reduce the fiber tensile strength (Gu, 2009). A mixture of caustic soda, washing soda, and sodium sulfide, as well as brief immersion procedures (approximately 2 hours), have been widely implemented on pure coconut fibers to reduce fiber degeneration, leading to a decrement in size, fineness, and flexure strength. The equivalent approach applied to kenaf: alkaline decantation produces tougher, elastic, and less fragile fibers (Amel et al., 2013); however, the alkali content affects fiber surface morphology due to greater alkaline content enhances microvoid volume percentages, lowering the mechanical properties of the fibers (Konczewicz & Wojtysiak, 2015). Moderate circumstances reduce kenaf fiber deterioration: using a low concentration like sodium sulfite results in lighter fibers, finer and gentler structures, and greater tenacity (Umoru et al., 2014). Regarding hemp fibers, similar reagents were examined. It turned out that the mixture of caustic soda and sodium sulfite generates the chalky color and finest size fibers, but not the superior spinnability (acquired after acidification). Researchers emphasize the importance of using mild alkali pretreatment on flax fibers, demonstrating that immersion for 1 minute in a diluted caustic soda solution improves adherence property among flax fibers and epoxy resin, enhancing the tensile and flexural strengths of the produced materials (Van de Weyenberg et al., 2006). Another way to eliminate lignin is through oxidative deterioration, which is typically utilized when the objective is to completely degrade and consume the lignocellulosic substance for the purpose of recovering the residues for utilizing in the chemical sectors or as energy. In these circumstances, peroxides are used to achieve deterioration, and the raw natural fibers are highly diverse, including wheat straw (Klinke et al., 2002), mulberry (Cong & Dong, 2007), and poplar wood (Chang & Holtzapple, 2000). The modification of bagasse with hydrogen peroxide significantly enhances the fiber predisposition followed by enzymatic hydrolysis (Azzam, 1989): the identified findings demonstrate that 2% alkaline hydrogen peroxide absorbed approximately half of the lignin content and complete hemicellulose constituent of bagasse within 8 hours. A comparable bleaching procedure on kenaf fibers (Amel et al., 2013) eliminates the hemicellulose in 60 seconds (delignification was previously done by alkali maceration) while keeping the fiber width and enhancing crystalline nature. A detailed examination into which oxidizing agent, including peroxide, sodium hypochlorite, and caustic soda is effective for delignifying the coconut fiber. According to the earlier research, peroxide is perhaps more efficient in separating wax, impurities, and fatty acids from the fiber coverings. In contrast, sodium hypochlorite/caustic soda treatment is the most efficient at eliminating hemicellulose and, as a result, exposing cellulose (which decides the intrinsic hydrophilicity and is helps in functionalization) (Thomsen et al., 2006). A novel way for preparing fibers for subsequent modifications is to use acidic chemical treatment. In this scenario, the possibility of degrading the cellulosic portion is significant, and managing the production waste is also tricky. As a result, such a technique is uncommon, with just a few works describing it. When the primary objective is total cellulose breakdown, the use of extremely strong acids such as sulfuric and hydrochloric acids to modify lignocellulosic fiber is extensively described (Sun & Cheng, 2002). Strong acids were utilized in varying concentrations more commonly in methods that exploited fewer severe circumstances, although they usually resulted in glucose recovery (Zhang et al., 2008). Chemical retting is undoubtedly a viable alternative to microbiological water retting, subject to meteorological concerns and significant yield reductions. The former is unaffected by climate variations and can maintain fiber quality. Therefore, due to ongoing waste problems and a high risk of fiber degradation, such chemical treatments are today less appealing than in the earlier days.

1.7 HISTORY AND PRESENT SCENARIO OF NATURAL FIBER POLYMER COMPOSITES

Approximately 7,000 years past, three Egyptian technologists joined each other and blended straw and clay to build bricks, which gave birth to the theory of composites. Mongols in Iraq placed pieces of wood on their side jointly at different angles to make plywood at roughly the same time. Traditional Mongolian shafts were gradually becoming supplanted by composite wood, tendon, and timber. Flash forward to industrialization when polymers and plastics became popular. Because of its great enhanced power tolerance, it has a wide range of functions in different fields. Glass fiber polymers, which were first developed in the 1930s, are still widely used in the modern composite industry. Carbon fiber's development in the 1960s opened up a slew of new possibilities for this composite industry (Pickering et al., 2007). Academicians and researchers have produced and classed composites on a variety of grounds throughout the past several decades. Unfortunately, in the early 1920s, owing to inherent non-biodegradability, the utilization and destruction of synthetic fibers, particularly glass or carbon-based composites, has become increasingly crucial in the face of rising ecological awareness and governmental requirements for their reprocessing (Konczewicz & Wojtysiak, 2015). The removal of non-disposable composite (NDC) materials following their specified service life span has become a serious and costly concern as fabrication and massive quantity use has increased. The majority of composites are subsequently burnt and disposed of in the environment. Combustion, on the other hand, is both costly and troublesome. The emitted gases may result in more contamination. Individuals must be motivated to experience a different efficient method of disposing of since landfill regions are shrinking and additional environmental concerns arise due to these existing disposal methods. Furthermore, because composites mix two incompatible elements, they are challenging to reutilize and reprocess, despite various efforts to provide it. Wood filler and natural fibers were employed as strengthening phases in initial composites to minimize composite manufacture and produce lighter products.

Composite materials consisting of cellulosic fibers mixed with thermoset matrix material were first described in the late 1970s (McLaughlin & Tait, 1980; Zadorecki et al., 1986). Later, thermoplastic matrix materials reinforced with wood filler were observed in the late 1980s, and this was expanded to greater mechanical strength natural-fiber-strengthened thermoplastic composite materials (Flix & Gatenholm, 1991; Myers et al., 1991). The scientific research of entirely green composite materials has been sponsored by authorities in India, Germany, and Japan. Companies have increased their attempt to include green material into their offerings. Toyota Motor Company, for instance, has employed kenaf fiber-based green composite materials to substitute plastic and metal-supported vehicle components. Significant collaboration between industry, research centers, and ministries is required to promote green material–based composites. Green production composites are projected to have mechanical qualities equivalent to and sometimes better than glass fiber reinforcement, demonstrate morphological and chemical integrity throughout retention and usage, and helps in effortless microbiological and ecological decomposition following the burial, all while posing no environmental risk (Reddy et al., 2009). The construction of natural fiber polymer composites at affordable rates, which can also acquire social and media acceptability, is one of the ongoing progress trends in entirely sustainable composite materials. The implementation price will most likely fall as the invention of inexpensive and better biodegradable polymers, market conditions, and mass manufacturing continue. But, on the contrary, existing and anticipated research should focus on creating new multifaceted biodegradable composites made up of diverse polymers and fibers in different types, such as woven and non-woven fabrics. Optimal information of visualization and interpretation of mechanical characteristics in current biodegradable composites should be summarized to a file with the goal of comprehensive strategy deployment and manufacturing. However, since natural fibers have such a complicated structure, additional research on fiber and composite characteristics is needed to create assurance in their possible development (Sheng et al., 2014). Biodegradable composites with superior physio-mechanical qualities are no longer a general idea. Biomaterials with better qualities can be made using correct molding procedures and fiber modifications to meet various needs. Chemical treatments, cross-linking, binding compounds, and other fiber modification procedures can increase fiber characteristics and fiber/resin compatibility. Adding nanoparticle and micro- and nano-crystalline cellulose to polymeric materials is another way that can improve the mechanical characteristics of composite material by increasing the functional groups of fibers and polymers. This process also enhances the thermal attributes of some polymers and, in some situations, generates moisture (Zhang et al., 2008). In order to create higher strength of the composite materials in the context of the difficulty of substituting traditional carbon or glass fibers, certain techniques have been beneficial: effective natural fiber pretreatment, polymers modification, and determination of optimized process parameters. Covering the fibers with plasma polymerization and many other processes, for instance, is found to reduce moisture susceptibility as enhance tenacity. Furthermore, the eventual manufacture of sophisticated biodegradable composites will be aided by the rapid growth of advanced fibers with greater strength based on cellulose. Nexia

Biosystems conducted the most important study, in which spider silk proteins were implanted in sheep (Turner & Karatzas, 2004). After that, the spider silk is produced in soy milk, refined, and woven into fibers.

Hybridizing fibers in the present scenario most efficient method to compensate for one fiber's deficit to accomplish natural fiber polymer composite sustainability, low cost, and increased productivity (Guna et al., 2018; Swolfs et al., 2014). Hybridization technologies allow for greater freedom in selecting fibers for the most significant manufacturing qualities based on application objectives. Hybridized fibers polymer composites are often made by combining natural fibers with some other natural or human-made fiber(s) with outstanding properties, including physio-mechanical performance, chemical constancy, low toxicity, non-computability, heat resistance, and thermally or acoustic insulation. Fiber hybridization is typically achieved by mixing multiple types of fibers together or interlaminating components of various fibers before adding them to the polymer (Sanjay & Yogesha, 2017). The physio-mechanical properties of natural fiber reinforced polymer composites can be considerably improved by combining natural fibers with human-made fibers such as glass, aramid, nylon, Kevlar, carbon fibers, and nanotubes. Hybrid composites with jute fibers with glass fiber (1:1 w/w proportion) enhanced the tenacity of jute fiber polyester by three times. Similarly, hybridizing jute fibers with carbon fibers enhanced the tenacity by ten times and impact by five times of jute fiber reinforced polyester resin, respectively. The volume and inter breakage estimates of the sisal and glass fibers reinforced with polypropylene hybrid composite can be configured by regulating the injection molding operating conditions to mitigate composite inconsistencies because of shrinkage and voids. Thus, interviewing the material's optimum deformation specifications for below the lid parts used in the automobile sector (Sezgin & Berkalp, 2017). Fiber loading, hybrid proportion, fiber alignment (Senthilkumar et al., 2018; Shahzad, 2012), layering patterns (Yahaya et al., 2016), and inherent fiber properties all play a role in the hybrid composite material enhanced functionality. The hybrid-reinforced polymer composites' tensile and fracture strength increased by 2% for flax strengthened composites and 3% for glass fiber–strengthened composites. Hybridization has the potential to mitigate the inherent drawbacks of particular bioproducts. Yusoff et al., for instance, found that varied layering arrangements of the hybrid kenaf, bamboo, and coir reinforcement in the polylactic acid polymer produced a wide variety of mechanical strengths (Yusoff et al., 2016).

Because of the large definite surface areas, greater crystal hardness, toughness, and preserved volume fraction, cellulosic nanoparticles such as nano-crystalline cellulose (NCCs) and cellulose nanofibrils (CNFs) in the present scenario is preferred as promising filler materials (Abitbol et al., 2016). The tenacity of 8.5 GPa and tensile modulus of 210–320 GPa were observed for cellulose nanocrystals (Moon et al., 2011). Micro- and nano-ascended fillers, like inorganic fibers, include a larger volume for bonding and adhering to the polymer, which is advantageous for physiological bonding at the junction, load shifting, and the composite's overall tensile and flexural strengths (Saba et al., 2014). On the other hand, inconsistency and agglomeration of nanocellulose in the polymer represent the most significant obstacles to producing superior-performance nano cellulosic material. Excellent thermo-mechanical and barrier performances can be attained with nano-sized fillers at low strengthening concentrations, often less than 10%. For instance, at 0.8 wt% fiber content, cellulose nanofibrils from softwood/ epoxy composite exhibit the highest mechanical properties (Saba et al., 2017). The use of a greater proportion of nanocellulose filler particles causes a number of issues, including nanofiber agglomeration, decreased fiber-matrix adherence, constrained composite integration throughout the crosslinking technique, deformities in the composite materials, inefficient load shifting, and, inevitably, breached composite productivity. Benhamou et al. (2015) discovered that the presence of hemicellulose, waxy, pectin, and lignin residues in cellulose nanocrystals improved the polymer's interfacial and barrier consistency. Likewise, Graupner et al. (2014) demonstrated that lignin loading improved Lyocell fiber adhesion and perceived tensile interface strengths in polylactic acid and polypropylene matrices. The advantages of using nanoparticles are: a) The inclusion of lignin decreased the viscous nature of the polylactic acid matrix; b) cleansed lignin particulates resulted in greater coverage of fiber sites for interfacial adhesion; c) chemical linkages between the fiber and polymer were enhanced through the van der Waals interactions (Swolfs et al., 2014).

Some other typical applications in the present scenario of natural fiber-reinforced composites (NFRPC) are: Flame-resistant material such as fiber-strengthened inorganic polymer composites have been broadly utilized to reinforce concrete for several decades. The current studies have added inorganic polymers with cementitious ingredients to build composites. The application of phosphorus cement reinforced composites substitute epoxy in natural fiber-reinforced composite structures improves structural strength (Komuraiah et al., 2014). NFRP films attached to the tension applied of concrete beams show a considerable increase in strength properties and stress-transfer capability, even though exposed to the aggressive environment (Manimaran et al., 2019). There has been a significant rise in the need for lightweight composites in the vehicle sector in order to improve fuel economy while lowering emissions. For instance, NFRPC is employed in the secure and convenient handling, and distribution of gaseous fuels like hydrogen, natural gas, and pressure vessels made up of natural fiber polymer composites are required (Omar et al., 2019). A discard truck utilizes a hydraulic structure comprising an actuator constructed of a telescoping pneumatic pump for the movement of earth particles. When a steel barrel is substituted with a carbon fiber–based epoxy polymer composite, the mass is reduced by 95%. When this telescoping cylinder was added, the entire hydraulic system was reduced by 50%. Because of its superior structural, optical, mechanical, and electrical features, NFRPC materials are being used in the aerospace industry to provide very resilient, heat-resistent, lighter materials for airframes structures (Barile & Casavola, 2019; Chand & Fahim, 2008). Natural fiber thermoset and thermoplastic coverings have the features needed for airframe sheets, such as fire and light resistance, simple reprocessing, and product discharge less expensive and lighter than standard face sheets (Booth et al., 2004). Although NFRPC materials have a wide range of usage in the aviation industry because of their superior physio-mechanical qualities and lightweight design, reprocessing is a challenge. Also, with biodegradability and reduced price, biocomposite products have opened up new opportunities in the aerospace sector (Rowell & Stout, 2007). The use of conductive fiber-based composites arrangement eliminates the need for distinctive wires for interconnected network transmitters for transmitting wireless signals. When power is added to the matrix, the fibers carry electrical energy to specific electronics equipment (Kelse & Thompson, 1989). NFRPC materials are utilized in dental and orthodontics in the biomedical sector because of their strength and nontoxicity. The development of lower-limb sporting prostheses has experienced remarkable scientific advancements. A novel NFRPC biomaterial substitutes the materials utilized during handmade cranial implants to restore bone of the skull and face deformities (Lazar et al., 2016). Protein immobilization, therapeutic implants and gadgets, current orthopedic therapy, and antibacterial material are examples of aramid fibers used in biomedical applications. Polyamide, often known as nylon, is a human-made polymer with outstanding mechanical properties utilized in implantation, and fibrous composites are used to make prostheses and sutures equipment. Chitosan and aramid hybrid increase the permeability of constructed composites for antibacterial usage (Kim & Lee, 2013). In tissue building, collagen and silk hybrid composites have a prospective use in regenerative medicine to repair lesioned cells after electrospun the composites, the toughness, the flexibility of the material increase, along with a rise in silk proportion (Kapoor et al., 2001). A fiber composites framework comprised of biodegradable synthetic polymers, poly (glycolic acid) (PGA), polylactic co-glycolic acid (PLGA), polyesters, polyurethane, gelatin, polyanhydrides, and elastin (PGE) can sustain dense cellular proliferation and supply massive amounts of cells. These have a wide range of applications in tissue regeneration, such as meeting design requirements for creating naturally derived biomaterial frameworks (Mengyan et al., 2006; Zhu et al., 2014).

1.8 ADVANTAGES AND DISADVANTAGES OF NATURAL FIBER–REINFORCED COMPOSITES

Natural fiber–reinforced polymer composites (NFRPC) have many benefits. They are superior to standard composite material in many ways, including higher specific toughness and specific mechanical properties, better wear resilience, improved reusability, high fatigue resistance, lower overall costs, increased impact uptake endurance, as well as being less poisonous (Mohammed et al., 2019). In particular, the benefits of their ingredients, especially natural fibers, manifested in such uses. Furthermore, the features and efficiency of goods made from biocomposite are primarily impacted by the components' specific characteristics, adaptability, and the fiber-polymer interfacial properties, which expand the possibilities for generating various products. Increased use of cellulose-based composites instead of synthetic fibers would have various advantages in terms of structural management, general durability, and an environmentally friendly manufacturing theme (Arun et al., 2021). In addition to their benefits, natural fiber polymer composites have some notable drawbacks. When contrasted with ordinary plastics, NFRPC materials have low water resilience, dimensional constancy, degradation heat, flame, ultraviolet, and microbiological resistance, limiting their potential uses (Vinod et al., 2020; Zhang et al., 2008). Natural fibers boost the polymer composite's mechanical performance, which improves as the fiber loading rises. However, as the proportion of fiber rises, the flexibility increases proportionally, and so does the chance of water uptake and unpleasant odor (Mallick, 2012). The poor interaction of natural fibers with polymers is another major restraining issue for using natural fiber–based polymer composites in high load-carrying, internal, and non-structural activities (Baillie, 2004). The degree of interaction among natural fibers and polymers impacts the composites' tensile strength, endurance, and efficiency. The long-term features of natural fiber polymer composites, such as slip and wear behaviors, persistence, and life span estimation, are unclear and a key issue (Nishino, 2004). As a result, various novel approaches and experimental procedures should be used to investigate the long-term effectiveness of NFRPC. The NFRPC materials are also developed for commercialization by fiber modification, polymer mixing, softener, and other filling additives, coating, and other sophisticated fabrication processes. Although natural fibers and polymers are less expensive than conventional counterparts, specific fabrication procedures necessitate a significant capital outlay as well as power consumption (Simsek et al., 2019). To overcome these restricting issues, specific essential characteristics and suggestions for future natural fiber polymer composite materials development should be stated. These are a) investigation, classification, and characterization of novel natural fibers are necessary for natural fibers to substitute dangerous synthetic fibers in a broader range of applications. Although animal fibers are considered to have greater mechanical qualities over plant fibers, there have been few types of research on animal fiber–based biopolymer composites; b) no separation technique has been recognized as a standard approach for a particular variety of fiber among the currently employed fiber separation techniques. It's also worth noting that there isn't a single standard procedure for making biopolymer hybrids (Omar et al., 2019). Setting up consistent techniques for composite processing reduces labor, energy, and input materials usage; c) because of its widely distributed sources and varied qualities, quality verification and the development of optimum natural fibers are required for the use of natural fiber polymer composites in many domains such as the transport and manufacturing fields; d) it's also crucial to learn more about the effects of strengthening enhanced natural fibers in polymer composites to forecast the best application and functionality (Jankauskiene et al., 2015). In addition, new surface treatment technologies must be investigated to implement economic and environmentally benign approaches; e) although NFRPC materials have demonstrated their use in automobile, aviation, digital, and biomedical applications, additional investigation in healthcare frameworks for wound healing, immune functions, and regenerative medicine is required. Natural fiber polymer composites have created significant commercialization for value-added commodities in the automobile industry; f) one of the most significant disadvantages of bio-based products is the uncertainty surrounding their endurance. The uncertain persistence of natural fiber–based polymer composite materials, combined with their low structural rigidity, is a significant barrier to their adoption in the automobile industry (Réquilé et al., 2018). The majority of knowledge about the endurance of composites comes from small-scale trials. The behavior of composite materials in situ has not been extensively studied to confirm laboratory results and analysis. Newly developed strategies for monitoring, evaluating, and controlling the endurance and degradability of natural fiber polymer composites should be investigated; g) composites' reusability is a potential subject to investigate. Future research on the reusability of NFRPC is required to recycle the composites in some form instead of allowing them to biodegrade after their intended usage; h) the application of nano-technology to NFRPC materials has shown a plethora of possibilities for improving their properties. Nanocrystalline cellulose is tougher than aluminum and harder than steel, according to investigations. As a result, the synthesis of nanofibers from natural fibers is a fascinating platform for natural fiber–based polymer nanocomposite construction materials (Malkapuram et al., 2009).

CONCLUSION

There has been a significant enhancement in industrial awareness in polymer composites over the past few years. Natural fibers, when modified, have demonstrated their capacity to outperform synthetic fibers due to their vast accessibility, relatively inexpensive, and environmentally beneficial properties. Kenaf fiber has proved to have the most durable mechanical qualities of all-natural fibers, whereas jute fiber has been shown to have the most strength and excellent durability. Because natural fibers are utilized for composite fabrication and other reasons, it encourages cultivation and horticulture, which minimizes air pollution by lowering greenhouse gas emissions and improves soil health. Natural fiber–reinforced polymer composites are recyclable, non-corrosive, environmentally benign, compostable, light, economical, and simple to make, with greater mechanical strength and toughness that can be adjusted to meet a variety of performance needs. Chemical constituents, physicochemical state of fiber and polymer, fiber treatment techniques, fiber length, the amount of additives and softener, the strategy of reinforcements, topology and alignment of fiber in polymer, structural issues, the connection among fiber and polymer, and ecological conditions all influence the functioning of natural fiber–reinforced polymer composites (NFRPC). Integrating several fibers in a homogeneous material to create greatly evaluated polymer composites is another frequently applied composite fabrication strategy. The composites were found to be consistent with the automobiles, aviation, and structural, pharmaceutical, and food industries and have a wide range of uses. Additional investigation into natural fiber–reinforced polymer composites can improve petroleum-based polymers in the coming years. When these composites grow more robust and perform better, there is the potential for emerging businesses to emerge. When a natural fiber is strengthened to a polymer matrix, several criteria must be recognized, including the objective, design phase, and climatic circumstances to which the product is subjected. Researches into functionality and life-cycle evaluations are required to determine the natural fiber reinforced polymer composites suitability.

REFERENCES

Abhemanyu, P. C., Prassanth, E., Kumar, T. N., Vidhyasagar, R., Marimuthu, K. P., & Pramod, R. (2019). Characterization of natural fiber reinforced polymer composites. *AIP Conference Proceedings; AIP Publishing: Melville*, NY, USA, 2019; *Volume 2080*, p. 020005. 10.1063/1.5092888

Abitbol, T., Rivkin, A., Cao, Y., Nevo, Y., Abraham, E., Ben-Shalom, T., Lapidot, S., & Shoseyov, O. (2016). Nanocellulose, a tiny fiber with huge applications. *Current Opinion in Biotechnology, 39*, 76–88. 10.1016/j.copbio.2016.01.002

Adeniyi, A. G., Abdulkareem, S. A., Ighalo, J. O., Onifade, D. V., Adeoye, S. A., & Sampson, A. E. (2020). Morphological and thermal properties of polystyrene composite reinforced with biochar from elephant grass (Pennisetum purpureum). *Journal of Thermoplastic Composite Materials*. 10.1177/0892705720939169

Ahmed, Z., & Akhter, F. (2001). Jute retting: An overview. *Journal of Biological Sciences*, *1*(7), 685–688. 10.3923/jbs.2001.685.688

Akin, D. E., Condon, B., Sohn, M., Foulk, J. A., Dodd, R. B., & Rigsby, L. L. (2007). Optimization for enzyme-retting of flax with pectate lyase. *Industrial Crops and Products*, *25*, 136–146. 10.1016/j.indcrop.2006.08.003

Ali, A., Sanuddin, A. B., & Ezzeddin, S. (2010). The effect of aging on Arenga pinnata fiber-reinforced epoxy composite. *Materials and Design*, *31*(7), 3550–3554. 10.1016/j.matdes.2010.01.043

Ali, M., Razak, A., Latif, M. A., Khoo, K. C., & Kasim, J. (1993). Physical properties, fibre dimensions and proximate chemical analysis of Malaysian rattans. *Forest Research Institute Malaysia, Kuala Lumpur* (Malaysia), *1*, 59–70.

Amaducci, S., & Gusovious, H. J. (2010). Hemp cultivation, extraction and processing. In J. Mussig (Ed.), *Industrial Application of Natural Fibres: Properties and Technical Application* (pp.109–134). Wiley, UK. 10.1002/9780470660324.ch5

Amel, B. A., Paridah, M. T., Sudin, R., Anwar, U. M. K., & Hussein, A. S. (2013). Effect of fiber extraction methods on some properties of kenaf bast fiber. *Industrial Crops and Products*, *46*, 117–123. 10.1016/j.indcrop.2012.12.015

Anastas, P. T., & Farris, C. A. (1994). Benign by design: Alternative synthetic design for pollution prevention. *American Chemical Society Symposium Series nr. 577*, American Chemical Society, Washington DC.

Anastas, P. T., & Kirchhoff, M. M. (2002). Origin, current status and failure challenges of green chemistry. *Accounts of Chemical Research 2002*, *35*, 686–693. 10.1021/ar01 0065m

Anastas, P. T., Heine, L. G., & Williamson, T. C. (2002). *Green Chemical Syntheses and Processes* (pp. 364). American Chemical Society, Washington DC, *2000*. 10.1021/op010086w

Anjum, A., Suresha, B., Prasad, S. L. A., & Harshvardhan, B. (2020). Wear behaviour of sansevieria and carbon fiber reinforced epoxy with nanofillers: Taguchi method. *Journal of the Serbian Tribology Society*, *42*(3), 443–460. 10.24874/ti.861.03.20.06

Antonov, V., Marek, J., Bjelkova, M., Smirous, P., & Fisher, H. (2007). Easily available enzymes as natural retting agents. *Biotechnology Journal*, *2*, 342–346. 10.1002/biot.2 00600110.

Arun, I., Yuvaraj, C., Madhu, A., & Ramesh, T. (2021). A comparison on microstructure and mechanical properties of electric discharge metal matrix nickel and silica composite coating on duplex stainless steel. *Journal of Composite Materials*, *55*(4), 507–520. 10.1177/0021998320953882

Ashok, B., Obi Reddy, K., Madhukar, K., Cai, J., Zhang, L., & Rajulu, A. V. (2015). Properties of cellulose/Thespesia lampas short fibers bio-composite films. *Carbohydrate Polymers*, *127*, 110–115. 10.1016/j.carbpol.2015.03.054

Asim, M., Saba, N., Jawaid, M., & Nasir, M. (2018). Potential of natural fiber/biomass filler-reinforced polymer composites in aerospace applications. In M. Jawaid & M. Thariq (Eds.),*Woodhead Publishing Series in Composites Science and Engineering, Sustainable Composites for Aerospace Applications* (pp. 253–268). Woodhead Publishing. 10.1016/B978-0-08-102131-6.00012-8

ASTM D1166-84 (2001), Standard Test Method for Methoxyl Groups in Wood and Related Materials. ASTM International, West Conshohocken, PA. www.astm.org

ASTM E1721-01 (2020), Standard Test Method for Determination of Acid-Insoluble Residue in Biomass. ASTM International, West Conshohocken, PA.

ASTM E1755-01 (2020), Standard Test Method for Ash in Biomass. ASTM International, West Conshohocken, PA. www.astm.org

Athijayamani, A., Thiruchitrambalam, M., Natarajan, U., & Pazhanivel, B. (2009). Effect of moisture absorption on the mechanical properties of randomly oriented natural fibers/polyester hybrid composite. *Materials Science and Engineering, 517*, 344–353. 10.1016/j.msea.2009.04.027

Atiqah, A., Jawaid, M., Sapuan, S. & Ishak, M. (2018). Physical properties of silane-treated sugar palm fiber reinforced thermoplastic polyurethane composites. *IOP Conference Series: Materials Science and Engineering, 368*, 012047. 10.1088/1757-899X/368/1/012047.

Audsley, E., & Annetts, J. E. (2003). Modelling the value of a rural biorefinery – Part I: The model description. *Agricultural Systems, 76*(1), 39–59. 10.1016/S0308-521X(02)00038-0.

Azzam, A. M. (1989). Pretreatment of cane bagasse with alkaline hydrogen peroxide for enzymatic hydrolysis of cellulose and ethanol fermentation. *Journal of Environmental Science and Health, Part B, 24*(4), 421–433.

Azammi, A. M. N., Ilyas, R. A., Sapuan, S. M., Ibrahim, R., Atikah, M. S. N., Asrofi, M., & Atiqah, A. (2020). Characterization studies of biopolymeric matrix and cellulose fibres based composites related to functionalized fibre-matrix interface. In G. L. Goh, M. K. Aswathi, R. T. De Silva, & S. Thomas (Eds.), *Woodhead Publishing* Series in Composites Science and Engineering, *Interfaces in Particle and Fibre Reinforced Composites* (pp, 29–93). Woodhead Publishing. 10.1016/B978-0-08-102665-6.00003-0

Aznar-Cervantes, S. D., Pagan, A., Monteagudo Santesteban, B., & Cenis, J. L. (2019). Effect of different cocoon stifling methods on the properties of silk fibroin biomaterials. *Scientific Reports, 9*, 6703. 10.1038/s41598-019-43134-5

Azwa, Z. N., Yousif, B. F., Manalo, A. C., & Karunasena, W. (2013). A review on the degradability of polymeric composites based on natural fibres, *Materials and Design, 47*, 424–442. 10.1016/j.matdes.2012.11.025

Bacci, L., Di Lonardo, S., Albanese, L., Mastromei, G., & Perito, B. (2010). Effect of different extraction methods on fiber quality of nettle (Urtica Dioica L.). *Textile Research Journal, 81*, 827–837. 10.1177/0040517510391698

Bachtiar, D., Sapuan, S., Zainudin, E. S., Abdan, K., & Zaman, K. (2010). The tensile properties of single sugar palm (Arenga pinnata) fibre. *IOP Conference Series: Materials Science and Engineering, 11*, 012012. 10.1088/1757-899X/11/1/012012

Baillie, C. (2004). *Green Composites: Polymer Composites and the Environment*. Woodhead Publishing Ltd, Boca Raton.

Barber, E. J. W. (1991). *Prehistoric Textiles: The Development of Cloth in the Neolithic and Bronze Ages with Special Reference to the Aegean*. Princeton University Press, Princeton, NJ.

Bari, E., Morrell, J., & Sistani, A. (2018). Durability of natural/synthetic/biomass fiber-based polymeric composites: Laboratory and field tests. In M. Jawaid, M. Thariq, & N. Saba (Eds.), *Durability and Life Prediction in Biocomposites, Fibre-Reinforced Composites and Hybrid Composites* (pp. 15–26). Woodhead Publishing. 10.1016/B978-0-08-102290-0.00002-7

Barile, C., & Casavola, C. (2019). Mechanical characterization of carbon fiber-reinforced plastic specimens for aerospace applications. In M. Jawaid, M. Thariq, & N. Saba (Eds.), *Woodhead Publishing Series in Composites Science and Engineering, Mechanical and Physical Testing of Biocomposites, Fibre-Reinforced Composites and Hybrid Composites* (pp. 387–407). Woodhead Publishing. 10.1016/B978-0-08-102292-4.00019-9

Belouadah, Z., Ati, A., & Rokbi, M. (2015). Characterization of new natural cellulosic fiber from Lygeum spartum L. *Carbohydrate Polymers, 134*, 429–437. 10.1016/j.carbpol.2015.08.024.

Benhamou, K., Kaddami, H., Magnin, A., Dufresne, A., & Ahmad, A. (2015). Bio-based polyurethane reinforced with cellulose nanofibers: A comprehensive investigation on the effect of interface. *Carbohydrate Polymers*, *122*, 202–211. 10.1016/j.carbpol.2014.12.081

Benin, S. R., Kannan, S., Anbiah, J., & Bright, R. J. (2020). Mechanical characterization of prosopis juliflora reinforced polymer matrix composites with filler material. *Materials Today: Proceedings*, *33*(1), 1110–11115. 10.1016/j.matpr.2020.07.190.

Bhargava, A. K. (2010). *Engineering Materials: Polymers, Ceramics and Composites*. PHI learning Pvt. Ltd., New Delhi.

Bian, Y., Ni, J., Wang, C., Zhen, J., Hao, H., Kong, X., Chen, H., Li, J., Li, X., Jia, Z., Luo, W., & Chen, Z. (2021). Microstructure and wear characteristics of in-situ micro/nanoscale niobium carbide reinforced copper composites fabricated through powder metallurgy. *Materials Characterization*, *172*, 110847, ISSN 1044-5803, https://doi.org/10.1016/j.matchar.2020.110847

Blanco, P., Sieiro, C., & Villa, T. G. (1999). Production of pectic enzymes in yeast. *FEMS Microbiology Letters*, *175*, 1–9. 10.1111/j.1574-6968.1999.tb13595.x

Booth, I., Goodman, A. M., Grishanov, S. A., & Harwood, R. J. (2004). A mechanical investigation of the retting process in dew-retting hemp (Cannabis sativa). *Annals of Applied Biology*, *145*, 51–58. 10.1111/j.1744-7348.2004.tb00358.x

Botta, L., Fiore, V., Scalici, T., Valenza, A., & Scaffaro, R. (2015). New polylactic acid composites reinforced with Artichoke fibers. *Materials*, *8*, 7770–7779. 10.3390/ma8115422

Bozell, J. (2008). Feedstocks for the future-biorefinery production of chemicals from renewable carbon. *Clean-Soil Air Water*, *36*, 641–647. 10.1002/clen.200800100

Cabral, M. R., Nakanishi, E. Y., Mármol, G., Palacios, J., Godbout, S., Lagacé, R., Savastano, H., & Fiorelli, J. (2018). Potential of Jerusalem Artichoke (Helianthus tuberosus L.) stalks to produce cement-bonded particleboards. *Industrial Crops and Products*, *122*, 214–222. 10.1016/j.indcrop.2018.05.054

Caldeira-Pires, A., Luz, S., Palma-Rojas, S., Rodrigues, T., Silverio, V., Vilela, F., Barbosa, P., & Alves, A. (2013). Sustainability of the biorefinery industry for fuel production. *Energies*, *6*, 329–350. 10.3390/en6010329

Cantor, B., Dunne, F. P. E., & Stone, I. C. (2003). *Metal and Ceramic Matrix Composites* (pp. 430) (1st Edition). CRC Press, USA. 10.1201/9781420033977

Carmisciano, S., Rosa, I. M. D., Sarasini, F., Tamburrano, A., & Valente, M. (2011). Basalt woven fiber reinforced vinylester composites: Flexural and electrical properties. *Materials and Design*, *32*, 337–342. 10.1016/j.matdes.2010.06.042

Chand, N., & Fahim, M. (2008). Sisal reinforced polymer composites. In Woodhead Publishing Series in Composites Science and Engineering. In *Tribology of Natural Fiber Polymer Composites* (pp. 84–107). Woodhead Publishing. 10.1533/9781845695057.84

Chang, V. S., & Holtzapple, M. T. (2000). Fundamental factors affecting biomass enzymatic reactivity. *Applied Biochemistry and Biotechnology*, *84*(86), 5–37. 10.1385/ABAB:84-86:1-9:5

Chawla, K. K. (2013) *Ceramic Matrix Composites*. Springer Science & Business Media, University of Alabama at Birmingham, Birmingham, AL, USA.

Chollakup, R., Smitthipong, W., Kongtud, W., & Tantatherdtam, R. (2013). Polyethylene green composites reinforced with cellulose fibers (coir and palm fibers): Effect of fiber surface treatment and fiber content. *Journal of Adhesion Science and Technology*, *27*, 1290–1300. 10.1080/01694243.2012.694275

Christian, S. J. (2020). Natural fibre-reinforced noncementitious composites (biocomposites). In Kent A. Harries & B. Sharma (Eds.), *Woodhead Publishing Series in Civil and Structural Engineering, Nonconventional and Vernacular Construction Materials* (pp. 169–187) (Second Edition). Woodhead Publishing. 10.1016/B978-0-08-102704-2.00008-1

Chung, D. D. L. (2017). Introduction to carbon composites. In *Carbon Composites: Composites with Carbon Fibers, Nanofibers, and Nanotubes* (pp. 88–160). Elsevier Science, Amsterdam, The Netherlands. 10.1016/B978-0-12-804459-9.00002-6

Claudino, E. S., & Talamini, E. (2013). Life cycle assessment (LCA) applied to agribusiness – A review. *Revista Brasileira de Engenharia Agrícola e Ambiental, 17*(1), 77–85. 10.15 90/S1415-43662013000100011

Clyne, T. W., & Withers, P. J. (1993). *An Introduction to Metal Matrix Composites.* University Press, Cambridge. 10.1017/CBO9780511623080

Cong, R., & Dong, W. (2007). Structure and property of mulberry fiber. *Modern Applied Science, 1*, 14–17. 10.5539/mas.v1n4p14

Corrêa, A., Zuin, V., Ferreira, V. & Vazquez, P. (2013). Green chemistry in Brazil. *Pure and Applied Chemistry, 85*(8), 1643–1653. 10.1351/PAC-CON-12-11-16

Dalmis, R., Köktaş, S., Seki, Y., & Kilinç, A. C. (2020). Characterization of a new natural cellulose based fiber from Hierochloe Odarata. *Cellulose, 27*, 127–139. 10.1007/s105 70-019-02779-1

Daniyan, I. A., Mpofu, K., Adeodu, A. O., & Adesina, O. (2021). Development of carbon fibre reinforced polymer matrix composites and optimization of the process parameters for railcar applications. *Materials Today: Proceedings, 38*(2), 628–634. 10.1016/j.matpr.2020.03.480

Das, S., Singha, A. K., Chaudhuri, A., & Ganguly, P. K. (2019). Lengthwise jute fibre properties variation and its effect on jute–polyester composite. *The Journal of the Textile Institute, 110*(12), 1695–1702. 10.1080/00405000.2019.1613735

Dawit, J. B., Regassa, Y., Lemu H. G. (2020). Property characterization of acacia tortilis for natural fiber reinforced polymer composite. *Results in Materials, 5*, 100054. 10.1016/j.rinma.2019.100054.

De Resende, T. M., & Da Costa, M. M. (2020). Biopolymers of sugarcane. In F. Santos, S. C. Rabelo, M. De Matos, & P. Eichler (Eds.), *Sugarcane Biorefinery Technology and Perspectives* (pp. 229–254). Academic Press. 10.1016/B978-0-12-814236-3.00012-3

Deswarte, F. E. I., Clark, J. H., Hardy, J. J. E., & Rose, P. M. (2006). The fractionation of valuable wax products from wheat straw using CO_2. *Green Chemistry, 8*(1), 39–42. 10.1039/b514978a

Deyholos, M. K., & Potter, S. (2014). Engineering bast fiber feedstocks for use in composite materials. *Biocatalalysts and Agricultural Biotechnology, 3*, 53–57. 10.1016/j.bcab.2 013.09.001

Di Fidio, N., Raspolli Galletti, A. M., Fulignati, S., Licursi, D., Liuzzi, F., De Bari, I., & Antonetti, C. (2020). Multi-step exploitation of raw Arundo donax L. for the selective synthesis of second-generation sugars by chemical and biological route. Catalysts, *10*, 79. 10.3390/catal10010079

Ding, L., Han, X., Li, H., Han, J., Cao, L., Chen, Y., Ling, Z., He, S., & Jiang, S. (2021). Characterization of novel natural fiber from manau rattan (Calamus manan) as a potential reinforcement for polymer-based composites. *Research Square Reprints.* 10.212 03/rs.3.rs-463693/v1

Dittenber, D. B., & GangaRao, H. V. (2012). Critical review of recent publications on use of natural composites in infrastructure. *Composite Part A: Applied Science and Manufacturing, 43*(8), 1419–1429. 10.1016/j.compositesa.2011.11.019

Dodson, J. R., Hunt, A. J., Parker, H. L., Yang, Y., & Clark, J. K. (2012). Elemental sustainability: Towards the total recovery of scarce metals. *Chemical Engineering and Processing: Process Intensification, 51*, 69–78. 10.1016/j.cep.2011.09.008

El-Galy, I. M., Ahmed, M. H., & Bassiouny, B. I. (2017). Characterization of functionally graded Al-SiCp metal matrix composites manufactured by centrifugal casting. *Alexandria Engineering Journal, 56*(4), 371–381. 10.1016/j.aej.2017.03.009

Elnaggar, M. Y., Mazied, N. A., Hassan, M. M., & Fathy, E. S. (2020). Kevlar fiber re-inforced composites based on waste polyethylene: Impact of ethylene-vinyl acetate and gamma irradiation. *Journal of Vinyl and Additive Technology, 26* (4), 577–585. 10. 1002/vnl.21771

Evans, J. D., Akin, D. E., & Foulk, J. A. (2002). Flax-retting by galacturonase-containing enzyme mixtures and effects on fiber properties. *Journal of Biotechnology, 97,* 223–231. 10.1016/s0168-1656(02)00066-4

Faruk, O., Bledzki, A. K., Fink, H. P., & Sain, M. (2012). Biocomposites reinforced with natural fibers: 2000–2010. *Progress in Polymer Science, 37*(11), 1552–1596. 10.1016/ j.progpolymsci.2012.04.003

Fiore, V., Scalici, T., & Valenza, A. (2014). Characterization of a new natural fiber from Arundo donax L. as potential reinforcement of polymer composites. *Carbohydrate Polymers, 106,* 77–83. 10.1016/j.carbpol.2014.02.016

Flix, J. M., & Gatenholm, P. P. (1991). The nature of adhesion in composites of modified cellulose fibers and polypropylene. *Journal of Applied Polymer Science, 42,* 609–620. 10.1002/app.1991.070420307

Foulk, J. A., Akin, D. E., & Dodd, R. B. (2008). Influence of pectinolytic enzymes on retting effectiveness and resultant fiber properties. *BioResources, 3,* 155–169.

George, A., Sanjay, M. R., Srisuk, R., Parameshwaranpillai, J., & Siengchin, S. (2020). A comprehensive review on chemical properties and applications of biopolymers and their composites. *International Journal of Biolological Macromolecules, 154,* 329–338. 10.1016/j.ijbiomac.2020.03.120

Getme, A. S., & Patel, B. (2020). A review: Bio-fiber's as reinforcement in composites of polylactic acid (PLA). *Materials Today: Proceedings, 26*(2), 2116–2122. 10.1016/ j.matpr.2020.02.457

Giudicianni, P., Cardone, G., Sorrentino, G., & Ragucci, R. (2014). Hemicellulose, cellulose and lignin interactions on Arundo donax steam assisted pyrolysis. *Journal of Analytical and Applied Pyrolysis, 110,* 138–146. 10.1016/j.jaap.2014.08.014

Gominho, J., Curt, M. D., Lourenço, A., Fernández, J., & Pereira, H. (2018). Cynara car-dunculus L. as a biomass and multi-purpose crop: A review of 30 years of research. *Biomass and Bioenergy, 109,* 257–275. 10.1016/j.biombioe.2018.01.001

Goutianos, S., Peijs, T., Nystrom, B., & Skrifvars, M. (2006). Development of flax fibre based textile reinforcements for composite applications. *Applied Composite Materials, 13,* 199–215. 10.1007/s10443-006-9010-2

Graupner, N., Fischer, H., Ziegmann, G., & Müssig, J. (2014). Improvement and analysis of fibre/matrix adhesion of regenerated cellulose fibre reinforced PP-, MAPP-and PLA composites by the use of Eucalyptus globulus lignin. *Composite Part B: Engineering, 66,* 117–125. 10.1016/j.compositesb.2014.05.002

Gu, H. (2009). Tensile behaviours of the coir fibre and related composites after NaOH treatment. *Materials and Design, 30,* 3931–3934. 10.1016/j.matdes.2009.01.035

Gualtieri, A. F., & Tartaglia, A. (2000). Thermal decomposition of asbestos and recycling in traditional ceramics. *Journal of the European Ceramic Society, 20*(9), 1409–1418. 10.1 016/S0955-2219(99)00290-3

Guna, V., Ilangovan, M., Ananthaprasad, M., & Reddy, N. (2018). Hybrid biocomposites. *Polymer Composites, 39,* E30–E54. 10.1002/pc.24641

Gurunathan, T., Mohanty, S., & Nayak, S. K. (2015). A review of the recent developments in biocomposites based on natural fibres and their application perspectives. *Composites Part A: Applied Science and Manufacturing, 77,* 1–25. 10.1016/j.compositesa.2015. 06.007

Haim, A. (2017). Stability of composite stringer-stiffened panels. In *Stability and Vibrations of Thin Walled Composite Structures* (pp. 461–507). Woodhead Publishing, Sawston, UK; Cambridge, UK.

Haldar, D., & Purkait, M. K. (2020). Thermochemical pretreatment enhanced bioconversion of elephant grass (*Pennisetum purpureum*): Insight on the production of sugars and lignin. *Biomass Conversion and Biorefinery*, *12*, 345–567. 10.1007/s13399-020-00689-y

Hallad, S., Banapurmath, N., Dhage, V., Ajarekar, V., Godi, M., & Shettar, A. (2018). Kevlar reinforced polymer matrix composite for structural application. *IOP Conference Series: Materials Science and Engineering*, *376*, 012074. 10.1088/1757-899X/376/1/012074

Hänninen, T., Thygesen, A., Mehmood, S., Madsen, B., & Hughes, M. (2012). Mechanical processing of bast fibres: The occurrence of damage and its effect on fibre structure. *Industrial Crops and Products*, *39*, 7–11. 10.1016/j.indcrop.2012.01.025

Hinestroza, J., & Netravali, A. N. (2014) *Cellulose Based Composites: New Green Nanomaterials* (pp. 328). Wiley-VCH. 10.1002/9783527649440

Höfer, R., & Bigorra, J. (2008). Biomass-based green chemistry: Sustainable solutions for modern economies. *Green Chemistry Letters and Reviews*, *1*(2), 79–97. 10.1080/17518250802342519

Hosseini, S. B. (2020). Natural fiber polymer nanocomposites. In B. Han, S. Sharma, T. Nguyen, L. Longbiao, & K. S. Bhat (Eds.), *Fiber-Reinforced Nanocomposites: Fundamentals and Applications* (pp. 279–299). Elsevier.

Hyness, N. R. J., Vignesh, N. J., Senthamaraikannan, P., Saravanakumar, S. S., & Sanjay, M. R. (2018). Characterization of new natural cellulosic fiber from heteropogon contortus plant. *Journal of Natural Fibers*, *15*(1), 146–153. 10.1080/15440478.2017.1321516

Ilyas, R. A., Sapuan, S. M., Ibrahim, R., Abral, H., Ishak, M. R., Zainudin, E. S., Asrofi, M., Atikah, M. S. N., Huzaifah, M. R. M., Radzi, A. M., Azammi, A. M. N., Shaharuzaman, M. Nurazzi, A. N. M., Syafri, E., Sari, N. H., Norrrahim, M. N. F., & Jumaidin, R. (2019). Sugar palm (Arenga pinnata (Wurmb.) Merr) cellulosic fibre hierarchy: A comprehensive approach from macro to nano scale. *Journal of Materials Research and Technology*, *8*(3), 2753–2766. 10.1016/j.jmrt.2019.04.011

Indran, S., Raj, R. E., & Sreenivasan, V. S. (2014). Characterization of new natural cellulosic fiber from Cissus quadrangularis root. *Carbohydrate Polymers*, *110*, 423–429. 10.1016/j.carbpol.2014.04.051

Iqbal, T., Irfan, M., Ramay, S. M., Mahmood, A., Saleem, M., & Siddiqi, S. A. (2020). ZnO–PVA polymer matrix with transition metals oxide nano-fillers for high dielectric mediums. *Journal of Polymer and the Environment*, *28*, 2422–2432. 10.1007/s10924-020-01768-x

Islam, M., Isa, N., Yahaya, A., Beg, M., & Yunus, R. (2017). Mechanical, interfacial, and fracture characteristics of poly (lactic acid) and Moringa oleifera fiber composites. *Advances in Polymer Technology*, *37*, 1665–1673. 10.1002/adv.21823

Jankauskiene, Z., Butkute, B., Gruzdeviene, E., Ceseviciene, J., & Fernando, A. L. (2015). Chemical composition and physical properties of dew- and water-retted hemp fibers. *Industrial Crops and Products*, *75*, 206–211. 10.1016/j.indcrop.2015.06.044

Jawaid, M., & Khalil, H. P. S. A. (2011). Cellulosic/synthetic fibre reinforced polymer hybrid composites: A review. *Carbohydrate Polymers*, *86*(1), 1–18. 10.1016/j.carbpol.2011.04.043

Jawaid, M., Thariq, M., & Saba, N. (2019). *Mechanical and Physical Testing of Biocomposites, Fibre-Reinforced Composites and Hybrid Composites*. Woodhead Publishing, Sawston, UK; Cambridge, UK.

Jayani, R. S., Saxena, S., & Gupta, R. (2005). Microbial pectinolytic enzymes: A review. *Process Biochemistry*, *40*, 2931–2944. 10.1016/j.procbio.2005.03.026

Jayaramudu, J., Guduri, B. R., & Rajulu, A. V. (2010). Characterization of new natural cellulosic fabric Grewia tilifolia. *Carbohydrate Polymers*, *79*(4), 847–851. 10.1016/j.carbpol.2009.10.046

John, M. J., & Anandjiwala, R. D. (2008). Recent developments in chemical modification and characterization of natural fiber-reinforced composites. *Polymer Composites, 29,* 187–207. 10.1002/pc.20461

Jordá-Vilaplana, A., Carbonell, A., Samper, M., Pop, P., & Garcia-Sanoguera, D. (2017). Development and characterization of a new natural fiber reinforced thermoplastic (NFRP) with Cortaderia selloana (Pampa grass) short fibers. *Composites Science and Technology, 145,* 1–9. 10.1016/j.compscitech.2017.03.036

Jose, J. P. & Joseph, K. (2012). Advances in polymer composites: Macro- and microcomposites—state of the art, new challenges, and opportunities. *Polymer Composites, 1,* 1–16. 10.1002/9783527645213

Kabenge, I., Omulo, G., Banadda, N., Seay, J., Zziwa, A., & Kiggundu, N. (2018). Characterization of banana peels wastes as potential slow pyrolysis feedstock. *Journal of Sustainable Development, 11,* 14. 10.5539/jsd.v11n2p14

Kalapathy, U., Proctor, A., & Shultz, J. (2002). An improved method for production of silica from rice hull ash. *Bioresource Technology, 85*(3), 285–289. 10.1016/S0960-8524(02) 00116-5

Kandachar, P., & Brouwer, R. (2001). Applications of bio-composites in industrial products. *Material Research Society Proceedings, 702,* 101–112. 10.1557/PROC-702-U4.1.1

Kandpal, B. C., Chaurasia, R., & Khurana, V. (2015). Recent advances in green composites—A review. *Global Journalof Pharmacology, 9*(3), 267–271. 10.5829/ idosi.gjp.2015.9.3.94289

Kandpal, B. C., Kumar, J., & Singh, H. (2017). Fabrication and characterisation of Al2O3/ aluminium alloy 6061 composites fabricated by Stir casting. *Materials Today: Proceedings, 4*(2), 2783–2792. 10.1016/j.matpr.2017.02.157

Kapoor, M., Beg, Q. K., Bhushan, B., Singh, K., Dadich, K. S., & Hoondal, G. S. (2001). Application of alkaline and thermostable galacturonase from Bacillus sp. MG-cp-2 in degumming of ramie (Boehmeria nivea) and sunn hemp (Crotolaria juncia) bast fibers. *Process Biochemistry, 36,* 803–807. 10.1016/S0032-9592(00)00282-X

Kaserer, L., Braun, J., Stajkovic, J., Leitz, K. H., Singer, P., Letofsky-Papst, I., Kestler, H., & Leichtfried, G. (2020). Microstructure and mechanical properties of molybdenum-titanium-zirconium-carbon alloy TZM processed via laser powder-bed fusion. *International Journal of Refractory Metals and Hard Materials, 93,* 105369. 10.1016/ j.ijrmhm.2020.105369

Kelse, J. W., & Thompson, C. S. (1989). The Regulatory and mineralogical definitions of asbestos and their impact on Amphibole dust analysis. *AIHA Journal, 50*(11), 613–622. 10.1080/15298668991375245

Kengkhetkit, N., & Amornsakchai, T. (2012). Utilisation of pineapple leaf waste for plastic reinforcement: 1. A novel extraction method for short pineapple leaf fiber. *Industrial Crops and Products, 40,* 55–61. 10.1016/j.indcrop.2012.02.037

Khan, A., Vijay, R., Singaravelu, D. L., Sanjay, M. R., Siengchin, S., Verpoort, F., Alamry, K. A., & Asiri, A. M. (2020). Characterization of natural fibers from *Cortaderia Selloana* grass (Pampas) as reinforcement material for the production of the composites. *Journal of Natural Fibers, 11,* 1893–1901. 10.1080/15440478.2019.1709110

Khan, J. A., & Khan, M. A. (2015). The use of jute fibers as reinforcements in composites. In O. Faruk & M. Sain (Eds.),*Biofiber Reinforcements in Composite Materials* (pp. 3–34). Woodhead Publishing, Sawston, UK; Cambridge, UK. 10.1533/9781782421276.5.648

Khondker, M., Hassan, M., Alfasane, M., & Shahjadee, U. (2010). Flowering and fruiting characteristics and biochemical composition of an endangered palm species (*Corypha taliera* Roxb.). *Bangladesh Journal of Plant Taxonomy, 17*(1), 79–86. 10.3329/bjpt.v1 7i1.5393

Kian, L. K., Jawaid, M., Ariffin, H., & Karim, Z. (2018). Isolation and characterization of nanocrystalline cellulose from roselle-derived microcrystalline cellulose. *International Journal of Biological Macromolecules, 114*, 54–63. 10.1016/j.ijbiomac.2018.03.065

Kılınç, A. C., Köktaş, S., Seki, Y., Atagür, M., Dalmış, R., Erdoğan, U. H., Göktaş, A. A., & Seydibeyoğlu, M. O. (2018). Extraction and investigation of lightweight and porous natural fiber from Conium maculatum as a potential reinforcement for composite materials in transportation. *Composites Part B: Engineering, 140*, 1–8. 10.1016/j.compositesb.2017.11.059

Kim, S. S., & Lee, J. (2013). Miscibility and antimicrobial properties of m-aramid/chitosan hybrid composite. *Industrial and Engineering Chemistry Research, 52*, 12703–12709. 10.1021/ie400354b

Klinke, H. B., Ahring, B. K., Schmidt, A. S., & Thomsen, A. B. (2002). Characterization of degradation products from alkaline wet oxidation of wheat straw. *Bioresource Technology, 82*, 15–26. 10.1016/s0960-8524(01)00152-3

Komuraiah, A., Kumar, N. S., & Prasad, B. D. (2014). Chemical composition of natural fibers and its influence on their mechanical properties. *Mechanics of Composite Materials, 50*, 359–376. 10.1007/s11029-014-9422-2

Konczewicz, W., & Wojtysiak, J. (2015). The effect of physical factors on the process of physical-mechanical degumming of flax fibers. *Textile Research Journal, 85*, 391–403. 10.1177/0040517514547214

Krausmann, F., Simone, G., Nina, E., Karl-Heinz, E., Helmut, H., & Marina F. K. (2009). Growth in global materials use, GDP and population during the 20th century. *Ecological Economics, 68*(10), 2696–2705. 10.1016/j.ecolecon.2009.05.007

Kulandaisamy, S. V., & Govindasamy, P. (2020). Thermal analysis of Acacia Arabica and Pencil Cactus fiber hybrid polymer composite. *Research Journal of Chemistry and Environment, 24*, 52–55.

Kuo, C. C., Liu, L. C., Teng, W. F., Chang, H. Y., Chien, F. M., Liao, S. J., Kuo, W. F., & Chen, C. M. (2016). Preparation of starch/acrylonitrile-butadiene-styrene copolymers (ABS) biomass alloys and their feasible evaluation for 3D printing applications. *Composites Part B: Engineering, 86*, 36–39. 10.1016/j.compositesb.2015.10.005

La Mantia, F. P., & Morreale, M. (2011). Green composites: A brief review. *Composites Part A: Applied Science and Manufacturing, 42*(6), 579–588. 10.1016/j.compositesa.2011.01.017

Laban, O., & Mahdi, E. (2016). Energy absorption capability of cotton fiber/epoxy composite square and rectangular tubes. *Journal of Natural Fibers, 13*, 726–736. 10.1080/15440478.2015.1130003

Lazar, M. A., Rotaru, H., Bâldea, I., Boşca, A. B., Berce, C. P., Prejmerean, C., Prodan, D., & Câmpian, R. S. (2016). Evaluation of the biocompatibility of new fiber-reinforced composite materials for craniofacial bone reconstruction. *Journal of Craniofacial Surgery, 27*(7), 1694–1699. 10.1097/SCS.0000000000002925. PMID: 27763970.

Lee, S. A., & Eiteman, M. A. (2001). Ground kenaf core as a filtration aid. *Industrial Crops and Products, 13*(2), 155–161. 10.1016/S0926-6690(00)00062-5

Li, X., Du, G.,Wang, S., & Yu, G. (2014). Physical and Mechanical Characterization of Fiber Cell Wall in Castor (Ricinus communis L.) Stalk. *BioResources, 9*(1). 10.15376/biores.9.1.1596-1605.

Li, Y., Wang, S., & Wang, Q. (2017). A molecular dynamics simulation study on enhancement of mechanical and tribological properties of polymer composites by introduction of grapheme. *Carbon, 111*, 538–545. 10.1016/j.carbon.2016.10.039

Liu, Y., Ma, Y., Yu, J., Zhuang, J., Wu, S., & Tong, J. (2019). Development and characterization of alkali treated abaca fiber reinforced friction composites. *Composite Interfaces, 26*, 67–82. 10.1080/09276440.2018.1472456

Lodha, P., & Netravali, A. N. (2002). Characterization of interfacial and mechanical properties of "green" composites with soy protein isolate and ramie fiber. *Journal of Materials Science, 37*, 3657–3665. 10.1023/A:1016557124372

Loganathan, T. M., Sultan, M. T. H., Jawaid, M., Ahsan, Q., Naveen, J., & Perumal, V. (2020). Characterization of new cellulosic *Cyrtostachys renda* and *Ptychosperma macarthurii* fibers from landscaping plants. *Journal of Natural Fibers, 19*(2), 669–684. 10.1080/15440478.2020.1758865

Loureiro, N. C., & Esteves, J. L. (2019). Green composites in automotive interior parts: A solution using cellulosic fibers. In G. Koronis & A. Silva (Eds.), *Woodhead Publishing Series in Composites Science and Engineering, Green Composites for Automotive Applications* (pp. 81–97). Woodhead Publishing. 10.1016/B978-0-08-102177-4.00004-5

Madueke, C. I., Kolawole, F., & Tile, J. (2021). Property evaluations of coir fibres for use as reinforcement in composites. *SN Applied Sciences, 3*, 262. 10.1007/s42452-021-04283-3

Maepa, C. E., Jayaramudu, J., Okonkwo, J. O., Ray, S. S., Sadiku, E. R., & Ramontja, J. (2015). Extraction and characterization of natural cellulose fibers from maize tassel. *International Journal of Polymer Analysis and Characterization, 20*(2), 99–109. 10.1 080/1023666X.2014.961118

Maheshwaran, M. V., Hyness, N. R. J., Senthamaraikannan, P., Saravanakumar, S. S., & Sanjay, M. R. (2018). Characterization of natural cellulosic fiber from *Epipremnum aureum* stem. *Journal of Natural Fibers, 15*(6), 789–798. 10.1080/15440478.2017.13 64205

Malkapuram, R., Kumar, V., & Singh, Y. (2009). Recent development in natural fiber reinforced polypropylene composites. *Journal of Reinforced Plastics and Composites, 28*(10), 1169–1189. 10.1177/0731684407087759

Mallick, P. K. (2012). Advanced materials for automotive applications: An overview. In J. Rowe (Ed.), *Advanced Material in Automotive Engineering* (pp. 5–27). Woodhead Publishing. 10.1533/9780857095466.5

Mallick, P. K. (1997). *Composites Engineering Handbook.* CRC Press, New York.

Manimaran, P., Saravanakumar, S. S., Mithun, N. K., & Senthamaraikannan, P. (2016). Physicochemical properties of new cellulosic fibers from the bark of *Acacia Arabica*. *International Journal of Polymer Analysis and Characterization, 21*(6), 548–553. 10.1 080/1023666X.2016.1177699

Manimaran, P., Saravanan, S. P., Sanjay, M. R., Siengchin, S., Jawaid, M., & Khan, A. (2019). Characterization of new cellulosic fiber: Dracaena reflexa as a reinforcement for polymer composite structures. *Journal of Materials Research and Technology, 8*(2), 1952–1963. 10.1016/j.jmrt.2018.12.015

Manimaran, P., Saravanan, S. P., Sanjay, M. R., Jawaid, M., Siengchin, S., & Fiore, V. (2020). New lignocellulosic *Aristida adscensionis* fibers as novel reinforcement for composite materials: Extraction, characterization and weibull distribution analysis. *Journal of Polymer and the Environment, 28*, 803–811. 10.1007/s10924-019-01640-7

Manimaran, P., Senthamaraikannan, P., Muruganathan, K., & Sanjay, M. R. (2018). Physicochemical properties of new cellulosic fibers from *Azadirachta indica* plant. *Journal of Natural Fibers, 15*(1), 29–38. 10.1080/15440478.2017.1302388

Manjula, R., Raju, N. V., Chakradhar, R. P. S., Kalkornsurapranee, E., & Johns, J. (2018). Influence of chemical treatment on thermal decomposition and crystallite size of coir fiber. *International Journal of Thermophysics, 39*(3). 10.1007/s10765-017-2324-5

Marteel, A. E., Davies, J. A., Olson, W. W., & Abraham, M. A. (2003). Green chemistry and eningeering: Drivers, metrics, and reduction to practice. *Annual Review of Environment and Resources, 28*, 401–428. 10.1146/annurev.energy.28.011503.163459

McLaughlin, E. C., & Tait, R. A. (1980). Fracture mechanism of plant fibres. *Journal of Material Sciemce, 15*, 89–95. 10.1007/BF00552431

Mengyan, L., Mondrinos, M. J., Xuesi, C., & Lelkes, P. I. (2006). Electrospun blends of natural and synthetic polymers as scaffolds for tissue engineering. In *Proceedings of the 2005 IEEE Engineering in Medicine and Biology 27th Annual Conference*, Shanghai, China. 10.1109/IEMBS.2005.1615822

Mhatre, A. M., Raja, A. S. M., Saxena, S., & Patil, P. G. (2019). Environmentally benign and sustainable green composites: Current developments and challenges. In S. Muthu (Ed.), *Green Composites. Textie Science and Clothing Technology* (pp. 53–90). Springer, Singapore. 10.1007/978-981-13-1969-3_3

Mishra, K., & Sinha, S. (2021). Biodegradable green composite film developed from Moringa Oleifera (Sahajana) seed filler and PVA: Surface functionalization, characterization and barrier properties. *Journal of Thermoplastic Composite Materials.* 10.1177/08927057211007550

Mishra, K., & Sinha, Shishir. (2020). Development and assessment of Moringa oleifera (Sahajana) leaves filler/epoxy composites: Characterization, barrier properties and in situ determination of activation energy. *Polymer Composites*, *41*(12), 5016–5029. 10.1002/pc.25771

Mlik, Y. B., Jaouadi, M., Rezig, S., Khoffi, F., Slah, M., & Durand, B. (2018). Kenaf fibre-reinforced polyester composites: Flexural characterization and statistical analysis. *The Journal of The Textile Institute*, *109*(6), 713–722. 10.1080/00405000.2017.1365580

Mohamed, H., Laperrière, L., & Mahi, H. (2018). Replacing stitching and weaving in natural fiber reinforcement manufacturing, Part 2: Mechanical behavior of flax fiber composite laminates. *Journal of Natural Fibers*, *17*(3), 388–397. 10.1080/15440478.2018.1494079

Mohamed, S. A. N., Zainudin, E. S., Sapuan, S. M., Azaman, M. D., & Arifin, A. M. T. (2018). Introduction to natural fiber reinforced vinyl ester and vinyl polymer composites. In S. M. Sapuan, H. Ismail, & E. S. Zainudin (Eds.), Woodhead Publishing Series in Composites Science and Engineering, *Natural Fibre Reinforced Vinyl Ester and Vinyl Polymer Composites* (pp. 1–25). Woodhead Publishing. 10.1016/B978-0-08-102160-6.00001-9

Mohammed, H. A., Rashid, S. A., Bakar, M. H. A., Anas, S. B. A., Mahdi, M. A., & Yaacob, M. H. (2019). Fabrication and characterizations of a novel etched-tapered single mode optical fiber ammonia sensors integrating PANI/GNF nanocomposite. *Sensors and Actuators B: Chemical*, *287*, 71–77. 10.1016/j.snb.2019.01.115

Mohnen, D. (2008). Pectin structure and biosynthesis. *Current Opinion in Plant Biology*, *11*, 266–277. 10.1016/j.pbi.2008.03.006

Moon, R. J., Martini, A., Nairn, J., Simonsen. J., & Youngblood, J. (2011). Cellulose nanomaterials review: Structure, properties and nanocomposites. *Chemical Society Reviews*, *40*(7), 3941–3994. 10.1039/c0cs00108b

Munshi, T. K., & Chattoo, B. B. (2008). Bacterial population structure of the jute-retting environment. *Microbial Ecology*, *56*(2008), 270–282. 10.1007/s00248-007-9345-8

Myers, G. E., Chahyadi, I. S., Coberly, C. A., & Ermer, D. S. (1991). Wood flour/polypropylene composites: Influence of maleated polypropylene and process and composition variables on mechanical properties. *International Journal of Polymeric Materials and Polymeric Biomaterials*, *15*(1), 21–44. 10.1080/00914039108031519

Nagappan, S., Subramani, S., & Ta, S. (2016). Extraction and characterization of new cellulose fiber from the agrowaste of lagenaria sicararia (bottle guard) plant. *Journal of Advances in Chemistry*, *12*, 4382–4388. 10.24297/jac.v12i9.3991

Nair, A. B., & Joseph, R. (2014). Eco-friendly bio-composites using natural rubber (NR) matrices and natural fiber reinforcements. In S. Kohjiya & Y. Ikeda (Eds.), *Chemistry, Manufacture and Applications of Natural Rubber. Woodhead Publishing*, Sawston, UK; Cambridge, UK.

Naveen, J., Jawaid, M., Amuthakkannan, P., & Chandrasekar, M. (2019). Mechanical and physical properties of sisal and hybrid sisal fiber-reinforced polymer composites. In M. Jawaid, M. Thariq, & N. Saba (Eds.), *Woodhead Publishing Series in Composites Science and Engineering* (pp. 427–440). Woodhead Publishing. 10.1016/B978-0-08-102292-4.00021-7

Netravali, A. N., & Chabba, S. (2003). Composites get greener. *Materials Today, 6*(4), 22–29. 10.1016/S1369-7021(03)00427-9

Nishino, T. (2004). Natural fiber sources. In C. Baille (Ed.), Green Composites: Polymer Composites and the Environment (pp. 49–80). Woodhead Publishing.

Nobe, R., Qiu, J., Kudo, M., Ito, K., & Kaneko, M. (2019). Effects of SCF content, injection speed, and CF content on the morphology and tensile properties of microcellular injection-molded CF/PP composites. *Polymer Engineering and Science, 59*(7), 1371–1380. 10.1002/pen.25120

Ochi, S. (2008). Mechanical properties of kenaf fibers and kenaf/PLA composites. *Mechanics of Materials, 40*(4), 446–452. 10.1016/j.mechmat.2007.10.006.

Olcay, H., & Kocak, E. D. (2020). The mechanical, thermal and sound absorption properties of flexible polyurethane foam composites reinforced with artichoke stem waste fibers. *Journal of Industrial Textiles.* 10.1177/1528083720934193

Olesen, P. O., & Plackett, D. V. (1999). Perspectives on the Performance of Natural Plant Fibres. Natural Fibres Performance Forum, Copenhagen, 27-28 May 1999, Denmark, 1-7.

Oliveira, J. R., Kotzebue, L. R. V., Freitas, D. B., Mattos, A. L. A., Júnior, A. E. D. C., Mazzetto, S. E., & Lomonaco, D. (2020). Towards novel high-performance bio-composites: Polybenzoxazine-based matrix reinforced with Manicaria saccifera fabrics. *Composites Part B: Engineering, 194*, 108060. 10.1016/j.compositesb.2020.108060

Omar, M. F., Jaya, H., & Zulkepli, N. N. (2019). Kenaf fiber reinforced composite in the automotive industry. In S. Hashmi & I. A. Choudhury (Eds.), *Encyclopedia of Renewable and Sustainable Materials* (pp. 95–101). Elsevier. 10.1016/B978-0-12-803581-8.11429-8

Orden, D. R. V. (1964). Asbestos. In R. D. Morrison & B. L. Murphy (Eds.), *Environmental Forensics* (pp. 19–33). Academic Press. 10.1016/B978-012507751-4/50024-0

Ouajai, S., & Shanks, R. A. (2005). Morphology and structure of fiber after bioscouring. *Macromolecular Bioscience, 5*, 124–134. 10.1002/mabi.200400151

Panneerdhass, R., Gnanavelbabu, A., & Rajkumar, K. (2014). Mechanical properties of luffa fiber and ground nut reinforced epoxy polymer hybrid composites. *Procedia Engineering, 97*, 2042–2051. 10.1016/j.proeng.2014.12.447

Panthapulakkal, S., & Sain, M. (2007). Injection-molded short hemp fiber/glass fiber-reinforced polypropylene hybrid composites—Mechanical, water absorption and thermal properties. Journal of *Applied Polymer Science, 103*(4), 2432–2441. 10.1002/app.25486

Pereira, P. H. F., Rosa, M. D. F., Cioffi, M. O. H., Benini, K. C. C. D. C., Milanese, A. C., Voorwald, H. J. C., & Mulinari, D. R. (2015). Vegetal fibers in polymeric composites: A review. *Polímeros, 25*(1), 9–22. 10.1590/0104-1428.1722

Petkieva, D., Ozerin, A., Kurkin, T., Golubev, E., Kova, E. I., & Zelenetskii, A. (2020). Carbonization of oriented poly (vinyl alcohol) fibers impregnated with potassium bi-sulfate. *Carbon Letter, 30*, 637–650. 10.1007/s42823-020-00135-z

Pfaltzgraff, L. A., & Clark, J. H. (2014). Green chemistry, biorefineries and second generation strategies for re-use of waste: An overview. In K. Waldron (Ed.), *Advances in Biorefineries* (pp. 3–33). Woodhead Publishing. 10.1533/9780857097385.1.3

Pickering, K. L., Farrell, R. L., & Lay, M. C. (2007). Interfacial modification of hemp fiber reinforced composites using fungal and alkali treatment. *Journal of Biobased Material and Bioenergy, 1*, 109–117. 10.1166/jbmb.2007.012

Porras, A., Maranon, A., & Ashcroft, I. A. (2016). Thermo-mechanical characterization of Manicaria Saccifera natural fabric reinforced poly-lactic acid composite lamina. *Composites Part A: Applied Science and Manufacturing*, *81*, 105–110. 10.1016/j.compositesa.2015.11.008

Pradhan, S., Ghosh, S., Barman, T. K., & Sahoo, P. (2017). Tribological behavior of Al-SiC metal matrix composite under dry, aqueous and alkaline medium. *Silicon*, *9*, 923–931. 10.1007/s12633-016-9504-y

Prakash, S. (2019). Experimental investigation of surface defects in low-power CO2 laser engraving of glass fiber-reinforced polymer composite. *Polymer Composite*, *40*(12), 4704–4715. 10.1002/pc.25339

Quadri, A. I., & Alabi, O. (2020). Assessment of sponge gourd (Luffa Aegyptiaca) fiber as a polymer reinforcement in concrete. *Concrete*, *2*, 125–132. 10.22034/JCEMA.202 0.232358.1026

Rafiqah, S., Khalina, A., Harmaen, A., Zaman, K., Jawaid, M., & Ching Hao, L.(2018). Effect of modified tapioca starch on mechanical, thermal, and morphological properties of PBS blends for food packaging. *Polymers*, *10*, 1187. 10.3390/polym10111187

Ragauskas, A., Williams, C., Davison, B., Britovsek, G., Cairney, J., Eckert, C. A., Frederick Jr., Hallett, J., Leak, D., Liotta, C., Mielenz, J., Murphy, R. J., Templer, R., & Tschaplinski, T. (2006). The path forward for biofuels and biomaterials. *Science*, *311*, 484–489. 10.1126/science.1114736

Raj, S. S., Kannan, T., & Rathanasamy, R. (2020). Influence of Prosopis Juliflora wood flour in poly lactic acid – Developing a novel bio-wood plastic composite. *Polímeros*, *30*(1). 10.1590/0104-1428.00120

Rajak, D. K., Pagar, D. D., Kumar, R., & Pruncu, C. (2019). Recent progress of reinforcement materials: A comprehensive overview of composite materials. *Journal of Material Research and Technology*, *8*(6), 6354–6374. 10.1016/j.jmrt.2019.09.068

Ramamoorthy, S. K., Skrifvars, M., & Persson, A. (2015). A review of natural fibers used in biocomposites: Plant, animal and regenerated cellulose fibers. *Polymer Reviews*, *55*(1), 107–162. 10.1080/15583724.2014.971124

Rao, T. B. (2017). An experimental investigation on mechanical and wear properties of Al7075/SiCp composites: Effect of SiC content and particle size. *Journal of Tribology*, *140*, 31601–31608. 10.1115/1.4037845

Ratwani, M. M. (2010). Composite materials and sandwich structures—A primer. *Rto-En-Avt*, *156*, 1–16. 10.14339/RTO-EN-AVT-156

Razali, N., Sapuan, S., Jawaid, M., Ishak, M., & Lazim, Y. (2015). A study on chemical composition, physical, tensile, morphological, and thermal properties of roselle fibre: Effect of fibre maturity. *Bioresources*, *10*(1). 10.15376/biores.10.1.1803-1824

Reddy, K., Guduri, B., & Rajulu, A. V. (2009). Structural characterization and tensile properties of Borassus fruit fibers. *Journal of Applied Polymer Science*, *114*, 603–611. 10.1002/app.30584

Réquilé, S., Duigou, A., Bourmaud, A., & Baley, C. (2018). Peeling experiments for hemp retting characterization targeting biocomposites. *Industrial Crops and Products*, *123*, 573–580. 10.1016/j.indcrop.2018.07.012

Ribeiro, A., Pochart, P., Day, A., Mennuni, S., Bono. P., Baret, J. L., Spadoni, J. L., & Mangin, I. (2015). Microbial diversity observed during hemp retting. *Applied Microbiology and Biotechnology*, *99*(10), 4471–4484. 10.1007/s00253-014-6356-5

Ridzuan, M. J. M., Abdul Majid, M. S., Afendi, M., Aqmariah Kanafiah, S. N., Zahri, J. M., & Gibson, A. G. (2016). Characterisation of natural cellulosic fibre from Pennisetum purpureum stem as potential reinforcement of polymer composites. *Materials and Design*, *89*, 839–847. 10.1016/j.matdes.2015.10.052.

Rimašauskas, M., Kuncius, T., & Rimašauskiene, R. (2019). Processing of carbon fiber for 3D printed continuous composite structures. *Materials and Manufacturing Processes, 34*(13), 1528–1536. 10.1080/10426914.2019.1655152

Rohatgi, P. (2013). Metal matrix composites. *Defence Science Journal, 43*(4), 323–349. 10.14429/dsj.43.4336

Rosso, M. (2006). Ceramic and metal matrix composites: Routes and properties. *Journal of Materials Processing Technology, 175*(1–3), 364–375. 10.1016/j.jmatprotec.2005.04 .038

Rowell, R. M., & Stout, H. P. (2007). Jute and kenaf. In M. Lewin & M. Dekker (Eds.), Handbook of Fiber Chemistry (pp. 405–452) (3rd Edition). Taylor & Francis, Boca Raton; New York. 10.1201/9781420015270.ch7

Runcang, S., Lawther, J. M., & Banks, W. B. (1996). Fractional and structural characterization of wheat straw hemicelluloses. *Carbohydrate Polymers, 29*(4), 325–331. 10.1 016/S0144-8617(96)00018-5

Saba, N., Mohammad, F., Pervaiz, M., Jawaid, M., Alothman. O., & Sain, M. (2017). Mechanical, morphological and structural properties of cellulose nanofibers reinforced epoxy composites. *International Journal of Biological Macromolecules, 97*, 190–200. 10.1016/j.ijbiomac.2017.01.029

Saba, N., Tahir, P. M., & Jawaid, M. (2014). A review on potentiality of nano filler/natural fiber filled polymer hybrid composites. *Polymers, 6*(8), 2247–2273. 10.3390/ polym6082247

Saleh, H. E. D., & Koller, M. (2018). Introductory chapter: Principles of green chemistry. In *Green Chemistry: An Introductory Text* (3rd Edition). 10.5772/intechopen.71191

Sanjay, M. R., & Yogesha, B. (2017). Studies on natural/glass fiber reinforced polymer hybrid composites: An evolution. *Materials Today: Proceedings, 4*(2), 2739–2747. 10.1016/j.matpr.2017.02.151

Sanyang, M. L., Sapuan, S. M., Jawaid, M., Ishak, M. R., & Sahari, J. (2016). Development and characterization of sugar palm starch and poly (lactic acid) bilayer films. *Carbohydrate Polymers, 1*(146), 36–45. 10.1016/j.carbpol.2016.03.051

Saputro, A., Verawati, I., Ramahdita, G., & Chalid, M. (2017). Preparation of microfibrillated cellulose based on sugar palm ijuk (Arenga pinnata) fibres through partial acid hydrolysis. *IOP Conference Series: Materials Science and Engineering, 223*, 012042. 10.1088/1757-899X/223/1/012042.

Sathishkumar, T. P., Navaneethakrishnan, P., Shankar, S., & Rajasekar, R. (2013). Characterization of new cellulose sansevieria ehrenbergii fibers for polymer composites. *Composite Interfaces, 20*(8), 575–593. 10.1080/15685543.2013.816652

Sathishkumar, T., Naveen, J., & Satheeshkumar, S. (2014). Hybrid fiber reinforced polymer composites – A review. *Journal of Reinforced Plastics and Composites, 33*(5), 454–471. 10.1177/0731684413516393

Say, Y., Guler, O., & Dikici, B. (2020). Carbon nanotube (CNT) reinforced magnesium matrix composites: The effect of CNT ratio on their mechanical properties and corrosion resistance. *Materials Science and Engineering: A, 798*, 139636. 10.1016/ j.msea.2020.139636

Scalici, T., Fiore, V., & Valenza, A. (2016). Effect of plasma treatment on the properties of Arundo Donax L. leaf fibres and its bio-based epoxy composites: A preliminary study. *Composites Part B: Engineering, 94*, 167–175. 10.1016/j.compositesb.2016.03.053

Schoning, A. (1965). Absorptiometric determination of acid-soluble lignin in semichemical bisulfite pulps and in some woods and plants. *Svensk Papperstidning-nordisk Cellulosa, 68*.

Seki, Y., Sarikanat, M., Sever, K., & Durmuşkahya, C. (2013). Extraction and properties of Ferula communis (chakshir) fibers as novel reinforcement for composites materials. *Composites Part B: Engineering, 44*(1), 517–523. 10.1016/j.compositesb.2012.03.013

Senthilkumar, K., Saba, N., Rajini, N., Chandrasekar, M., Jawaid, M., Siengchin, S., & Alotman, O. Y. (2018). Mechanical properties evaluation of sisal fibre reinforced polymer composites: A review. *Construction and Building Materials, 174*, 713–729. 10.1016/j.conbuildmat.2018.04.143

Sezgin, H., & Berkalp, O. B. (2017). The effect of hybridization on significant characteristics of jute/glass and jute/carbon-reinforced composites. *Journal of Industrial Textiles, 47*(3), 283–296. 10.1177/1528083716644290

Shahinur, S., & Hasan, M. (2019). Jute/coir/banana fiber reinforced bio-composites: Critical review of design, fabrication, properties and applications. In S. Hashmi & I. A. Choudhury (Eds.), *Encyclopedia of Renewable and Sustainable Materials* (pp. 751–756). Elsevier. 10.1016/B978-0-12-803581-8.10987-7

Shahzad, A. (2012). Hemp fiber and its composites – A review. *Journal of Composite Materials, 46*, 973–986. 10.1177/0021998311413623

Sharma, A. K., Bhandari, R., & Aherwar, A. (2018). Mechanical aspects of metallic bio materials in human hip replacement. *International Journal of Research and Analytical Reiews, 5*, 363–368. 10.1729/Journal.18652

Sheng, W., Gao, J., Jin, Z., Dai, H., Zheng, L., & Wang, B. (2014). Effect of steam explosion on degumming efficiency and physicochemical characteristics of banana fiber. *Jornal of Applied Polymer Science, 131*, 40598–44606. 10.1002/app.40598

Shin, L. J., Dassan, E. G. B., Abidin, M. S. Z., & Anjang, A. (2020). Tensile and compressive properties of glass fiber-reinforced polymer hybrid composite with eggshell powder. *Arabain Journal of Science and Engineering, 45*, 5783–5791. 10.1007/s13369-020-04561-z

Silva, F., Ribeiro, C. E. G., Demartini, T., & Rodriguez, R. J. S. (2020). Physical, chemical, mechanical, and microstructural characterization of banana pseudostem fibers from Musa Sapientum. *Macromolecular Symposia, 394*, 2000052. 10.1002/masy.202000052

Silva, G. G., De Souza, D. A., Machado, J. C., & Hourston, D. J. (2000). Mechanical and thermal characterization of native Brazilian coir fiber. *Journal of Applied Polymer Science, 76*, 1197–1206. 10.1002/(SICI)1097-4628(20000516)76:7<1197::AID-APP23>3.0.CO;2-G

Simsek, D., Simsek, I. & Ozyurek, D. (2019). Relationship between Al2O3 content and wear behavior of Al+2% graphite matrix composites. *Science and Engineering of Composite Materials, 27*(1), 177–185. 10.1515/secm-2020-0017

Singh, J. K., Rout, A. K., & Kumari, K. (2021). A review on Borassus flabellifer lignocellulose fiber reinforced polymer composites. *Carbohydrate Polymers, 262*, 117929. 10.1016/j.carbpol.2021.117929

Singh, R., Singh, I., & Kumar, R. (2019). Mechanical and morphological investigations of 3D printed recycled ABS reinforced with bakelite–SiC–Al2O3. *Proceedings of the Institution of Mechanical Engineers, Part C: Journal of Mechanical Engineering Science, 233*(17), 5933–5944. 10.1177/0954406219860163

Singh, S., Ramakrishna, S., & Berto, F. (2019). 3D Printing of polymer composites: A short review. *Material Design and Processing Communications, 2*(2), 123–135. 10.1002/mdp2.97

Singh, T. J., & Samanta, S. (2015). Characterization of kevlar fiber and its composites: A review. *Materials Today: Proceedings, 2*(4–5), 1381–1387. 10.1016/j.matpr.2015.07.057

Sisti, L., Totaro, G., Vannini, M., Fabbri, P., Kalia, S., Zatta, A., & Celli, A. (2016) Evaluation of the retting process as a pre-treatment of vegetable fibers for the preparation of high-performance polymer biocomposites. *Industrial Crops and Products, 81*, 56–65. 10.1016/j.indcrop.2015.11.045

Stamboulis, A., Baillie, C. A., Garkhail, S. K., Van Melick, H. G. H., & Peijs, T. (2000).

Environmental durability of flax fibres and their composites based on polypropylene matrix. *Applied Composite Materials*, *7*, 273–294. 10.1023/A:1026581922221

Stevens, C. V., & Verhe, R. V. (2004). *Renewable Bioresources: Scope and Modification of Non-food Applications*. JohnWiley & Sons, Chichester.

Suaduang, N., Ross, S., Ross, G. M., Pratumshat, S., & Mahasaranon, S. (2019). Effect of spent coffee grounds filler on the physical and mechanical properties of poly(lactic acid) bio-composite films. *MaterialsToday Proceedings*, *17*(4), 2104–2110. 10.1016/j.matpr.2019.06.260

Sucinda, E. F., Abdul Majid, M. S., Ridzuan, M. J. M., Sultan, M. T. H., & Gibson, A. G. (2020). Analysis and physicochemical properties of cellulose nanowhiskers from Pennisetum purpureum via different acid hydrolysis reaction time. *International Journal of Biological Macromolecules*, *155*, 241–248. 10.1016/j.ijbiomac.2020.03.199

Sullins, T., Pillay, S., Komus, A., & Ning, H. (2017). Hemp fiber reinforced polypropylene composites: The effects of material treatments. *Composites Part B: Engineering*, *114*, 15–22. 10.1016/j.compositesb.2017.02.001

Sun, Y., & Cheng, J. (2002) Hydrolysis of lignocellulosic materials for ethanol production: A review. *Bioresource Technology*, *83*, 1–11. 10.1016/S0960-8524(01)00212-7

Svennerstedt, B., & Sevenson, G. (2006). Hemp (*Cannabis sativa L.*) trials in Southern Sweden 1999-2001. *Journal of Industrial Hemp*, *11*(1), 17–25. 10.1300/J237v11n01_03

Swolfs, Y., Larissa, G., & Verpoest, I. (2014). Fibre hybridisation in polymer composites: A review. *Composites Part A: Applied Science and Manufacturing*, *67*, 181–200. 10.1016/j.compositesa.2014.08.027

Tahir, M. P., Ahmed, A. B., SaifulAzry, S., & Ahmed, Z. (2011). Retting process of some bast plant fibres and its effect on fibre quality: A review. *BioResources*, *6*, 5260–5281.

Tamanna, T. A., Belal, S. A., Shibly, M. A. H., & Khan, A. N. (2021). Characterization of a new natural fiber extracted from *Corypha taliera* fruit. *Scientific Reports*, *11*, 7622. 10.1038/s41598-021-87128-8

Tamburini, E., León, A. G., Perito, B., Candilo, M. D., & Mastromei, G. (2004). Exploitation of bacterial pectinolytic strains for improvement of hemp water retting. *Euphytica*, *140*, 47–54. 10.1007/s10681-004-4754-y

Terzopoulou, Z. N., Papageorgiou, G. Z., Papadopoulou, E., Athanassiadou, E., Alexopoulou, E., & Bikiaris, D. N. (2015). Green composites prepared from aliphatic polyesters and bast fibers. *Industrial Crops and Products*, *68*, 60–79. 10.1016/j.indcrop.2014.08.034

Thomsen, A. B., Thygesen, A., Bohn, V., Vad Nielsen, K., Pallesen, B., & Jørgensen, M. S. (2006). Effect of chemical-physical pretreatment processes on hemp fibres for re-inforcement of composites and for textiles. *Industrial Crops and Products*, *24*, 113–118. 10.1016/j.indcrop.2005.10.003

Townsend, T. (2020). 1B - World natural fibre production and employment . In R. M. Kozłowski & M. Mackiewicz-Talarczyk (Eds.), *Woodhead Publishing Series in Textiles, Handbook of Natural Fibres* (pp. 15–36) (Second Edition). Woodhead Publishing. 10.1016/B978-0-12-818398-4.00002-5

Turner, J., & Karatzas, C. (2004). Advanced spider silk fibers by biomimicry. In F. T. Wallenberger & N. Weston (Eds.), *Natural Fibers, Plastics and Composites* (pp. 11–25). Kluwer Academic Publishers, Boston.

Uludag, S., Loha, V., Prokop, A., & Tanner, R. D. (1996). The effect of fermentation (re-tting) time and harvest time on kudzu (*Pueraria lobata*) fiber strength. *Applied Biochemistry and Biotechnology*, *57*, 75–84. 10.1007/BF02941690

Umashankaran, M., & Gopalakrishnan, S. (2021). Characterization of bio-fiber from *Pongamiapinnata L.* Bark as possible reinforcement of polymer composites. *Journal of Natural Fibers*, *18*(6), 823–833. 10.1080/15440478.2019.1658254

Umoru, P. E., Boryo, D. E. A., Aliyu, A. O., Adeyemi, O. O. (2014). Processing and evaluation of chemically treated kenaf bast (Hibiscus cannabinus). *International Journal
of Science Technology and Research*, *3*, 1–6.

Unterweger, C., Brüggemann, O., & Fürst, C. (2013). Synthetic fibers and thermoplastic
short-fiber-reinforced polymers: Properties and characterization. *Polymer Composites*,
35(2), 227–236. 10.1002/pc.22654

Van de Weyenberg, I., Truong, T. C., Vangrimde, B., & Verpoest, I. (2006). Improving the
properties of UD flax fibre reinforced composites by applying an alkaline fibre treatment.
Composite Part A-Applied Science, *37*, 1368–1376. 10.1016/j.compositesa.2005.08.016

Van der Werf, H. M. G., & Turunen, L. (2008). The environmental impacts of the production
of hemp and flax textile yarns. *Industrial Crops and Produsts*, *27*, 1–10. 10.1016/
j.indcrop.2007.05.003

Varghese, A. M., & Mittal, V. (2018). Polymer composites with functionalized natural fibers.
In N. G. Shimpi (Ed.), *Woodhead Publishing Series in Composites Science and
Engineering, Biodegradable and Biocompatible Polymer Composites* (pp. 157–186).
Woodhead Publishing. 10.1016/B978-0-08-100970-3.00006-7

Verma, A., Negi, P., & Singh, V. K. (2018). Experimental analysis on carbon residuum
transformed epoxy resin: Chicken feather fiber hybrid composite. Polymer Composite,
40(7), 2690–2699. 10.1002/pc.25067

Verma, D., Gope, P., & Maheshwari, M. K. (2012). Coir fiber reinforcement and application
in polymer composites: A Review. *Journal of Materials and Environmental Science*, *4*,
263–276.

Vijay, R., Vinod, A., Singaravelu, D. L., Sanjay, M. R., & Siengchin, S. (2021).
Characterization of chemical treated and untreated natural fibers from Pennisetum
orientale grass – A potential reinforcement for lightweight polymeric applications.
International Journal of Lightweight Materials and Manufacture, *4*(1), 43–49. 10.101
6/j.ijlmm.2020.06.008

Vinayaka, D. L., Guna, V. K., Madhavi, D., Arpitha, M., & Reddy, N. (2017). Ricinus
communis plant residues as a source for natural cellulose fibers potentially exploitable
in polymer composites. *Industrial Crops and Products*, *100*, 126–131. 10.1016/
j.indcrop.2017.02.019

Vinod, A., Gowda, T. G. Y., Vijay, R., Sanjay, M. R., Gupta, M. K., Jamil, M., Kushvaha,
V., & Siengchin, S. (2021). Novel Muntingia Calabura bark fiber reinforced green-
epoxy composite: A sustainable and green material for cleaner production. *Journal of
Cleaner Production*, *294*, 126337. 10.1016/j.jclepro.2021.126337

Vinod, A., Sanjay, M., Suchart, S., & Jyotishkumar, P. (2020). Renewable and sustainable
biobased materials: An assessment on biofibers, biofilms, biopolymers and biocomposites. *Journal of Cleaner Production*, *258*, 120978. 10.1016/j.jclepro.2020.12
0978

Wang, L., Tong, L., Zhu, S., Liang, J., & Zhang, H. (2020). Enhancing the mechanical
performance of glass fiber-reinforced polymer composites using MWCNTs. *Advanced
Engineering Materials*, *22*(8), 2000318. 10.1002/adem.202000318

Wang, Y., Wu, H., Zhang, C., Ren, L., Yu, H., Galland, M. A., & Ichchou, M. (2018).
Acoustic characteristics parameters of polyurethane/rice husk composites. *Polymer
Composite*, *40*(7), 2653–2661. 10.1002/pc.25060

Wei, J., Liao, M., Ma, A., Chen, Y., Duan, Z., Hou, X., Li, M., Jiang, N., & Yu, J. (2020).
Enhanced thermal conductivity of polydimethylsiloxane composites with carbon fiber.
Composites Communications, *17*, 141–146. 10.1016/j.coco.2019.12.004

Weiss, K. P., Bagrets, N., Lange, C., Goldacker, W., & Wohlgemuth, J. (2015). Thermal and
mechanical properties of selected 3D printed thermoplastics in the cryogenic temperature regime. *IOP Conference Series: Material Science and Engineering*, *102*,
012022. 10.1088/1757-899X/102/1/012022

Woodford, C. (2020). Cermets. Available at https://www.explainthatstuff.com/cermets.html.

Xiao, Z., Wang, S., Bergeron, H., Zhang, J., & Lau, P. C. (2008). A flax-retting endopolygalacturonase-encoding gene from Rhizopus oryzae. *Antonie Van Leeuwenhoek, 94*(4), 563–571. 10.1007/s10482-008-9274-7

Yadav, D., Amini, F., & Ehrmann, A. (2020). Recent advances in carbon nanofibers and their applications – A review. *European Polymer Journal, 138*, 109963. 10.1016/j.eurpolymj.2020.109963

Yahaya, R., Sapuan, S. M., Jawaid, M., Leman, Z., & Zainudin, E. S. (2016). Effect of fibre orientations on the mechanical properties of kenaf–aramid hybrid composites for spall-liner application. *Defence Technology, 12*(1), 52–58. 10.1016/j.dt.2015.08.005

Ye, N., Ren, X., & Liang, J. (2020). Microstructure and mechanical properties of the Ni/Ti/Nb multilayer composite manufactured by accumulative pack-roll bonding. *Metals, 10*(3), 354. 10.3390/met10030354

Yusoff, R. B., Takagi, H., & Nakagaito, A. N. (2016). Tensile and flexural properties of polylactic acid-based hybrid green composites reinforced by kenaf, bamboo and coir fibers. *Industrial Crops and Products, 94*, 562–573. 10.1016/j.indcrop.2016.09.017

Zadorecki, P., Karnerfors, H., & Lindenfors, S. (1986). Cellulose fibers as reinforcement in composites: Determination of the stiffness of cellulose fibers. *Composites Science and Technology, 27*(4), 291–303. 10.1016/0266-3538(86)90072-2

Zhang, L., Li, D., Wang, L., Wang, T., Zhang, L., Chen, K. D., & Mao, Z. (2008). Effect of steam explosion on biodegradation of lignin in wheat straw. *Bioresource Technology, 99*, 8512–8515. 10.1016/j.biortech.2008.03.028

Zhang, Y., Yu, C., Chu, P. K., Lv, F., Zhang, C., Ji, J., Zhang, R., & Wang, H. (2012). Mechanical and thermal properties of basalt fiber reinforced poly(butylene succinate) composites. *Materials Chemistry and Physics, 133*, 845–849. 10.1016/j.matchemphys.2012.01.105

Zhao, X., Wang, X., Wu, Z., Keller, T., & Vassilopoulos, A. P. (2018). Temperature effect on fatigue behavior of basalt fiber-reinforced polymer composites. *Polymer Composite, 40*, 2273–2283. 10.1002/pc.25035

Zhu, B., Li, W., Lewis, R. V., Segre, C. U., & Wang, R. (2014). E-Spun composite fibers of collagen and dragline silk protein: Fiber mechanics, biocompatibility, and application in stem cell differentiation. *Biomacromolecules, 16*, 202–213. 10.1021/bm501403f

2 Extraction of Natural Fibers

Deeksha Jaiswal and G. L. Devnani
Chemical Engineering, Harcourt Butler Technical University
Kanpur, India

INTRODUCTION

Natural fiber–based composites providing efficient utilization in almost all sectors in the current scenario, as it has no issues of degradability such as recycling or disposal problems. Fiber's properties vary on the account of what kind of fiber is used for application. One of the most important factors is that origin of plant, such as plant type or mineral or if it belongs to animal fiber. Plant fibers associated with higher amounts of cellulosic components on the other hand, protein is the main constituent of animal fiber. Characteristics of fiber also have dependency on the structure of plant and chemical components as well, which can be related directly to how the fiber has been extracted from its source, what kind of fiber it is, time of harvesting, conditions until plant maturity, and maturity and decortications as well (Senthilkumar et al., 2018). Generally, there are main types of natural fibers: plant, animal, and mineral fibers.

Plant fibers comprise of seed, leaf, stalk, bast, and fruit fiber, and in animal fibers, silk fibers and hair of animal are used, and amosite and chrysotile come under the category of minerals (Santhosh Kumar et al., 2019). Figure 2.1 shows various plant fibers, such as fiber properties that depend on extractions; various methods of extraction are available, with their pros and cons such as physical (steam explosion, retting, crushing, grinding, rolling mills), chemical (degumming, alkali retting, chemical retting), and physical and chemical (CMT, RMT) (Zakikhani et al., 2014). Figure 2.2 shows various extraction methods.

In the previous methods, retting is one of the most used method as it is environmentally friendly and economical as well. Using all of the above techniques, fibers are extracted. But the properties of these fibers play an important role and cannot be taken for granted, such as flexibility, mechanical properties, etc. and hence various methodologies are available for extraction of fibers. The technique is described as manually ripening of "pseudo-stem" extraction of fiber starts by subsequent use of outer sheath detachment. The barks were washed and the waste was repealed, and the surface was sticky with the remainder. For reducing remaining moisture, pristine bark is exposed in the environment or carried out for oven dry. According to requirements, these barks are treated with the retting method. In this section, various methods for extraction of some traditionally available fibers such as bamboo, banana, bagasse, sisal, jute, and

DOI: 10.1201/9781003201724-2

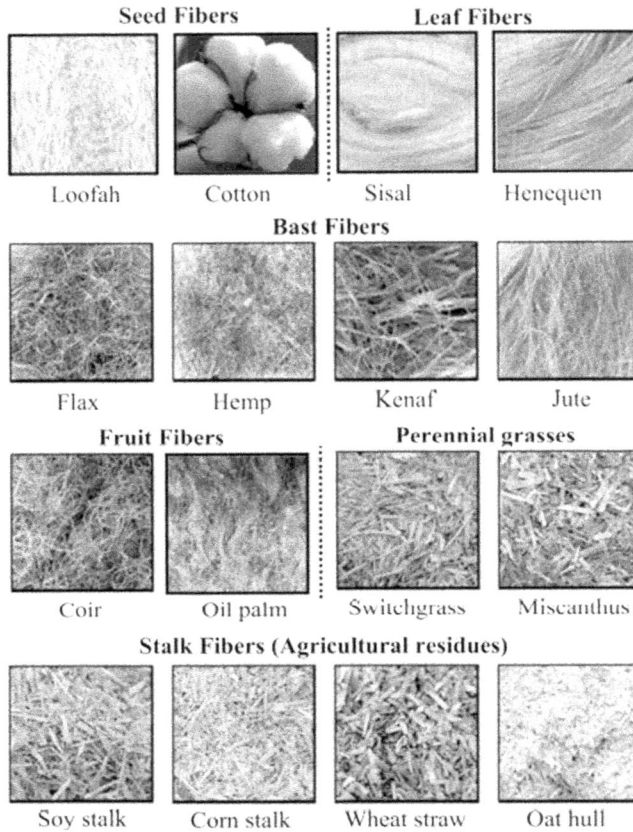

FIGURE 2.1 Types of plant fibers (Santhosh Kumar et al., 2019).

some other fibers as well are described. Banana fiber is a type of plant extraction fiber; generally it is known as pseudo-stem lignocellulosic fiber. Banana fiber consists of higher cellulose fraction but reports lower microfibrillar angle and shows better thermal properties and mechanical properties as compared to others (Biswal et al., 2012). Lignocellulosic fiber is constituent with cellulose, hemicellulose lignin, pectin wax, and some other component (Barreto et al., 2010). So, there are different techniques available for extraction of banana fiber according to the part of the plant and amount of the fiber that must be extracted (Paramasivam et al., 2020). Sisal fiber botanical name (*Agave sislana*) is categorized into hard fibers obtained from the sisal plant leaves. Because of ease in cultivation, this is the most utilized naturally occurring fiber (Li et al., 2000). Some other treatments are also there, such as enzymatic (Vardhini & Murugan, 2016) and chemical treatments for the production of fibers. KOH, H_2SO_4, and $NaClO_2$ chemicals are used for the separation of lignin non-cellulosic fractions from the fiber through chemical treatment.

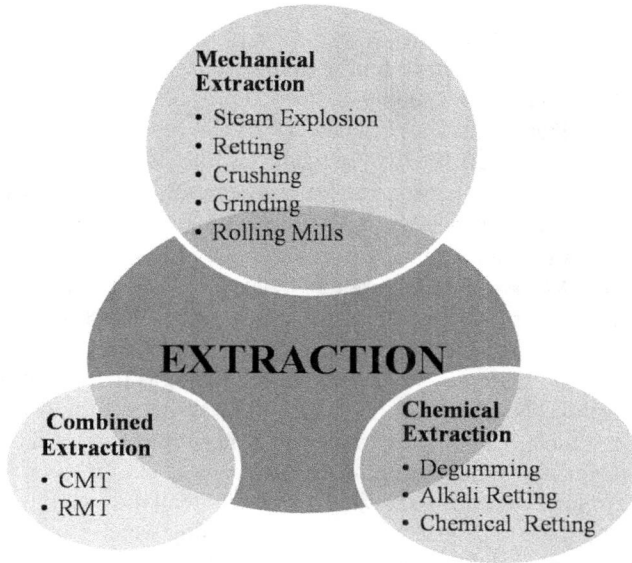

Mechanical Extraction
- Steam Explosion
- Retting
- Crushing
- Grinding
- Rolling Mills

EXTRACTION

Combined Extraction
- CMT
- RMT

Chemical Extraction
- Degumming
- Alkali Retting
- Chemical Retting

FIGURE. 2.2 Extraction methods of natural fibers.

2.1 PHYSICAL EXTRACTION

It involves various methods namely as grinding, crushing, rolling, steam explosion/ heat steam method, and retting. These methods are used for extracting fibers for successful application of natural fiber–based reinforced composites.

Mechanical retting, which is a conventional method used for extraction of fibers, is the same but the manual work is replaced by machines. Two kinds of machines are used for the process of extraction. In first kind, a roller mill is used for decortication by using rolling cylinders that are made of fiber, which leads to reduction in destruction of fiber. Hammer mill decortication is performed by rotating rolls in which straw is thrashed in hurds; mesh is provided in these machines for further fiber filtration and, according to requirements, these machines are used when high length fiber are required, fulfilled by a hammer mill (Brindha et al., 2017). Steam explosion associated with high pressure steam, due to this cellulosic and non-cellulosic content, is broken though compression; bark of the fibers is distorted because of an increase in temperature and production of fiber caused by mechanical loads. Most of the time short strands of fiber are obtained by decompression through elevation (Sheng et al., 2014). Chemical retting is also one of the most-used techniques for the extraction of fiber; mainly NaOH is used for the chemical retting. Besides retting, other methods also exist for the extraction of fiber such as manual extraction. It is performed by scalping and sickling of bark sheath.

2.1.1 RETTING

In this method, as its name suggests, retting means soaking. This method is associated with soaking fiber in water for a couple of days and then the fibers are separated from the cell wall and fiber strips are obtained by combining, beating, and scrapping. Retting is a long-established method; even today this method is in use for extraction of fibers. Acquiring fiber from a banana tree trunk is generally termed "pseudo-stem," as it constitutes a stalk that is compactly packed in a cylindrical way (Pappu et al., 2015). The outer part of a stem yields a baggy fiber that can be taken apart easily even when load is very light. As these fibers do not associate with better quality, they are not interesting at all; even the main part of the pseudo-stem, which contains miscible content, is not suitable as well. So, the intercede part is the subject of attentiveness (Vardhini & Murugan, 2016). For extracting fibers, retting is familiar even today as a conventional method for extracting fiber. Fibers are developed by primeval micro-organisms. In microbiological retting, dew and water retting are small parts of this method (Kengkhetkit & Wongpreedee, 2018; Zuluaga et al., 2004). It is found that fiber extracted from water retting shows better quality than dew retting. Atmospheric conditions are the basic dependency of dew types retting which is advantageous for multiplication of fungi. Quality of fiber is more affected in the case of scraping, resulting in less broken fibers in length. In some studies, it was found that they did not use the above-described method; they just simply cut bamboo in parts and cleaned them and subjected them to fermentation with water for a period of 2 months at normal temperature. In these two methods, aerobic and anaerobic retting was introduced. Authors reported that this method provides good separation of fiber bundles and from this bundle any length of fiber can be obtained. In a study, Vakka and date fiber and bamboo are extracted by the physical retting method. In this, Vakka, which is a verdure of a tree, falls when ripened on the ground. It is a sheath that includes leaves and stem. This fiber was soaked in water about 15 days. After 15 days, when the first layer was removed, it was again soaked in water for another 3 days. Then, the fiber was washed and rubbed by hand for extraction of fibers. The authors reported that this technique of extraction is environmentally friendly and economical as well. In the case of dates, the stem containing surface layers and leaves is cut by using a knife and subjected to shade drying for 5 days. Later, these stems, followed by beating, are put in water for 3 days. Then a sharp knife was used for scraping so that other material can be removed. In a similar way, bark attached with bamboo fibers is removed; the culm portion was used for extraction of fibers. By this, a culm strip was made around 10 mm wide and 1.5 mm thick strips. These strips are immersed in water for 72 hours. Followed by beating, these retted fibers are scraped and combed for separation of fibers (Rao & Rao, 2007). Pineapple leaf fiber was extracted by separating the leaves by scraping and retting and the author reported 1.8% and 1.4% yield for retting and scraping, respectively (Kengkhetkit & Amornsakchai, 2012). Jowar fiber harvested by base cutting trimmings of culms was done and then shade dried for 7 days. The culms were cut and separated. For making strips, pieces are peeled longitudinally. Then, the strips were soaked in water for a time duration of 3 hours; then, by beating with a mallet, plastic fibers were separated followed by scraping and combing (Prasad & Rao, 2011).

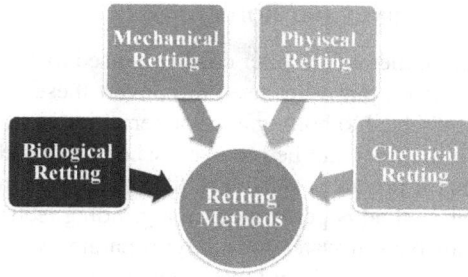

FIGURE 2.3 Types of retting methods.

Properties hurt when retting does not happen in a proper manner, such as some time involved with higher retting termed over-retting and sometimes lower, called under-retting; for overcoming this problem, a supervision is required during the retting process (Brindha et al., 2017). Sometimes a whole field is spoiled because of the brown stains left by microbiological retting, so some development is considered, like the use of seawater because of huge water body pollution. To reduce pollution and the time of the water retting process, some advancement are made; enzymatic retting is a possible alternate method method. In water tanks, enzymes are streamed directly and provide results within a day and this method is also nature friendly (Paridah et al., 2011). Figure 2.3 shows various retting methods.

2.1.2 STEAM EXPLOSION METHOD

In 1962, steam explosion was introduced for detaching the walls of cells for making pulp as a less energy-consuming method. In a study of bamboo fiber it was found that there is not an effective separation of fiber. By using a sifter machine along with a mesh fiber diameter, 125 to 210 µm fiber bundles were obtained. These fibers dried for a period of 120 minutes at a temperature of 120°C (Zakikhani et al., 2014). In another study, pineapple leaf fiber, jute fiber, and banana fiber, followed by mercerization, were subjected to a steam explosion method at 137 pascal pressure. Lignocellulosic material was directly loaded in a steam gun and at treatment was done at a higher pressure steam within a temperature range of 200°C to 250°C. Then the fibers were washed with water. For this method, autoclave was used, which works with the previously proposed pressure (Abraham et al., 2011). As this method did not reach the desired separation of lignin, a mixing machine helped to fulfill this and from this machine bamboo fiber cotton was produced (Okubo et al., 2004). For extracting banana fibers through steam explosion, small pellets of banana fibers near 10 cm in size are prepared. At a temperature of 115°C and at a pressure of about 20 lb, these fibers are immersed in a NaOH solution of 2% concentration after treatment of fibers subjected to drawing and bleaching by using NaOH and acetic acid; afterward, oxalic acid used for hydrolyization and some mechanical treatment performed at the end of the process (Deepa et al., 2011).

2.1.3 Crushing, Grinding, and Rolling Mill

In the crushing extraction method, a roller crusher is used to chop the raw fiber into smaller fractions and then a pin roller used to convert these fibers. Before drying these fibers, they are subjected to boiling at some temperature for a time duration of 10 hours so that fat in the fiber can be removed. The only disadvantage associated with this method is that it produces small size fibers and leads to a powdered form when another physical method is performed. In a grinding technique, fibers are cut in strip forms and immersed in water for a day; separated strips are chopped into small pieces manually using a knife and the wider strips are converted into smaller chips by passing in extrudes. Now these fibers are blended for a half-hour in a grinder with the help of sieves of various sizes of fields obtained. Then the fibers were oven dried (Thwe & Liao, 2002). In studies it was found that fibers that were extracted by this method possess good tensile load and modules but are not less effective in tensile strength and enhanced crystallization was found as well (Yingchen et al., 2010). In the rolling mill technique, small pieces of fibers were soaked in water for 60 minutes and then subjected to a rolling mill at a lower speed; obtained strips were immersed in water for half an hour and then fibers were separated with a razor blade. In roller mill bonding, the strength of fiber is reduced. Chopped pieces of pineapple leaf fiber containing about 80% water content are subjected to a ball mill. Ball mills are made of rigid PVC pipes and balls (50 in number) made of steel. Chopped 240 g of leaves of pineapples, along with 150 mL water, was put in a ball mill. Then the mill was set to rotate with 60 rpm in a shaft for a time duration of 6 hours (Kengkhetkit & Amornsakchai, 2012).

2.2 CHEMICAL EXTRACTION

In chemical extraction, as its name specifies, its meaning is using chemicals for extraction. Various chemicals are used; available methods with categories under chemical extraction are acid retting or alkali retting, CAN method (chemical assisted natural), degumming, and chemical retting. This method signifies its advantage as it reduces lignin components and another microstructures as well.

2.2.1 Degumming

In this process, gummy and pectin fractions are removed from fibers. In an study, the outer covering of bamboo was chopped off manually and exposed to the sun for drying. The fiber contains a gummy material and tissues; these are removed by degumming, which is a chemical method of extraction. The author used a degumming method and reported a 33% yield of fiber on the basis of weight (Rao & Rao, 2007).

2.2.2 Chemical Retting

Fibers are soaked in acidic or in alkaline solution for some time and press machined, and a steel nail is used to separate the fibers. Various chemical solutions are used: $Zn(NO_3)_2$ sodium silicate, sodium sulphite, sodium polyphosphate. Chemical retting is a technique that uses chemicals for the extraction of fiber. It comprises good-quality fiber production and less time for the process as well. Uses of alkali are more common for this method, such as NaOH (Parre et al., 2019).

2.3 PHYSICAL AND CHEMICAL EXTRACTION

Chemically treated fiber is subjected to compression moulding technique (CMT) and roller mill technique (RMT) for extraction of fiber. Bamboo fiber was extracted by cutting into small fractions and then grounded and dried in air. Extraction of bamboo fiber was done using toluene/ethanol volume/volume in a Soxhlet apparatus and then oven dried and finally these fibers were subjected to ball milling with 4 rpm of 500. After that, these fibers were used for further investigation (Yang et al., 2013). In another study, steam explosion fiber was applied for bleaching with a chemical such as chlorine dioxide for removal of lignin 5% oxalic acid treatment and again subjected to steam explosion. Now these fibers were cleaned with water and sent to a mechanical stirrer and then further characterized (Abraham et al., 2011).

2.4 EFFECT OF EXTRACTION TECHNIQUE ON QUALITY OF FIBERS

The availability of different kinds of extraction techniques leaves various significant effects on the properties of fibers. These properties are investigated by convenient characterization methods according to requirements such as FTIR, SEM, EDX, TGA, and DSC, etc. These techniques not only provide properties but also provide a clear idea about their compatibility with a polymer matrix. SEM (scanning electron microscopy) is method for defining the morphology of the fiber. It gives a clear view on how effectively fiber can be used for reinforcement, and provides information on whether the fiber's surface requires any modification or not. FTIR leads to proof the pressure of cellulosic or non-cellulosic component. TGA gives an idea of the thermal behavior of fibers. In some studies, sound absorption behavior was also determined; increment in thickness of fiber leads to improvement in acoustic properties (Malawade & Jadhav, 2019). Similarly, other techniques also provide significant information in their respective areas. Bamboo fibers were extracted with steam explosion and alkaline methods and with other physical and chemical methods; the authors reported that a reduction in lignin content was observed in the case of steam explosion and chemical methods; removal of lignin from the surface of bamboo fibers made more compatible with a polymer matrix (Zakikhani et al., 2014). Figure 2.4 is a SEM image of pineapple fiber by different extraction methods (Table 2.1).

2.5 NEW AND UNCOMMON NATURAL FIBERS

Reinforced polymer associated with excellent performance can be produced by inclusion of fibril, such as carbon and glass fibers. But when our objective lies in making lower-weight products, through natural fiber reinforcement, the requirement of lightweight material can be achieved. Natural fiber–reinforced composite can be

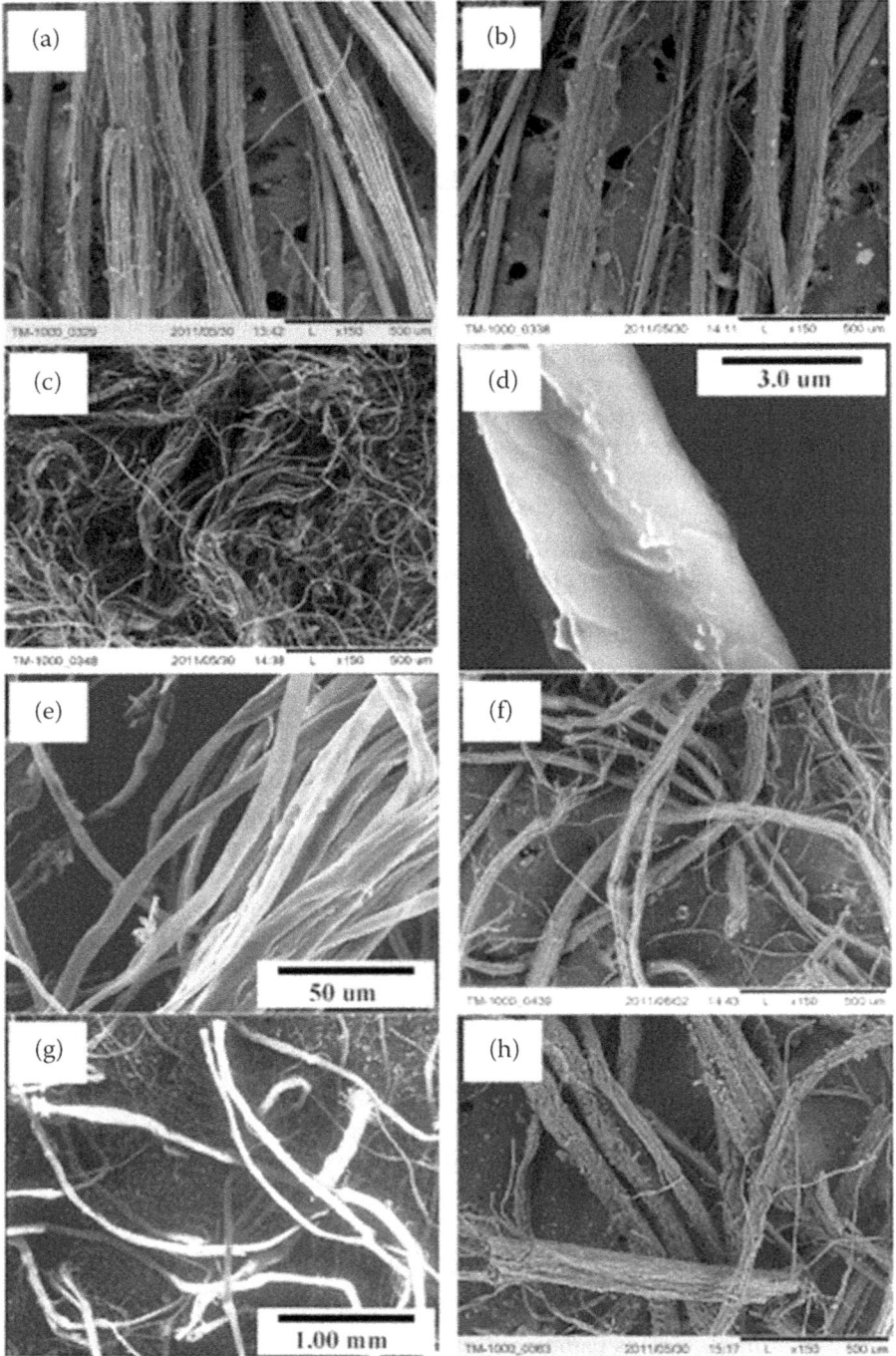

FIGURE 2.4 SEM images of pineapple leaf fiber (PALF) of different extraction methods (a) scraping – PALF, (b) retting – PALF, (c, d, e), ball milling – PALF, (f, g) milling – PALF, (h) milling of dry leaves – PALF (Kengkhetkit & Amornsakchai, 2012).

TABLE 2.1
Effect of Extraction Technique on Traditional Fibers

Fiber Name	Method of Fiber Extraction	Physical Properties (Density, Diameter)	Mechanical Properties (Tensile Strength, Tensile Modulus)	Characteristics of Fiber (FTIR, SEM, TGA, and Some Other Techniques)	References
Pineapple Leaf Fiber	Retting Scraping Milling Ball milling Dried leaves milling	Diameter – 5–166 5–129 3–68 3–95 5–194	Improvement in tensile properties was observed on increasing content of pineapple leaf fiber	Lower weight loss was observed in second step of degradation in TGA in case of dried leaves milling as compared to other extraction methods	(Kengkhetkit & Amornsakchai, 2012)
Banana fiber	Manual extraction Chemical retting	Density – 760–960 Kg/cm^3 diameter – 50–250 μm diameter – 50–250 μm	Tensile strength – 110–400 MPa Tensile modulus – 20–25 GPa Tensile strength – 700–800 MPa Tensile modulus – 24-32 GPa	Water absorption percentage in manually extracted fiber was 55%–60% XRD reveals peak at 5.99 and 5.33, according to TGA moisture removal was occurred at 100°C	(Guimarães et al., 2009; Tadasse et al., 2018)
Bamboo fiber	Steam explosion Alkaline Chemical +compression Chemical+ roller mill	Diameters – 15–210 μm Diameters – 230 ± 180 μm Density – 0.8–0.9 g/cm^3 diameter – 50–400 μm Diameter – 50–100	Tensile strength – 441 ± 220 MPa, Tensile modulus – 36 ± 13 GPa Tensile strength – 395 ± 155 MPa, Tensile modulus – 26.1 ± 14.5 GPa Tensile strength – 645 MPa Tensile strength – 379 MPa	In removal of lignin, steam explosion technique and chemical method performed well in order to find better adhesive with polymer matrix, short fiber were obtained through these methods but long fibers were obtained in case of rolling milling	(Zakikhani et al., 2014)
Bagasse	Crushing Ground in knife mill	Thickness of 10, 20, 30 mm. Diameters – 0.125 m–0.355 m 0.355 m–0.500 m 0.500 m–0.850 m	–	Increment in thickness of fiber leads to improvement acoustic properties. As particles size increase true density decreases highest density recorded for 0.125 m–0.355 m is 1.4989 for dried sample r	(Cardoso et al., 2013; Malawade & Jadhav, 2019)

applied in various sectors, one of them being food packaging (Sydow & Bieńczak, 2018). Traditionally available natural fibers, as described in an earlier section, such as banana, bamboo, sisal, jute, and bagasse, etc. are the materials with tremendous properties. Studies of these traditionally available fibers have been done in a very vast manner and found that these fibers can replace man-made fibers. As a environmental point of view, sole dependency on traditional fibers is not worthy so researches are keen on exploring new natural fibers as they reduce dependency on natural fibers (Youssef et al., 2019). Exploring new fibers provides a way towards independent and eco-friendly products. As natural fibers are associated with enormous properties, in order to sustain these properites it is required to use these fibers in an efficient way. Extraction of natural fibers is an important factor in order to preserve the quality of fibers; in this context, various extraction methods are available as described before. With the help of these methods, various new and uncommon natural fibers are extracted and their properties are studied. In this section, new and uncommon natural fibers are explained with their extraction method and their corresponding properties that have been explored in recent years.

2.5.1   FIBERS EXPLORED IN YEAR 2018

Area palm leaf stalk fibers are separated manually and then a water retting technique was used for extraction of fibers and then cleaned and dried. Authors reported that impurities were found in untreated fibers removed after alkali treatment and also physical and chemical property improvements (Shanmugasundaram et al., 2018). Albiazia amara fibers were extracting using retting method fiber density was lower than other fibers; the authors suggested that lower density fiber will lead to preparation of a lightweight composite (Senthamaraikannan et al., 2019). Furcraea foetide leaf fibers were extracted by immersing fibers in water for 14 days. Afterwards, the fibers were seperated using a metal tooth comb. Higher amounts of cellulosic content 68.35 wt% was determined and found that oxygen and carbon were present in higher amount by EDX analysis (Manimaran et al., 2018).

Phaseolous vulgaris fibers were extracted by peeling and found 43.07 crystallinity and higher amounts of cellulose 62.14 wt% and thermal stability at 150°C (Gurukarthik Babu et al., 2019). Figure 2.5 shows the extraction process of fibers.

The bark of pigeon pea was used as a natural fiber source and extracted by immersing in water so that the outermost layer and inner part are separated by virtue of microbial degradation (Kulandaivel et al., 2018). African teff straw, which was collected from Ethiopia, was cleaned using water in order to reduce impurities of the surface and termed untreated teff straw and it was found that this fiber contains good thermal and mechanical properties when compared to other available natural fibers (Devnani & Sinha, 2018). Onion and garlic skin and stalk were used as source of natural fiber in a study. These discarded wastes were cleaned and dried and then subjected to grinding for making a powdered form of fiber cellulosic component of onion skin and stalk was found at 41.1% and 45.5%, respectively. The authors suggested using this fiber as a potential reinforced composite that can be utilized in various sectors, such as food packaging applications (Reddy & Rhim, 2018) (Table 2.2).

Areca palm leaf stalk tree   Areca palm leaf stalk with leaves   Manual removal of leaves from the stalk

Areca palm leaf stalks   stalks were immersed in the water   Areca palm leaf stalk fibers

FIGURE 2.5   Extraction process of Areca palm leaf stalk fibers (Shanmugasundaram et al., 2018).

2.5.2   FIBERS EXPLORED IN YEAR 2019

Tridax fibers were seperated manually by retting process and then washed and dried and carried out for further treatment and characterization it was reported that physical and chemical properties were enhanced, and the roughness of the surface has become more favorable on alkali treatment for composite reinforcement (Vijay et al., 2018). For Saccharum Bengalense, the fiber retting process was used for extraction and then dried under sunlight; a higher amount of lignin was reported that led to fractures on the surface at the time of tensile testing but found rougher surfaces with cracks that would be suitable for the reinforcement. The authors suggested the applications of this fiber in household furniture and some utilization in the automobile sector (Vijay et al., 2019). Pigeon pea fibers were obtained by mechanical process (R. S. Kumar et al., 2019). Ficus Religiosa tree fiber was extracted by the method of microbial degradation for a couple of weeks, and then extracted fibers were carried out for further characterization. The density of this fiber was found comparable to other available traditional natural fibers such as banana, sisal, hemp, amd kenaf fiber; smoothness was observed on the surface and Moshi et al., 2019, using a Briodio plot, activation energy of the fiber was reported at 68.02 kJ/mol (Moshi et al., 2019). Eleusine Indica grass and Elettaria card amomum fibers were obtained by using the retting process and then cleaned and subjected to sunlight for drying (Ahmed et al., 2019; Khan et al., 2019). The water retting process is used for separation of fibers from Eichhornia Crassipes; the fibers

TABLE 2.2

Properties of Extracted Fibers in the Year 2018

Fiber Name	Method of Extraction	Composition (%)	Density (g/cm⁻³) and Diameters (μm)	Mechanical Properies	FTIRWavenumber (cm⁻¹) and Corresponding Functional Groups	Thermal Properties (°C)	Crystallinity (%)	References
Furcracea foetida	*Retting*	Cellulose – 68.35 Hemicellulose – 11.46 Lignin – 12.32	Density – 0.778 Diameter – 12.8 μm	Tensile strength – 623.52 ± 45 Young's Modulus – 6.52 ± 1.9	$3,800–3,100$ cm⁻¹ – OH stretching $2,935$ cm⁻¹ – CH stretching $2,844$ cm⁻¹ – CH₂,1,647 cm⁻¹ – OH bending $1,023$ cm⁻¹ – CO and OH stretching	320.5	52.6	(Manimaran et al., 2018)
Pigeon pea	Water retting	Cellulose – 55.03 Lignin – 18.32	Density – 1.728	Tensile strength – 131 Young's Modulus – 2.1	$3,417$ cm⁻¹ – OH stretching $2,916$ and $2,846$ cm⁻¹ – CH stretching $1,704$ and $1,627$ cm⁻¹ – C-O group $1,056$ cm⁻¹ – C-O bond bending of alkoxy	225	65.89	(Kulandaivel et al., 2018)
Areca palm leaf stalk fiber		Cellulose – 57.49 Hemicellulose – 18.34 Lignin – 7.26	Density – 1.09 Diameter – 85–330 μm	Tensile strength – 334.66 ± 21.46 Young's Modulus – 7.64 ± 1.13	$3,872$ cm⁻¹ – OH stretching 2954 cm⁻¹ – CH stretching of cellulose and CH₂ of hemicellulose $1,634$ cm⁻¹ – C=O group $1,054$ cm⁻¹ – CO and OH stretching attribute to polysaccharides in cellulose	279		(Shanmugasundaram et al., 2018)
Albizia amara	Water retting	Cellulose – 64.54 Hemicellulose – 14.32 Lignin – 15.61	Density – 1.043	Tensile strength 640 ± 13.4	$3,448$ cm⁻¹ – OH stretching $2,918$ cm⁻¹ – CH stretching (cellulose) $2,361$ cm⁻¹ – CH₂ streching (wax) $1,647$ cm⁻¹ – C=O of α keto carboxylic acid in lignin $1,023$ cm⁻¹ – CO stretching attribute to lignin content	330.6	63.78	(Senthamaraikannan et al., 2019)
African teff straw	-		Density – 1.15	Tensile strength – 280–326 Young's Modulus –9.2–10.7	$3,328$ cm⁻¹ – OH stretching $2,918$ and $2,850$ cm⁻¹ – CH stretching (cellulose) and CH₂ stretching (hemicellulose)	260	54	(Devnani & Sinha, 2018)

Fiber	Method	Chemical composition	Properties	FTIR analysis			Reference
Tridax procumbens	*Retting*	Cellulose – 57.49, Hemicellulose – 18.34, Lignin – 7.26	Density – 1.16, Diameter – 233.1 μm	Tensile strength – 25.57 ± 2.45, Young's modulus – 0.94 ± 0.09	1,738 and 1,226 cm^{-1} – attributing to vibration of aromatic skeletal 1,010 cm^{-1} – CO stretching 3,296 cm^{-1} – OH stretching 2,932 cm^{-1} – aldehyde group 2,894 cm^{-1} – carboxylic acid 1,590 cm^{-1} – = O linkage 1,375 cm^{-1} – C-O group	250 · 34.46	(Vijay et al., 2018)
Onion skin	Grinding	Cellulose – 41.1 ± 1.1, Hemicellulose – 16.2 ± 0.6, Lignin – 38.9 ± 1.3	–		Significant peak was observed at 1,738 cm^{-1}, 1,731 cm^{-1}, 1,735 cm^{-1}, 1,732 cm^{-1} for onion skin, onion stalk, garlic skin, garlic stalk, respectively, attributing presence of C=O acetyl ester functional group and carbonyl aldehyde group significant to hemicellulose and lignin content	240	(Reddy & Rhim, 2018)
Onion stalk		Cellulose – 45.5 ± 0.9, Hemicellulose – 25.5 ± 0.3, Lignin – 26.2 ± 1.7				245	
Garlic skin		Cellulose – 41.7 ± 2.1, Hemicellulose – 20.8 ± 1.6, Lignin – 34.6 ± 2.4				235	
Garlic stalk		Cellulose – 49.8 ± 1.8, Hemicellulose – 17.5 ± 0.9, Lignin – 28.6 ± 0.7				257	

possessed a higher content of cellulose but lower fraction of hemicellulose was found and activation energy of this fiber was 66.324 KJ/mol, and surface modification of fiber aws recommended for better adhesion towards a polymer matrix (Palai et al., 2019). Extraction of (Ampelodesmos Mauritanicus) Diss fiber was performed using the retting process; wetted fibers were scraped and fiber threads were produced and exposed to surface modification in a NaOH solution; the authors evaluated the density of fiber as 0.93 g/cm^3, which is quite low and suitable for lightweight application (Nouri et al., 2019). The retting method was also used in extraction of Cereus Hildmannianus fiber and Albizia Lebbeck Bark fiber; these fibers showed porous and rougher surfaces and exhibited cellulosic content 58.40% ± 0.56% for Cereus Hildmannianus fiber and higher cellulosic fraction of 72.36 wt% was determined in the case of albizia lebbeck (Manimaran, Kumar, et al., 2019; Subramanian et al., 2019).

The retting method was applied for the extraction of date palm tree fiber; different parts of the date palm tree were taken for study, namely the trunk of the tree, leaf stalk, leaf sheath, and fruit bunch of palm tree, which were further termed DPTRF, DPLST, DPLSH, and DPFBS; in all above parts, the date palm fruit bunch showed better properties compared to other fiber parts (Alotaibi et al., 2019). Celosia Argentea fiber extraction was performed manually; higher amounts of cellulose 64.34 wt% were reported and less amounts of waxy material were found on fiber surfaces (Manimaran, Sanjay, et al., 2019). Pongmia Pinnata l., which has a height of 50 to 60 ft, the fully grown leaves are taken for study and fiber was extracted by the water retting process. The mechanical processing was done with a metal comb and a higher fraction of cellulose was reported 62.34 wt% and XRD crystallinity of 45.31%. An SEM revealed that the surface was smooth and the authors suggested some chemical modifications of the fibers (Umashankaran & Gopalakrishnan, 2019). The retting technique was used for extraction of Cardiospermum Halicababum; findings revealed that fibers had less amounts of waxy content and lower moisture content, which would be favorable for reinforcement (Vinod et al., 2019). Mechanical decorticator was used for extraction of fibers from the aerial root of the banyan tree (by stripping the skin) (Ganapathy et al., 2019). The retting method was also used for extracting Acacia Tortills bark fiber, kans grass, and Furcracea Foetida fiber (Dawit et al., 2019; Devnani & Sinha, 2019; Manimaran et al., 2018). Figure 2.6 shows the extraction of Kans grass fiber (Table 2.3).

2.5.3   FIBERS EXPLORED IN YEAR 2020

Acacia Nilotica L. was extracted using the retting method; chemical composition of cellulose was 56.46%, and 8.33% of hemicellulose was determined. Fiber was degraded at a maximum at 339°C temperature and 69.739 kJ/mol of activation energy was evaluated (R. Kumar et al., 2020). Sesbania Rostrata plants were collected from the field of turmeric, and the fibers were extracted manually from the Sesbania Rostrata plant's stem, initially fiber was covered with impurities waxy materials which were reduced, and roughness was produced an surface modification of fiber, chemical constituent of fiber; cellulose − 72.75, hemicellulose − 8.01, lignin − 15.91 and in physical properties density (1.482 g/cm^3 was determined, mechanical properties such as tensile strength (439 ± 26.31 MPa), Young's

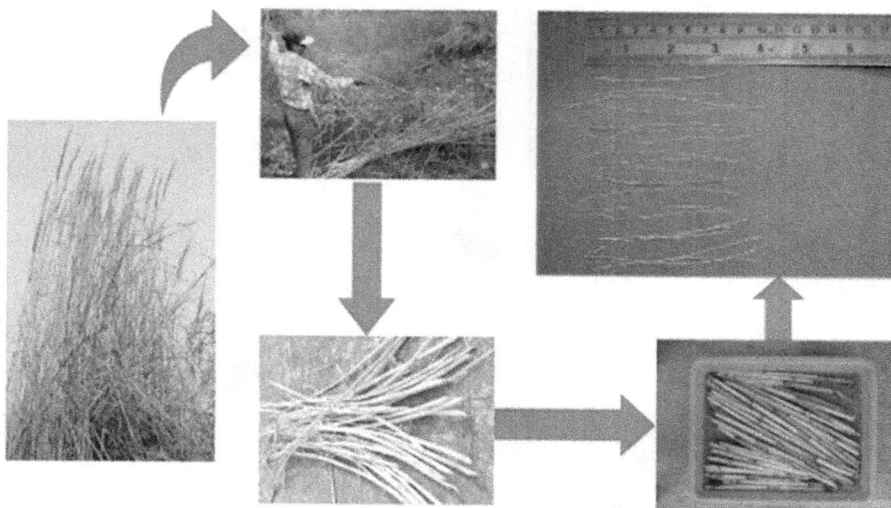

FIGURE 2.6 Extraction of Kans grass fiber (Devnani & Sinha, 2019).

Modulus (42.83 ± 6.92 Gpa) was investigated. Crystallinity was determined as 69.71% (Raja, Senthilkumar, et al., 2020). Leaf of the Yankee pineapple was extracted mechanically, and the fibers exhibited good mechanical and chemical properties and higher crystallinity was observed at 55.22%, which is desirable for application of composites (Najeeb et al., 2020). Leucas aspera fibers were extracted by the retting process manually; observed reduction in diameter and increment in density on treatment with silane by the authors suggested that leucas aspera fiber treated with silane can be utilized in composite application significantly (Vijay, Manoharan, et al., 2020). Extraction of Adansonia Digitata L. (Baobab) is done manually by the retting process of fibers was extracted from the bast covering of Baobab tree stem the stem, chemical components of fiber were 60.70%, 21.8%, and 5.91% for cellulose, hemicellulose, and lignin, respectively. Fiber density was 1.1041 g/cm^3, which is desired applications where light weight is demanded, such as the automobile sector and textile industry as well (Eltahir et al., 2020). Shwetark stem fiber is extracted by peeling the fibrous material from the stem, and cellulose content was estimated at 69.65% and determined density of fiber 1.364 g/cm^3 favorable for reinforcement, good mechanical strength was reported for this fiber, roughness was observed at the surface of fiber through AFM and no degradation was seen up to 225°C (Raja, Prabu, et al., 2020). Fiber extraction of red banana peduncle is done by manually and reported that the alkali treatment improved the crystallinity and mechanical properties as well (Pillai et al., 2020).

Fiber of Cortaderia Selloana grass was separated from the fiber by the retting method; fiber stability was observed upto 320°C temperature, which more than thermoplastic polymerization (Khan et al., 2020). Retting method was used for extraction of Vernonia elaeagnifolia fiber, as shown in Figure 2.7, thermal, physical and chemical properties was analyzed using TGA, FTIR, and SEM; fiber exhibit

TABLE 2.3

Properties of Extracted Fibers in the Year 2019

Fiber Name	Method of Extraction	Composition	Density and Diameters	Mechanical Properites	FTIR	Thermal Properties	Crystallinity	References
Phaseolus vulgaris	Peeling	Cellulose – 62.17 Hemicellulose – 7.04 Lignin – 9.13	Density – 0.852	–	3,748 cm^{-1} – OH-stretching 2,922 cm^{-1} – C≡C stretching 1,745 cm^{-1} – C=O 1,517 cm^{-1} – C=C stretching	150	43.01	(Gurukarthik Babu et al., 2019)
Saccharum bengalense	Retting	Cellulose – 53.45 Hemicellulose – 31.45 Lignin – 1.7	Density 1.165	Tensile strength – 33 ± 1.54	3,369 cm^{-1} – OH stretching vibration 2,915 cm^{-1} – C-H stretching vibration 2,850 cm^{-1} – CH$_2$ group 1,729 cm^{-1} – C=O linkage 1,575 cm^{-1} – C=C group 1,024 cm^{-1} – C-O group	–	44.02	(Vijay et al., 2019)
Ficus religiosa tree	Microbial Degradation	Cellulose – 55.58 Hemicellulose – 13.86 Lignin – 10.13	Density – 1.246 Diameter – 25.62 μm	Tensile strength – 433.32 ± 44 Young's modulus – 5.42 ± 2.6	3,326 cm^{-1} – OH group 2,922–2,857 cm^{-1} – C≡C1, 612 cm^{-1} – C=O stretching 1,037 cm^{-1} – = C-O and O-H stretching	325	42.92	(Moshi et al., 2019)
Eleusien indica grass	Retting	Cellulose – 61.3 Hemicellulose – 14.7 Lignin – 11.12	Density – 1.143	Tensile strength – 22 ± 1.0 Young's modulus – 10.75 ± 0.5	3,301 and 2,916 cm^{-1} – indicating presence of cellulose 2,842 cm^{-1} – showing presence of wax 1,723 cm^{-1} – C=O linkage 1,039 cm^{-1} – C=C group	205	45	(Khan et al., 2019)
Elettaria cardamomum	Retting	Cellulose – 63.12 Hemicellulose – 13.7 Lignin – 16.5	Density – 1.470	Tensile strength – 294 ± 1.62 Young's modulus – 7.63 ± 2.1	3,329 cm^{-1} – OH stretching 2,918 cm^{-1} – CH$_2$ group 1,640 cm^{-1} – C=C linkage 1,427 and 1,023 cm^{-1} – correspond to C-H and C-O group	230	36.84	(Ahmed et al., 2019)
Eichhornia crassipes	Biological retting	Cellulose – 59.86 Hemicellulose – 9.65 Lignin – 13.97	Density – 1.350	–	3,333 cm^{-1} – OH stretching 2,924 cm^{-1} – – CH group 1,738 cm^{-1} – C=O linkage 1,027 cm^{-1} – correspond to C-OH group	200	44.32	(Palai et al., 2019)

Fiber	Extraction method	Chemical composition	Physical property	Mechanical property	FTIR	Value	Reference
Cereus hildmannianus	Cutting, spin removal and retting	Cellulose – 58.40 Hemicellulose – 17.14 Lignin – 10.36	Density – 1.364	Tensile strength – 2013.98 ± 48–2897.47 ± 23 Modulus – 1.64 ± 0.4 GPa–2.98 ± 0.2	3,332 cm⁻¹ – OH stretching 2,897.63 cm⁻¹ – C-H group 1,731.84 cm⁻¹ – C=O linkage 1,640 cm⁻¹ – correspond to C-O group (carbonyl or acetal group)	285.9	(Subramanian et al., 2019)
Date palm fruit bunch stalk	Water Retting followed by crushing and chemical extraction (using ethanol benzene solution)	Cellulose – 44 Hemicellulose – 26.00 Lignin – 11.45	Diameter – 294.56 µm	–	–	245.3	(Alotaibi et al., 2019)
Albizia Lebbeck Bark	Retting	Cellulose – 72.59 Hemicellulose – 9.69 Lignin – 10.08	Density – 0.905	Tensile strength – 270 Young's modulus – 67.45	3,427 cm⁻¹ – OH stretching 2,915 cm⁻¹ – CH₂ group 2,719 cm⁻¹ CH₂ group in hemicellulose 1,742 cm⁻¹ – C-O stretching of hemicellulose 1,043 cm⁻¹ – C-O group of lignin	353.37	(Manimaran, Kumar, et al., 2019)
Celosia argentea	Manual extraction	Cellulose – 63.34 Hemicellulose – 12.61 Lignin – 8.99	Density – 0.843		3,392 cm⁻¹ – OH stretching 2,922 cm⁻¹ – CH group 1,631 cm⁻¹ – C=O stretching 1,039 cm⁻¹ – C-O group	324	(Manimaran, Sanjay, et al., 2019)
Pongamia pinnata l	Retting followed by mechanical processing	Cellulose – 62.34 Hemicellulose – 14.57 Lignin – 12.54	Density – 1.345	Tensile strength – 322 Young's modulus – 9.67	3,387 cm⁻¹ – OH stretching 2,927 cm⁻¹ – CH group 1,632 cm⁻¹ – acetyl group stretching hemicellulose 1,249 cm⁻¹ – acetyl group in lignin1,013 cm⁻¹ – C-OH group	210	(Umashankaran & Gopalakrishnan, 2019)
Cardiospermum halicababum	Retting	Cellulose – 59.82 Hemicellulose – 16.75 Lignin – 9.3	Density – 1.141 Diameter – 315.4 µm	Tensile strength – 20.7 ± 1.0	3,380 cm⁻¹ – OH stretching 2,916 and 2,859 cm⁻¹ – CH and CH₂ stretching (sp³ hybridized) 2,279 cm⁻¹ – C≡C 1,020 cm⁻¹ – C-O group	–	(Vinod et al., 2019)

(Continued)

TABLE 2.3 (Continued)

Properties of Extracted Fibers in the Year 2019

Fiber Name	Method of Extraction	Composition	Density and Diameters	Mechanical Properites	FTIR	Thermal Properties	Crystallinity	References
Aerial root of banyan tree fiber	Mechanical Decorticator followed by retting	Cellulose – 67.32 Hemicellulose – 13.46 Lignin – 15.62	Density – 1.234 Diameter – 0.09–0.14 μm	Tensile strength – 19.37 ± 7.72 Young's modulus – 1.80 ± 0.40	3,280 cm^{-1} – OH stretching 2,909 cm^{-1} – CH stretching 2,336 cm^{-1} – C≡C 1,013 cm^{-1} – C-OH vibration for lignin 649 cm^{-1} – C-S stretching	Thermal stability upto 230°C	72.47	(Ganapathy et al., 2019)
Acacia tortills bark fiber	Manually separation followed by retting	Cellulose – 61.89 Lignin – 21.26	Density – 0.906 Diameter – 0.48 μm	Tensile strength – 71.63 Young's modulus – 4.21	–	–	–	(Dawit et al., 2019)

FIGURE 2.7 Extraction of Vernonia elaeagnifolia fiber (a) plant, (b) stems, (c) soaked in water, (d) retting process (Shaker et al., 2020).

density of 1.22 to 1.43 g/cm^3 and tensile properties 37.75 GPa (tensile strength) and 259.6 MPa (tensile modulus) was determined. Fiber was able to bear temperatures up to 200°C; the authors remarked that Vernonia fiber can be used for outdoor applications as it is comparable to other available fibers (Shaker et al., 2020). Bauhinia Vahlii was extracted by beating the stem of Bauhinia vahlii bark; afterwards, fibers were separated manually by hand. Chemical composition for this fiber was cellulose (61%–72%), hemicellulose (11%–14%), and lignin (10%–13%) and tensile strength and Young's Modulus was 38.96–91.28 MPa and 3.08–6.9 GPa, respectively. FTIR recorded wave numbers at 3,600 cm^{-1}–3850 cm^{-1}, which corresponded to OH stretching, 2,800 cm^{-1}–3,000 cm^{-1} attributed to CH stretching, 2,250 cm^{-1}–2,390 cm^{-1} C≡C alkynes stretching was observed at 2,250 cm^{-1}–2,390 cm^{-1} and 1,455 cm^{-1}–1,900 cm^{-1} wave number signifying the presence of C=C of aromatic ring of lignin. Fiber thermal stability was observed up to a 240°C temperature (Patel et al., 2020) (Table 2.4).

CONCLUSION

Several methods are available for the extraction of natural fibers. Every method has their own demerits, merits, and identification that make them a unique and different technique. The water retting method is one of the most-used methods in all available extraction techniques as it is economical; also, significant properties were found in studies after applying the retting method as an extractive method of fibers,

Natural Fiber Composites

TABLE 2.4
Properties of Newly Extracted Fiber in the Year 2020

Fiber Name	Method of Extraction	Composition %	Density and Diameters	Mechanical Properties	FTIR	Thermal Properties	Crystallinity	References
Bark of Vachellia	Water retting followed by peeling manually	Cellulose – 38.3 Hemicellulose – 12.1 Lignin – 9.2	Density – 1.270	Tensile Strength – 33.075 ± 1.3 MPa	$3,317\ cm^{-1}$ represent stretching of OH $2,922-2,853\ cm^{-1}$ correspond to CH and CH_2 stretching $1,051\ cm^{-1}$ C-OH stretching corresponding to presence of lignin	345°C	13%	(Vijay, Daniel, et al., 2020)
Bauhinia vahii	Beating	Cellulose – 61–72 Hemicellulose – 11–14 Lignin – 10–13	–	Tensile Strength – 38.96–91.28 MPa Young's Modulus – 3.08–6.9 GPa	$3,600\ cm^{-1}-3,850\ cm^{-1}$ – OH stretching $3,350\ cm^{-1}-3,500\ cm^{-1}$ – OH stretching $2,800\ cm^{-1}-3,000\ cm^{-1}$ – CH stretching $2,250\ cm^{-1}-2,390\ cm^{-1}$ – C≡C alkynes stretching $1,455\ cm^{-1}-1,900\ cm^{-1}$ – C=C Aromatic ring of lignin $930\ cm^{-1}-1,210\ cm^{-1}$ – Alkyl amine $1,000\ cm^{-1}-1,100\ cm^{-1}$ – C-O stretching	Thermal stability – 240°C		(Patel et al., 2020)
Vernonia elaeaignifolia	Retting	–	Density – 1.43 g/cm^3	Tensile Strength – 259.62 MPa Young's Modulus – 37.25 GPa	$3,320\ cm^{-1}$ – OH group $2,852\ cm^{-1}$ – CH stretching $1,596\ cm^{-1}$ – C-O vibrations $1,028\ cm^{-1}$ – CO and OH stretching	Thermal stability – 200°C	–	(Shaker et al., 2020)
Cyrtostachys renda	Grinding	Cellulose – 21.48 Hemicellulose – 18.41 Lignin – 23.63	Density – 0.90	Tensile Strength – 51.82 MPa Young's Modulus – 0.69 ± 0.18GPa	$3,283\ cm^{-1}$ – OH stretching $2,917\ cm^{-1}$ – CH stretching $1,732\ cm^{-1}$ – C≡C stretching $1,603\ cm^{-1}$ – C=O stretching	125–295	44.72	(Loganathan et al., 2020)

Fiber	Extraction	Composition	Density	Mechanical	FTIR			Reference
Psychosperma macarthurii		Cellulose – 32.62 Hemicellulose – 21.64 Lignin – 13.13	Density – 0.94	Tensile strength – 42.00	3,288 cm^{-1} – OH stretching 2,917 cm^{-1}– CH stretching 1,734 cm^{-1} – C≡C stretching 1,605 cm^{-1} – C=O stretching		42.00	(R. Kumar et al., 2020)
Acacia nilotica L.	Retting	Cellulose – 56.46 Hemicellulose – 14.14 Lignin – 8.33	Density – 1.165	–	3,297 cm^{-1} – OH stretching of cellulose and hemicellulose 2,927 cm^{-1} – C≡C stretching (wax) 1,607 cm^{-1} C-O stretching of lignin 1,444 cm^{-1} – CH$_2$ stretching 0f cellulose 1,022 cm^{-1} – CO stretching of lignin	339	44.82	
Kigelia africana plant	Crushing	Cellulose – 55.1 ± 0.2 Hemicellulose – 9.34 ± 0.08 Lignin – 11.7 ± 0.2		Tensile strength – 379.28 Young's Modulus 15.68 ± 2.92	3,424 cm^{-1} – OH group stretching 2,922 cm^{-1} – CH stretching 1,734 cm^{-1} – C=O vibrations 1,637 cm^{-1} – C≡C stretch 1,028 cm^{-1} – CO stretching	300	59	(Ilangovan et al., 2020)
Kigeliaa fricana fruit	Retting	Cellulose – 61.5 Hemicellulose – 12.42 Lignin – 20.94	1.316	Tensile strength – 15.68 ± 2.92 Young's Modulus – 50.31 ± 24	3,751 cm^{-1} – OH group 3421.25 and 2,923.99 cm^{-1} – OH stretch and CH stretching of cellulose 2,131.16 cm^{-1} – C≡C of alkynes 1,734 cm^{-1} – C=O ester group (hemicellulose) 1,652 cm^{-1} – C=O stretch (lignin) 1,428.36 cm^{-1} – C=C (aromatic)		57.38	(Siva et al., 2020)
Sesbania rostrate	Manually	Cellulose – 72.75 Hemicellulose – 8.01 lignin – 15.91	Density – 1.482	Tensile strength – 439 ± 26.31 Young's Modulus – 42.83 ± 6.92	3,347 cm^{-1} – OH stretching 2,904 cm^{-1} – CH stretching	375	69.71	(Raja, Senthilkumar, et al., 2020)

(*Continued*)

TABLE 2.4 (Continued)
Properties of Newly Extracted Fiber in the Year 2020

Fiber Name	Method of Extraction	Composition %	Density and Diameters	Mechanical Properties	FTIR	Thermal Properties	Crystallinity	References
Yankee pineapple leaf	Mechanically	Cellulose – 47.74 Hemicellulose – 15.98 Lignin – 2.44	–	Tensile strength – 420.3	1,635 cm⁻¹ – C=C stretching; 1,376 cm⁻¹ – CH bending; 1,058 cm⁻¹ – CO stretching; 3,307.15 cm⁻¹ – OH stretching (cellulose); 2,849.30 cm⁻¹ – CH₂ stretching (hemicellulose)1,431 cm⁻¹ – O-CH₃ methoxyl (lignin); 1,376 cm⁻¹ – C-O-C bond (lignin)	340	55.22	(Najeeb et al., 2020)
Leucas aspera	Retted manually	Cellulose – 58.3 Hemicellulose – 8.9 Lignin – 4.5	Density – 1.268 ± 0.0218	Tensile strength – 43	3,303 cm⁻¹ – OH stretching (cellulose); 2,914 and 2,854 cm⁻¹ – CH and CH₂ stretching (cellulose and hemicellulose); 1,431 cm⁻¹ – O-CH₃ methoxyl (lignin); 1,376 cm⁻¹ – C-O-C bond (lignin)	325	22.04	(Vijay, Manoharan, et al., 2020)
Shwetark stem	Retting	Cellulose – 69.65 Hemicellulose – 0.2 Lignin – 16.82	Density – 1.364	Tensile strength – 110-533	3,446 cm⁻¹ – OH stretching; 2,919 cm⁻¹ – CH and CH₂ stretching; 1,720 cm⁻¹ – CO (aldehyde stretching); 1,588 cm⁻¹ – C=C (lignin); 1,022 cm⁻¹ – C-O acetyl group (lignin)	225	72.81	(Raja, Prabu, et al., 2020)

but sometimes it creates pollution because of the anaerobic process. Researchers identified the problems and started to use some other methods. The steam explosion and chemical extraction processes are short lengths of fibers but other methods such as grinding produce long fibers. Retting can manage the fiber length. These extraction methods can be applied according to the requirements of various applications. Extraction in a significant manner leads to potential reinforcement in polymer composites and a good reinforced composite can be used as an alternative to synthetic materials and can be utilized in various demanding applications as well.

REFERENCES

Abraham, E., Deepa, B., Pothan, L. A., Jacob, M., Thomas, S., Cvelbar, U., & Anandjiwala, R. (2011). Extraction of nanocellulose fibrils from lignocellulosic fibers: A novel approach. *Carbohydrate Polymers*, *86*(4), 1468–1475. 10.1016/j.carbpol.2011.06.034

Ahmed, J., Balaji, M. A. S., & Saravanakumar, S. S. (2019). A comprehensive physical, chemical and morphological characterization of novel cellulosic fiber extracted from the stem of Elettaria cardamomum plant. *Journal of Natural Fibers*, 1–12. 10.1 080/15440478.2019.1691121

Alotaibi, M. D., Alshammari, B. A., Saba, N., Alothman, O. Y., Sanjay, M. R., Almutairi, Z., & Jawaid, M. (2019). Characterization of natural fiber obtained from different parts of date palm tree (Phoenix dactylifera L.). *International Journal of Biological Macromolecules*, *135*, 69–76.

Barreto, A. C. H., Costa, M. M., Sombra, A. S. B., Rosa, D. S., Nascimento, R. F., Mazzetto, S. E., & Fechine, P. B. A. (2010). Chemically modified banana fiber: Structure, dielectric properties and biodegradability. *Journal of Environmental Polymer Degradation*, *18*(4), 523–531. 10.1007/s10924-010-0216-x

Biswal, M., Mohanty, S., & Nayak, S. K. (2012). Banana fiber-reinforced polypropylene nanocomposites: Effect of fiber treatment on mechanical, thermal, and dynamic-mechanical properties. *Journal of Thermoplastic Composite Materials*, *25*(6), 765–790. 10.1177/0892705711413626

Brindha, R., Narayana, C. K., Vijayalakshmi, V., & Nachane, R. P. (2017). Effect of different retting processes on yield and quality of banana pseudostem fiber. *Journal of Natural Fibers*, 1–10. 10.1080/15440478.2017.1401505

Cardoso, C. R., Oliveira, T. J. P., Junior, J. A. S., & Ataíde, C. H. (2013). Physical characterization of sweet sorghum bagasse, tobacco residue, soy hull and fiber sorghum bagasse particles: Density, particle size and shape distributions. *Powder Technology*, *245*, 105–114. 10.1016/j.powtec.2013.04.029

Dawit, J. B., Regassa, Y., & Lemu, H. G. (2019). Property characterization of Acacia tortilis for natural fiber reinforced polymer composite. *Results in Materials*, 100054. 10.1016/j.rinma.2019.100054

Deepa, B., Abraham, E., Mathew, B., Bismarck, A., Blaker, J. J., Pothan, L. A., Lopes, A., Ferreira, S., Souza, D., & Kottaisamy, M. (2011). Bioresource technology structure, morphology and thermal characteristics of banana nano fibers obtained by steam explosion. *Bioresource Technology*, *102*(2), 1988–1997. 10.1016/j.biortech.2010.09.030

Devnani, G. L., & Sinha, S. (2018). African teff straw as a potential reinforcement in polymer composites for light-weight applications: Mechanical, thermal, physical, and chemical characterization before and after alkali treatment. *Journal of Natural Fibers*, 1–15. 10.1080/15440478.2018.1546640

Devnani, G. L., & Sinha, S. (2019). Extraction, characterization and thermal degradation kinetics with activation energy of untreated and alkali treated Saccharum spontaneum

(Kans grass) fiber. *Composites Part B*, *166*(February), 436–445. 10.1016/j.compositesb.2019.02.042

Eltahir, H. A., Xu, W., Lu, X., Li, C., Ren, L., & Liu, J. (2020). Prospect and potential of Adansonia digitata L. (Baobab) bast fiber in composite materials reinforced with natural fibers. Part 1: Fiber characterization. *Journal of Natural Fibers*, 1–11. 10.1 080/15440478.2020.1724234

Ganapathy, T., Sathiskumar, R., Senthamaraikannan, P., Saravanakumar, S. S., & Khan, A. (2019). Characterization of raw and alkali treated new natural cellulosic fibers extracted from the aerial roots of banyan tree. *International Journal of Biological Macromolecules*, *138*, 573–581. 10.1016/j.ijbiomac.2019.07.136

Guimarães, J. L., Frollini, E., Silva, C. G., Wypych, F., & Satyanarayana, K. G. (2009). Characterization of banana, sugarcane bagasse and sponge gourd fibers of Brazil. *Industrial Crops and Products*, *30*(3), 407–415. 10.1016/j.indcrop.2009.07.013

Gurukarthik Babu, B., Prince Winston, D., SenthamaraiKannan, P., Saravanakumar, S. S., & Sanjay, M. R. (2019). Study on characterization and physicochemical properties of new natural fiber from Phaseolus vulgaris. *Journal of Natural Fibers*, *16*(7), 1035–1042. 10.1080/15440478.2018.1448318

Ilangovan, M., Guna, V., Prajwal, B., Jiang, Q., & Reddy, N. (2020). Extraction and characterisation of natural cellulose fibers from Kigelia Africana. *Carbohydrate Polymers*, *236*, 115996. 10.1016/j.carbpol.2020.115996

Kengkhetkit, N., & Amornsakchai, T. (2012). Utilisation of pineapple leaf waste for plastic reinforcement: 1. A novel extraction method for short pineapple leaf fiber. *Industrial Crops and Products*, *40*, 55–61. 10.1016/j.indcrop.2012.02.037

Kengkhetkit, N., Wongpreedee, T., & Amornsakchai, T. (2018). Pineapple leaf fiber: From waste to high-performance green reinforcement for plastics and rubbers. In S. Kalia (Ed.), *Lignocellulosic Composite Materials* (pp. 271–291). Springer International Publishing.

Khan, A., Vijay, R., Singaravelu, D. L., Sanjay, M. R., Siengchin, S., Verpoort, F., Alamry, K. A., & Asiri, A. M. (2019). Extraction and characterization of natural fiber from Eleusine indica grass as reinforcement of sustainable fiber-reinforced polymer composites. *Journal of Natural Fibers*, 1–9. 10.1080/15440478.2019.1697993

Khan, A., Vijay, R., Singaravelu, D. L., Sanjay, M. R., Siengchin, S., Verpoort, F., Alamry, K. A., & Asiri, A. M. (2020). Characterization of natural fibers from Cortaderia selloana grass (Pampas) as reinforcement material for the production of the composites. *Journal of Natural Fibers*, 1–9. 10.1080/15440478.2019.1709110

Kulandaivel, N., Muralikannan, R., & Kalyanasundaram, S. (2018). Extraction and characterization of novel natural cellulosic fibers from pigeon pea plant. *Journal of Natural Fibers*, 1–11. 10.1080/15440478.2018.1534184

Kumar, R. S., Balasundar, P., Al-Dhabi, N. A., Prithivirajan, R., Bhat, K. S., Senthil, S., Narayanasamy, P., Kumar, R. S., Balasundar, P., Al-Dhabi, N. A., & Prithivirajan, R. (2019). A new natural cellulosic pigeon pea (Cajanus cajan) pod fiber characterization for bio-degradable polymeric composites. *Journal of Natural Fibers*, 1–11. 10.1 080/15440478.2019.1689887

Kumar, R., Sivaganesan, S., Senthamaraikannan, P., Saravanakumar, S. S., Khan, A., Daniel, S. A. A., Loganathan, L., Sivaganesan, S., Senthamaraikannan, P., & Saravanakumar, S. S. (2020). Characterization of new cellulosic fiber from the bark of Acacia nilotica L. plant. *Journal of Natural Fibers*, 1–10. 10.1080/15440478.2020. 1738305

Li, Y., Mai, Y., & Ye, L. (2000). Sisal fiber and its composites: A review of recent developments. *Composites Science and Technology*, *60*(11), 2037–2055.

Loganathan, T. M., Thariq, M., & Sultan, H. (2020). Characterization of new cellulosic Cyrtostachys renda and Ptychosperma macarthurii fibers from landscaping plants. *Journal of Natural Fibers*, 1–16. 10.1080/15440478.2020.1758865

Malawade, U. A., & Jadhav, M. G. (2019). Investigation of the acoustic performance of bagasse. *Journal of Materials Research and Technology*, 1–8. 10.1016/j.jmrt.2019.11.028

Manimaran, P., Kumar, K. S. S., & Prithiviraj, M. (2019). Investigation of physico chemical, mechanical and thermal properties of the Albizia lebbeck bark fibers. *Journal of Natural Fibers*, 1–12. 10.1080/15440478.2019.1687068

Manimaran, P., Sanjay, M. R., Senthamaraikannan, P., Saravanakumar, S. S., Siengchin, S., Pitchayyapillai, G., Khan, A., Sanjay, M. R., Senthamaraikannan, P., & Saravanakumar, S. S. (2019). Physico-chemical properties of fiber extracted from the flower of Celosia Argentea plant. *Journal of Natural Fibers*, 1–10. 10.1080/15440478.2019.1629149

Manimaran, P., Senthamaraikannan, P., Sanjay, M. R., Marichelvam, M. K., & Jawaid, M. (2018). Study on characterization of Furcraea foetida new natural fiber as composite reinforcement for lightweight applications. *Carbohydrate Polymers*, *181*, 650–658. 10.1016/j.carbpol.2017.11.099

Moshi, A. A. M., Ravindran, D., Bharathi, S. R. S., et al., (2019). Characterization of a new cellulosic natural fiber extracted from the root of Ficus religiosa tree, International Journal of Biological Macromolecules. 10.1016/j.ijbiomac.2019.09.094

Najeeb, M. I., Sultan, M. T. H., Andou, Y., Shah, A. U. M., Eksiler, K., Ariffin, A. H., Sultan, M. T. H., Andou, Y., Shah, A. U. M., & Eksiler, K. (2020). Characterization of lignocellulosic biomass from Malaysian's Yankee pineapple AC6 toward composite application. *Journal of Natural Fibers*, 1–13. 10.1080/15440478.2019.1710655

Nouri, M., Griballah, I., Tahlaiti, M., & Grondin, F. (2019). Plant extraction and physico-chemical characterizations of untreated and pretreated Diss fibers (Ampelodesmos mauritanicus). *Journal of Natural Fibers*, *18*, 1083–1093. 10.1080/15440478.2019.1687062

Okubo, K., Fujii, T., & Yamamoto, Y. (2004). Development of bamboo-based polymer composites and their mechanical properties. *Composites Part A: Applied Science and Manufacturing*, *35*(3), 377–383. 10.1016/j.compositesa.2003.09.017

Palai, B. K., Sarangi, S. K., Mohapatra, S. S., & Fibers, E. C. (2019). Investigation of physiochemical and thermal properties of Eichhornia Crassipes fibers. *Journal of Natural Fibers*, *18*(9), 1–12. 10.1080/15440478.2019.1691110

Pappu, A., Patil, V., Jain, S., Mahindrakar, A., & Haque, R. (2015). Advances in industrial prospective of cellulosic macromolecules enriched banana biofiber resources: A review. *International Journal of Biological Macromolecules*, *79*, 449–458. 10.1016/j.ijbiomac.2015.05.013

Paramasivam, S. K., Panneerselvam, D., & Sundaram, D. (2020). Extraction, characterization and enzymatic degumming of banana fiber. *Journal of Natural Fibers*, 1–10. 10.1080/15440478.2020.1764456

Paridah, M. T., Basher, A. B., SaifulAzry, S., & Ahmed, Z. (2011). Retting process of some bast plant fibers and its effect on fiber quality: A review. *BioResources*, *6*(4), 5260–5281.

Parre, A., Karthikeyan, B., Balaji, A., & Udhayasankar, R. (2019). Investigation of chemical, thermal and morphological properties of untreated and NaOH treated banana fiber. *Materials Today: Proceedings*. 10.1016/j.matpr.2019.06.655

Patel, U., Ray, R., Mohapatra, A., & Das, S. N. (2020). Effect of different chemical treatments on surface morphology, thermal and tensile strength of Bauhinia Vahlii (BV) stem fibers. *Journal of Natural Fibers*, 1–12. 10.1080/15440478.2020.1739591

Pillai, G. P., Manimaran, P., & Vignesh, V. (2020). Physico-chemical and mechanical properties of alkali-treated red banana peduncle fiber. *Journal of Natural Fibers*, 1–10. 10.1080/15440478.2020.1723777

Prasad, A. V. R., & Rao, K. M. (2011). Mechanical properties of natural fiber reinforced polyester composites: Jowar, sisal and bamboo. *Materials and Design*, *32*(8–9), 4658–4663. 10.1016/j.matdes.2011.03.015

Raja, K., Prabu, B., Ganeshan, P., Sekar, V. S. C., Nagarajaganesh, B., Prabu, B., Ganeshan, P., Sekar, V. S. C., Nagarajaganesh, B., Raja, K., Prabu, B., Ganeshan, P., Sekar, V. S. C., & Nagarajaganesh, B. (2020). Characterization studies of natural cellulosic fibers extracted from Shwetark stem. *Journal of Natural Fibers*, 1–12. 10.1080/15440478.2 019.1710650

Raja, K., Senthilkumar, P., Nallakumarasamy, G., & Natarajan, T. (2020). Effect of eco-friendly chemical treatment on the properties of Sesbania Rostrata fiber. *Journal of Natural Fibers*, 1–13. 10.1080/15440478.2020.1725712

Rao, K. M. M., & Rao, K. M. (2007). Extraction and tensile properties of natural fibers: Vakka, date and bamboo. *Composite Structures*, *77*(3), 288–295. 10.1016/j.compstruct.2005. 07.023

Reddy, J. P., & Rhim, J. W. (2018). Extraction and characterization of cellulose microfibers from agricultural wastes of onion and garlic. *Journal of Natural Fibers*, *15*(4), 465–473. 10.1080/15440478.2014.945227

Santhosh Kumar, S., & Hiremath, S. S. (2019). Natural fiber reinforced composites in the context of biodegradability: A review. In *Encyclopedia of Renewable and Sustainable Materials* (Vol. 322, Issue 1567). Elsevier Ltd. 10.1016/B978-0-12-803581-8.11418-3

Senthamaraikannan, P., Sanjay, M. R., Bhat, K. S., Padmaraj, N. H., & Jawaid, M. (2019). Characterization of natural cellulosic fiber from bark of Albizia amara. *Journal of Natural Fibers*, *16*(8), 1124–1131. 10.1080/15440478.2018.1453432

Senthilkumar, K., Saba, N., Rajini, N., Chandrasekar, M., Jawaid, M., Siengchin, S., & Alotman, O. Y. (2018). Mechanical properties evaluation of sisal fiber reinforced polymer composites: A review. *Construction and Building Materials*, *174*, 713–729. 10.1016/j.conbuildmat.2018.04.143

Shaker, K., Muhammad, R., Ullah, W., Jabbar, M., Umair, M., Tariq, A., Kashif, M., & Nawab, Y. (2020). Extraction and characterization of novel fibers from Vernonia elaeagnifolia as a potential textile fiber. *Industrial Crops and Products*, *152*(April), 112518. 10.1016/j.indcrop.2020.112518

Shanmugasundaram, N., Rajendran, I., & Ramkumar, T. (2018). Characterization of un-treated and alkali treated new cellulosic fiber from an Areca palm leaf stalk as potential reinforcement in polymer composites. *Carbohydrate Polymers*, *195*, 566–575. 10.101 6/j.carbpol.2018.04.127

Sheng, Z., Gao, J., Jin, Z., Dai, H., Zheng, L., & Wang, B. (2014). Effect of steam explosion on degumming efficiency and physicochemical characteristics of banana fiber. *Journal of Applied Polymer Science*, *131*(16), 1–9. 10.1002/app.40598

Siva, R., Valarmathi, T. N., Palanikumar, K., & Samrot, A. V. (2020). Study on a novel natural cellulosic fiber from Kigelia africana fruit: Characterization and analysis. *Carbohydrate Polymers*, *244*(20), 116494. 10.1016/j.carbpol.2020.116494

Subramanian, S. G., Rajkumar, R., & Ramkumar, T. (2019). Characterization of natural cellulosic fiber from Cereus hildmannianus. *Journal of Natural Fibers*, 1–12. 10.1 080/15440478.2019.1623744

Sydow, Z., & Bieńczak, K. (2018). The Overview on the use of natural fibers reinforced composites for food packaging. *Journal of Natural Fibers*, 1–12. 10.1080/15440478.2 018.1455621

Tadasse, S., Abdellah, K., Prasanth, A., & Goytom, D. (2018). Mechanical characterization of natural fiber reinforced composites: An alternative for rural house roofing's. *Materials Today: Proceedings*, *5*(11), 25016–25026. 10.1016/j.matpr.2018.10.302

Thwe, M. M., & Liao, K. (2002). Effects of environmental aging on the mechanical properties of bamboo–Glass fiber reinforced polymer matrix hybrid composites. *Composites Part A: Applied Science and Manufacturing*, *33*(1), 43–52.

Umashankaran, M., & Gopalakrishnan, S. (2019). Characterization of bio-fiber from Pongamiapinnata L. bark as possible reinforcement of polymer composites. *Journal of Natural Fibers*, 1–11. 10.1080/15440478.2019.1658254

Vardhini, K. J. V., & Murugan, R. (2016). Effect of Laccase and Xylanase enzyme treatment on chemical and mechanical properties of banana fiber. *Journal of Natural Fibers*, 1–11. 10.1080/15440478.2016.1193086

Vijay, R., Daniel, J., Dhilip, J., Gowtham, S., Harikrishnan, S., Chandru, B., Amarnath, M., Khan, A., Daniel, J., Dhilip, J., Gowtham, S., & Harikrishnan, S. (2020). Characterization of natural cellulose fiber from the barks of Vachellia farnesiana. *Journal of Natural Fibers*, 1–10. 10.1080/15440478.2020.1764457

Vijay, R., Manoharan, S., Arjun, S., Vinod, A., & Singaravelu, D. L. (2020). Characterization of Silane-treated and untreated natural fibers from stem of Leucas aspera. *Journal of Natural Fibers*, 1–17. 10.1080/15440478.2019.1710651

Vijay, R., Singaravelu, D. L., Vinod, A., Raj, I. D. F. P., & Sanjay, M. R. (2019). Characterization of novel natural fiber from Saccharum Bengalense grass (Sarkanda). *Journal of Natural Fibers*, 1–9. 10.1080/15440478.2019.1598914

Vijay, R., Singaravelu, D. L., Vinod, A., Sanjay, M. R., Siengchin, S., Jawaid, M., Khan, A., & Parameswaranpillai, J. (2018). Characterization of raw and alkali treated new natural cellulosic fibers from Tridax procumbens. *International Journal of Biological Macromolecules*, *125*, 99–108. 10.1016/j.ijbiomac.2018.12.056

Vinod, A., Vijay, R., Singaravelu, D. L., Sanjay, M. R., & Siengchin, S. (2019). Extraction and Characterization of natural fiber from stem of Cardiospermum halicababum. *Journal of Natural Fibers*, 1–11. 10.1080/15440478.2019.1669514

Yang, D., Zhong, L., Yuan, T., Peng, X., & Sun, R. (2013). Studies on the structural characterization of lignin, hemicelluloses and cellulose fractionated by ionic liquid followed by alkaline extraction from bamboo. *Industrial Crops and Products*, *43*, 141–149. 10.1016/j.indcrop.2012.07.024

Ying-Chen, Z., Hong-Yan, W., & Yi-Ping, Q. (2010). Bioresource Technology morphology and properties of hybrid composites based on polypropylene/polylactic acid blend and bamboo fiber. *Bioresource Technology*, *101*(20), 7944–7950. 10.1016/j.biortech.201 0.05.007

Youssef, A. M., Hasanin, M. S., El-Aziz, M. E. A., & Darwesh, O. M. (2019). Green, economic, and partially biodegradable wood plastic composites via enzymatic surface modification of lignocellulosic fibers. *Heliyon*, *5*(3), e01332. 10.1016/j.heliyon.201 9.e01332

Zakikhani, P., Zahari, R., Sultan, M. T. H., & Majid, D. L. (2014). Extraction and preparation of bamboo fiber-reinforced composites. *Materials and Design*, *63*, 820–828. 10.1016/ j.matdes.2014.06.058

Zuluaga, R., Velez, J. M., & Gan, P. (2004). Biological natural retting for determining the hierarchical structuration of banana fibers. *Macromolecular Bioscience*, *4*(10), 978–983. 10.1002/mabi.200400041

3 Traditional and Advance Characterization Techniques for Natural Fibers

G. L. Devnani
Chemical Engineering, Harcourt Butler Technical University
Kanpur, India

Shishir Sinha
Chemical Engineering, Indian Institute of Technology
Roorkee, India

Dileep Kumar
Biochemical Engineering, Harcourt Butler Technical
University Kanpur, India

Shailendra Kumar Pandey
Chemical Engineering, Harcourt Butler Technical University
Kanpur, India

3.1 PROPERTIES AND EVALUATION OF NATURAL FIBERS

The manufacturing sector and researchers are exploring various environmentally friendly and renewable materials that are cost effective and energy efficient to produce a variety of products for sustainable development. In last few decades, agricultural-based materials have the drawn attention of researchers to replace existing nonrenewable materials. Environmental issues are also perturbing for the scientific community to concentrate on alternative technologies and materials, lignocellulosic natural fibers and fillers derived from various parts of the plant are offering numerous advantages like light weight, sustainable, environmentally friendly, cost effective, and having good mechanical properties and becoming an alternate of their synthetic counterpart like glass, aramid, as reinforcement in various polymer matrix. These lignocellulosic materials are basically made of cellulose, hemicellulose, and lignin (Kaushik et al., 2010). A good number of reviews are available on extraction and characterization of traditional natural fibers and fillers like coir, bagasse, bamboo, banana, pineapple, wheat straw, etc. and their reinforcement in different polymer matrix

DOI: 10.1201/9781003201724-3

like polypropylene (PP), high density polyethylene (HDPE), low density polyethylene (LDPE), epoxy, etc. (Faruk et al., 2012; Mittal et al., 2016). The major issues addressed by the scientists are the poor adhesion and the compatibility of these natural fibers and fillers with a different polymer matrix. To deal with these issues, different surface treatment methodologies have been suggested (G. L. Devnani, 2021). The addition of some additives like nanofillers have also been experimented to improve the quality of product (G. L. Devnani & Sinha, 2019a). The properties of these novel natural fibers depend on compositional analysis, geometry, and various climatic conditions. An efficient characterization of natural fiber is very important for their application in reinforced polymer composites. Table 3.1 summarizes the research work carried out by different researchers on these novel materials and their reinforced polymer composites.

3.2 COMPOSITIONAL ANALYSIS OF NATURAL FIBERS

The composition determination of lignocellulosic fiber is one of the most important analysis in order to determine the overall quality of reinforced polymer composite material. Different standard methods have been routinely explored by academicians and researchers. The mechanical strength of natural fibers depends largely on its composition. Lignocellulosic fibers mainly consist of three components (Table 3.2) that are cellulose, hemicellulose, and lignin, which are bonded because of covalent

TABLE 3.1
Compilations on Work Done on Natural Fibers in the Last Decade

S. No.	Work Done	Reference
1.	Processing, characterization of different natural fibers	(Faruk et al., 2012)
2.	Automotive application of natural fiber composites	(Koronis et al., 2013)
3.	Fabrication techniques and properties of natural fibers and their composites	(La Mantia & Morreale, 2011)
4.	Recent developments and specifically mechanical performance of these green composites	(Pickering et al., 2016)
5.	Comprehensive review on properties and characterization of natural fiber–reinforced polymer composites	(Sanjay et al., 2018)
6.	Food packaging applications	(Sydow & Bieńczak, 2019)
7.	Extraction, characterization, chemical and physical analysis traditional fibers like bagasse, jute along with some less common natural fibers like acacia leucophloea, elephant grass, etc	(Madhu et al., 2018, 2019)
8.	Different surface treatment methodologies of natural fibers like physical treatment, chemical treatment, and enzymatic treatment	(G. L. Devnani, 2021)

TABLE 3.2

Lignocellulosic Composition of Common Natural Fibers (Faruk 2012)

Fiber	Cellulose (wt%)	Hemicellulose (wt%)	Lignin (wt%)	Waxes (wt%)
Bagasse	55.2	16.8	25.3	–
Bamboo	26–43	30	21–31	–
Flax	71	18.6–20.6	2.2	1.5
Kenaf	72	20.3	9	–
Jute	61–71	14–20	12–13	0.5
Hemp	68	15	10	0.8
Ramie	68.6–76.2	13–16	0.6–0.7	0.3
Abaca	56–63	20–25	7–9	3
Sisal	65	12	9.9	2
Coir	32–43	0.15–0.25	40–45	–
Oil palm	65	–	29	–
Pineapple	81	–	12.7	–
Curaua	73.6	9.9	7.5	–
Wheat straw	38–45	15–31	12–20	–
Rice husk	35–45	19–25	20	14–17
Rice straw	41–57	33	8–19	8–38

bonds, different intermolecular interactions, and van der Wall's forces (Ayeni et al., 2015; Ioelovich, 2015).

3.2.1 CELLULOSE

The main component, cellulose, is a glucose consisting of a linear chain of several hundred to many thousands of glucose molecules connected by a beta acetal linkage. Cellulose is the basic structural component of the primary cell wall of vegetable plants. It is the most abundant organic polymer on Earth. Cellulose has a strong characteristic to form intra- and inter-molecular hydrogen bonding by the OH groups available on linear cellulose chains, which gives stiffness to the straight chain and induces aggregation to a crystalline structure and gives cellulose a nature of partially crystalline fiber structures (A. Sluiter et al., 2013; J. B. Sluiter et al., 2010).

3.2.2 HEMICELLULOSE

Hemicellulose is basically noncellulose polysaccharides. Its composition varies from plant to plant. Geographical and climatic factors are also responsible for variation of its percentage. Hemicelluloses are basically hydrophilic and amorphous in nature and ester bonds are formed between hemicelluloses and lignin in addition to physical bonds with cellulose. In the chemical structure chain of acetylated links of pentoses and hexoses are there. Various plant and vegetable fibers and their tissues contain hemicellulose.

3.2.3 Lignin

Amorphous, rigidity, and hydrophobicity are the key characteristics of lignin. It is aromatic in nature. Lignin is a complex polymeric structure of phenyl propane units cross linked with each other. One part of lignin in the plant is located inside the cell in the shape of hydrophobic nano layers that surround the cellulose fibrils and other parts of lignin are located between the cells.

3.2.4 Standard Methods for Estimation of Cellulose Hemicelluloses and Lignin

Conventional chemical analysis methods can be applied for determination of chemical composition of different types of plant natural fibers. Different researchers from different countries have developed different standard methods for the compositional analysis of natural fibers. TAPPI and NREL methods are frequently used for the estimation. However, a few researchers also developed their own methods similar to these standard guidelines (Kale et al., 2018; Varma & Mondal, 2016; Vinayaka et al., 2017).

3.2.4.1 Ash Content

The natural fiber used to be for 3 hours in muffle furnace at temperature of 540°C and the difference of weight between the original sample and the weight of cooled and dried samples gives the amount of ash.

3.2.4.2 Extractive Content

The natural straw fiber (w_0 g) can be leached with a benzene and ethanol mixture having a composition 2:1 by volume for the duration of 3 hours by keeping the temperature fixed at 60°C. After this process, the residue is dried to a constant weight, keeping the temperature at 105°C. The dried sample is now put in a desiccator and allowed to cool and then the weight is taken (w_1 g). The mass of extractives can be estimated by difference in weights w_0 and w_1 and so on.

3.2.4.3 Hemicellulose Content

The amount of hemicellulose can be calculated by boiling extracted (w_1) residue in a 0.5 nolar solution of NaOH for the duration of 3.5 hours; after that, the residue is washed and neutralized and dried to a constant weight. The sample that is dried is put again in the desiccator and allowed to cool; after that the weight is (w_2 g). The difference in weight of w_1 and w_2 gives the amount of hemicelluloses.

3.2.4.4 Lignin Content

The lignin amount is determined by digesting the fibers in 72% H_2SO_4, after refluxing for 5 hours in boiling water. The residue is washed after filtering lignin-containing water. The total amount of acid-insoluble lignin and ash is estimated based on the weight difference before digestion and weight of residue after the extraction of lignin.

3.2.4.5 Cellulose Content

The cellulose content and its percentage can be estimated by the difference in weight of the original sample w_0 and the total mass contributed by hemicellulose, lignin ash, and extractives. Figure 3.1 shows the diagrammatic representation of lignocellulosic estimation of natural fibers.

3.3 MECHANICAL PROPERTIES OF NATURAL FIBERS

The tensile strength of natural fiber is one of the most important characteristics that helps in the selection of fiber to be reinforced in a polymer matrix. ASTM D3822 is the standard that is normally used in the analysis of tensile strength of both natural and synthetic fibers that have sufficient length so that they can be mounted on a tensile tester. Normally, the single fibers are very delicate and fragile. Utmost care is very important for the analysis of the sample. Samples are usually tested dry or wet (with or

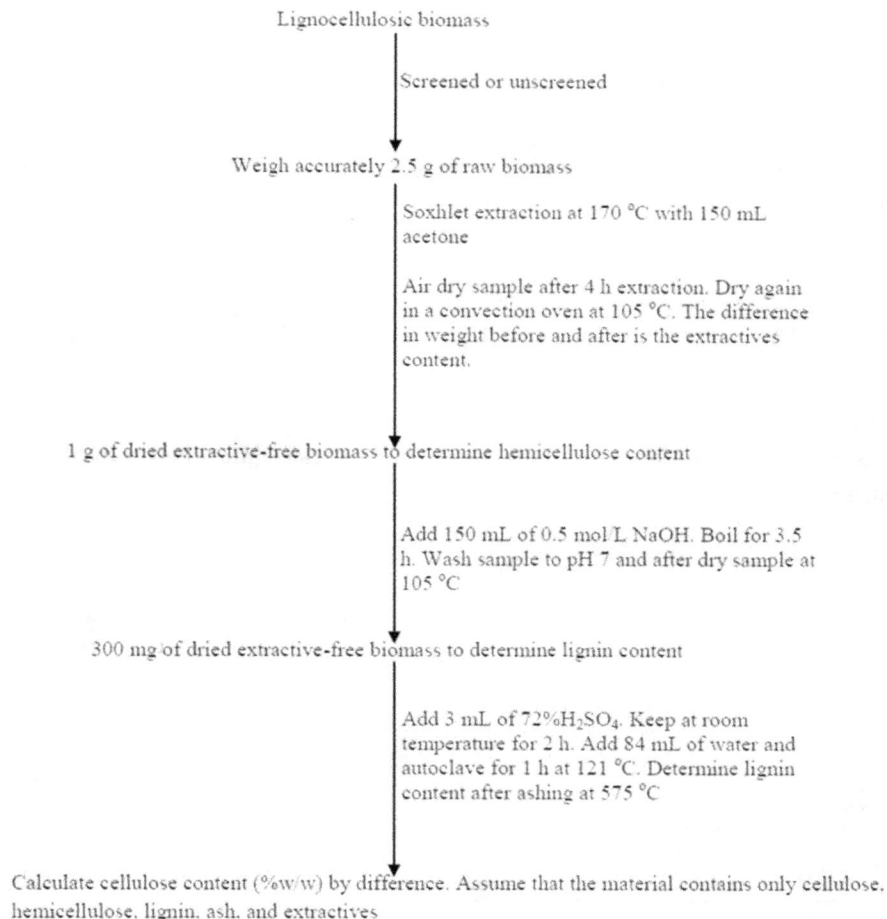

Lignocellulosic biomass

Screened or unscreened

Weigh accurately 2.5 g of raw biomass

Soxhlet extraction at 170 °C with 150 mL acetone

Air dry sample after 4 h extraction. Dry again in a convection oven at 105 °C. The difference in weight before and after is the extractives content.

1 g of dried extractive-free biomass to determine hemicellulose content

Add 150 mL of 0.5 mol/L NaOH. Boil for 3.5 h. Wash sample to pH 7 and after dry sample at 105 °C

300 mg of dried extractive-free biomass to determine lignin content

Add 3 mL of 72% H_2SO_4. Keep at room temperature for 2 h. Add 84 mL of water and autoclave for 1 h at 121 °C. Determine lignin content after ashing at 575 °C

Calculate cellulose content (%w/w) by difference. Assume that the material contains only cellulose, hemicellulose, lignin, ash, and extractives

FIGURE 3.1 Compositional analysis of natural fibers (Ayeni et al., 2015).

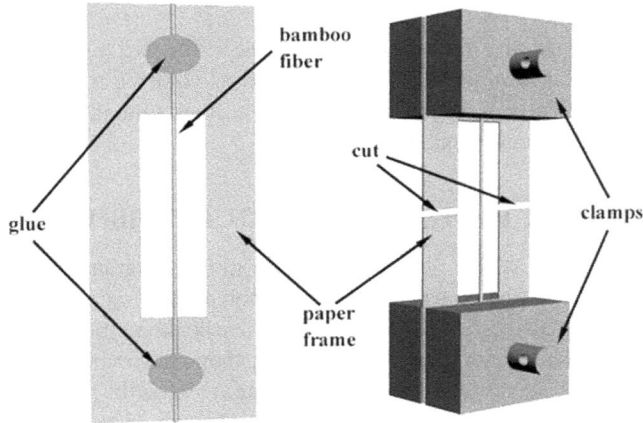

FIGURE 3.2 Tensile testing of a single fiber (Wang & Shao, 2014).

without immersion). The constant rate of extension is applied so that the fiber can be broken on a tensile tester. A force elongation curve is used to calculate the breaking force and the value of elongation at this point. Since the fibers have to be used ultimately for reinforcement purposes in different polymer matrices in making composites, the breaking tenacity and elongation are important factors to be analyzed properly. Figure 3.2 shows the tensile testing of a single fiber.

Table 3.3 presents the mechanical properties of some commonly available natural fibers that are frequently used as a reinforcing fiber in different thermosets, thermoplastics, and naturally occurring polymer matrices.

TABLE 3.3

Mechanical Properties of Traditional Natural Fibers (Djafari Petroudy, 2017; Faruk et al., 2012)

Fiber	Tensile Strength (MPa)	Young's Modulus (GPa)	Elongation at Break (%)	Density (g/cm^3)
Bamboo	140–230	11–17	–	0.6–1.1
Abaca	400	12	3–10	1.5
Hemp	690	70	1.6	1.48
Bagasse	20–290	17	–	1.25
Kenaf	930	53	1.6	–
Jute	393–773	26.5	1.5–1.8	1.3
Sisal	511–635	9.4–22	2.0–2.5	1.5
Flax	345–1035	27.6	2.7–3.2	1.5
Oil palm	248	3.2	25	0.7–1.55

3.4   DENSITY AND DIAMETER MEASUREMENT

3.4.1   Diameter Measurement

Normally the thickness or diameter of the natural fiber is measured with the help of a digital micrometer or by using an optical/scanning electron microscope. The accuracy of measurement using a digital micrometer is 0.001 mm. It is well known that the diameter measurement of natural fibers is a difficult task because of the irregular shape of fibers and variation of thickness at different cross sections. For consistency, 5–10 samples are use to be tested at different locations across the length and average value of diameter is calculated. Image analysis software is used for this estimation of diameter (Sanjay et al., 2019).

3.4.2   Density Measurement of Natural Fibers

Density is another important physical parameter of natural fiber. Low density and light weight are advantageous properties of natural fibers when we compare them to synthetic fibers. When we talk about quality of composites, the void volume fraction is an important parameter to evaluate. The fiber density is used to calculate accurate fiber and void fractions in the composites as per ASTM: D2374 year 1994 and ASTM: D3171 year 1990.

Percentage of fiber volume fraction in composite is given as

$$[(M_f/\rho_f)/(M_c/\rho_c)] - 100 \tag{3.1}$$

where M_f = mass of fiber, M_f = mass of composite sample, ρ_f = density of fiber, and ρ_c = density of composite. The void fraction in a composite % is given by

$$100 - \rho_c[(R/\rho_m) + (R/\rho_f)] \tag{3.2}$$

where ρ_m is the density of matrix material, R is the weight percent of matrix, and r is the weight percent of the fiber. There are several methods that have been used by researchers and academicians for the calculation of density of natural fibers that follow (Truong et al., 2009).

3.4.2.1   Linear Density Calculations

This is the easiest method to evaluate the density of fiber in which the large fiber is cut into small pieces and its diameter is calculated with the help of a microscope. Approximately 100 samples are analyzed to evaluate the average diameter and roughly 10 fiber specimens are used to measure the density of the fiber. The fiber density is calculated with the simple formula of mass/volume

$$\rho_f = W/(0.785d^2l)$$

where W is the mass of fiber, d is the diameter, and l is the length of fiber.

3.4.2.2 Buoyancy Method (Archimedes Principle)

ASTM D 3800-99 can be used for the evaluation of density of different natural fibers. Different liquids like benzene, canola oil, etc. whose density is less than the fiber can be used as immersion liquids and a high accuracy electronic weighing balance supplied with density determination kit, having a least count of 0.1 mg is required for this procedure. The following equation was used for the calculation of density of fibers:

$$\rho_f = \frac{\rho_w \cdot W_1}{W_1 - W_2}$$

where ρ_f is the density of fiber, ρ_w is the density of different liquids, and W_1 and W_2 are the weights of fiber in air and that liquid, respectively (G. L. Devnani & Sinha, 2019c).

3.4.2.3 Pycnometery (Helium)

In this method, at least 10 samples are used and the mean value is calculated. The principle is almost the same as Archimedes's principle, but liquid gas is used. Helium is the gas that is used to evaluate the fiber volume through what was called the helium gas pycnometer. The weight of the fiber sample is measured on a weighing balance, and the density of the fiber is determined by dividing the mass by the volume. Table 3.4 shows the density and diameter values newly explored natural fibers.

3.5 FOURIER TRANSFORM INFRARED SPECTROSCOPY (FTIR)

FTIR stands for Fourier transform infrared. It is the preferred method of infrared spectroscopy. When infrared radiation is passed through a specimen, some radiation is absorbed by it and some amount passes through (is transmitted). The resulting signal obtained through the detector is a spectrum representing a molecular "fingerprint" of the given specimen. The utility of infrared spectroscopy arises due to different spectral fingerprints produced by different chemical structures.

Figure 3.3 shows the schematic diagram of a FTIR instrument in which a coherent light source in which a sample is amplified and converted into a digital signal. The signal is then transferred to computer where the Fourier transform takes place. Table 3.5 shows the frequency range of various functional groups present in the sample (Mishra, 2020; Nascimento et al., 2018). With the help of these values, one can observe the change in functional groups before and after surface modifications of natural fibers.

3.6 X-RAY DIFFRACTION

X-ray diffraction analysis is a famous analysis technique to examine the crystalline and amorphous tendency of the given specimen. Fundamentally, it is based on constructive interference that occurs between monochromatic rays and of a

TABLE 3.4

Density and Diameter of New Natural Fibers (G. L. Devnani & Sinha, 2019c)

Fiber	Density (g/cm³)	Diameter (µm)
Kans grass fiber	1.188 ± 0.008	354 ± 25
Sansevieria	1.41	–
Century		210 ± 50
Gomuti (*Arenga pinnata*)	1.40	81–313
Prosopis juliflora	0.58	
Acacia leucophloea	1.382	
Nigerian coir	1.097 ± 0.0125	369 ± 5
Pennisetum purpureum		240 ± 20
Phoenix sp.	1.2576 ± 0.062	576 ± 204
Corn husks	0.34	186 ± 20
Juncus effusus L.	1.139	280 ± 56
Saharan Aloevera	1.325	91.15
Salago	1.023	6.23
Windmill palm		225.24 ± 97.77
Luffa sponge		590 ± 60
Areca palm leaf stalk	1.09 ± 0.024	285–330
Coccinia grandis L	1.243 ± 0.022	27.33 ± 0.3789
Conium maculatum	–	–
Furcraea foetida	0.778	12.8
Arundo donax	1.168	
Cellulosic Chloris barbata	0.634	
Calotropis gigantea Bast Fibers	1.324	

crystalline sample. X-rays produced through a cathode-ray tube with filters to make it monochromatic can be concentrated on the sample. Atoms of the given fiber sample produce and interface a pattern with incoming X-rays. Various applications of X-ray diffraction are characterization of crystalline material, identification, and analysis of fine-grained materials, etc.

The CrI (crystallinity index) that gives an idea about the fraction of crystalline material was calculated as

$$CrI = \frac{(I_{Total} - I_{Amorphous})}{I_{Total}}$$

CrI can be calculated using intensity at the main peak (~21.9° for cellulose) and because of amorphous peak intensity (estimated at the minimum ~18° between the main peak and the secondary peak at ~16°) (G. L. Devnani et al., 2018). Figure 3.4 shows the X-ray diffraction pattern of untreated and chemically treated bagasse fiber. Untreated bagasse fiber has three different peaks at 15.2°, 21.9°, and 29°. The

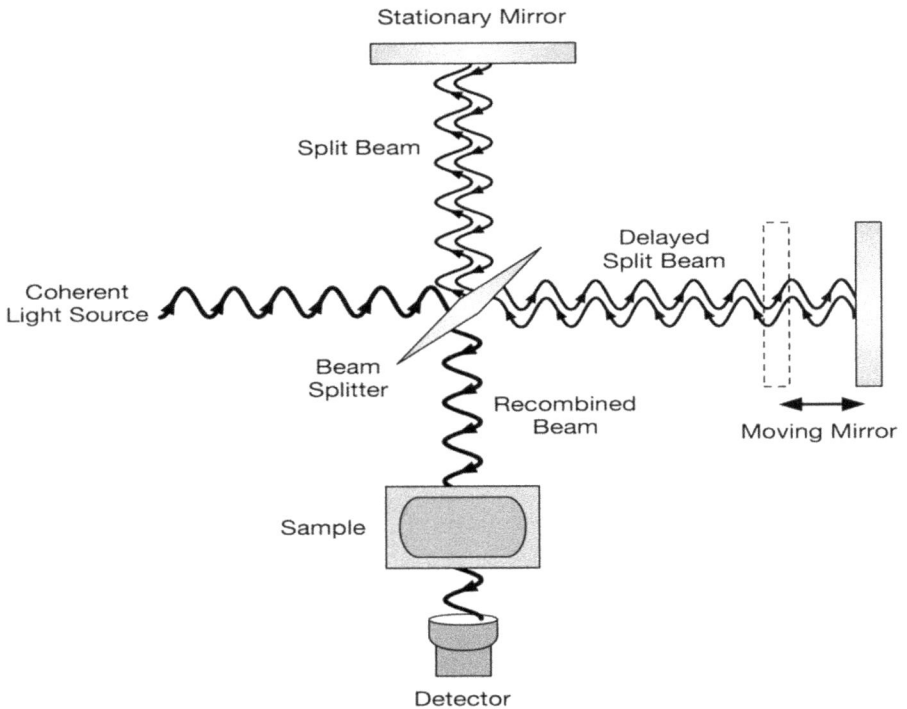

FIGURE 3.3   Line diagram of Fourier transform infrared spectroscopy (FT-IR).

first two peaks at 15.2° and 21.9° are the characteristic peaks of cellulose and along that there is an amorphous zone of lignin, too. It can be observed in the chemically treated X-ray spectra of bagasse fiber that the crsytallinity value has been increased. A sharp peak at 21.9° is visible in the figure and attenuation of amorphous peaks justify the outcomes that are primarily because of partial removal of amorphous lignin and hemicellulose because of chemical treatment steps. The crystallinity index value increased from 34% to 50% because of these treatment steps (G. L. Devnani et al., 2018).

3.7   X-RAY PHOTOELECTRON SPECTROSCOPY (XPS) IS ANOTHER SURFACE-SENSITIVE QUANTITATIVE

A spectroscopic technique that is based on the principle of the photoelectric effect is the technique useful in identifying the elements present within a material (compositional analysis of elements) or that are covering its surface, as well as their chemical nature and the complete electronic structure. The importance of XPS as a technique is enormous because it not only tells what elements are present, but also what other elements they are bonded to. The technique is beneficial for line profiling of the elemental composition across the surface. For getting the elemental

TABLE 3.5

Frequency Range of Various Functional Groups Present in the Sample (Nascimento et al., 2018)

S. No.	Frequency (cm^{-1})	Functional Groups
1.	3,200–3,500	Hydroxyl group and bonded OH stretching
2.	2,905	CH$_2$ and CH$_3$ stretching
3.	2,805	CH aliphatic and aromatic
4.	1,740	Carbonyl stretching vibrations (Ester groups)
5.	1,650	H-O-H boding of absorbed water
6.	1,616	Benzene stretching (lignin)
7.	1,500	Benzene stretching (lignin)
8.	1,440	CH$_2$ bending of lignin
9.	1,411	CH$_3$ bending of lignin
10.	1,310	Alcohol group of cellulose
11.	1,159	Non-symmetric bridge C-O-C
12.	1,070	Skeletal vibrations of CO group
13.	1,040	C-O stretching of lignocellulosic component
14.	879	Beta glucosidic linkage
15.	600	Out-of-phase bending

composition of natural fiber, this technique can be used; moreover, a C/O ratio of the fiber also tells about the hydrophilic or hydrophobic tendency, so after chemical or physical modifications, change in hydrophilic nature can also be estimated by this technique (Fuentes et al., 2013, 2015). Figure 3.5 shows the XPS graphs of conium maculatum fibers.

G. L. Devnani & Sinha (2021) obtained X-ray photoelectron spectroscopy (XPS) graphs of untreated and optimally treated (with 5% NaOH) African Teff straw fiber. The characteristics peaks were obtained at 284 and 531 eV, which ensures the presence of C-C/CH and C=O groups (Kılınç et al., 2018). The O/C ratio for untreated fiber is 0.16, while for 5% alkali-treated fiber the value is 0.23. A low O/C ratio signifies the hydrophobic nature of fiber, which is favorable for the fabrication of composites.

3.8 AFM ANALYSIS

3.8.1 Optical Microscopy

Optical microscopy is an analysis method that is applied for closely analyzing a sample with the support of the magnifying lens; a visible light is used for this purpose. Academicians are using this in traditional and conventional microscopy frequently, which was first explored before the eighteenth century and is still widely applicable in the present era, too. Different types of microscopes are in

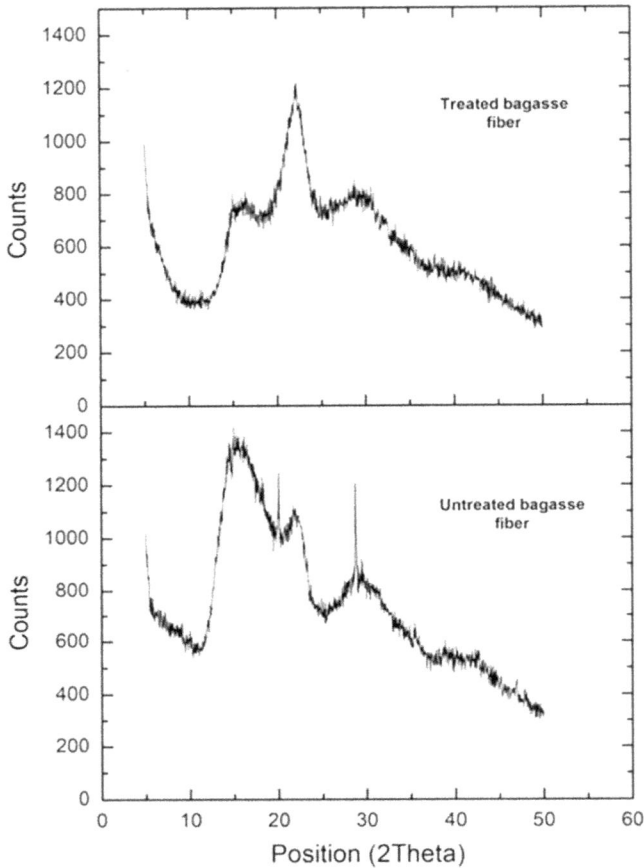

FIGURE 3.4   XRD spectra of untreated and treated bagasse fiber.

practice. The range and variants varying from an available basic design model to a very high complex design instrument offer a high level of contrast and high resolution power also. The categories of the optical microscopes are (i) a binocular stereoscopic microscope that helps in clear observation of three-dimensional (3D) images; (ii) a polarizing microscope (uses varying light transmission properties); (iii) a fluorescence microscope is another special type of biological microscope that records the emission fluorescence of the sample by applying mercury lamps, which is a required source of light; (iv) a laser microscope is in the category of advance microscope which is useful for the observation of even thicker samples having variable focal lengths by the use of laser beam. Other less common microscopes are multiphoton excitation microscope, structured illumination microscope, etc. With the help of optical microscopy, it was possible to get magnifying micrographs of the filament with the corresponding measurements of the diameter and length, as shown in Figure 3.6, where the filament of banana fibers is shown (Vasconcelos et al., 2014).

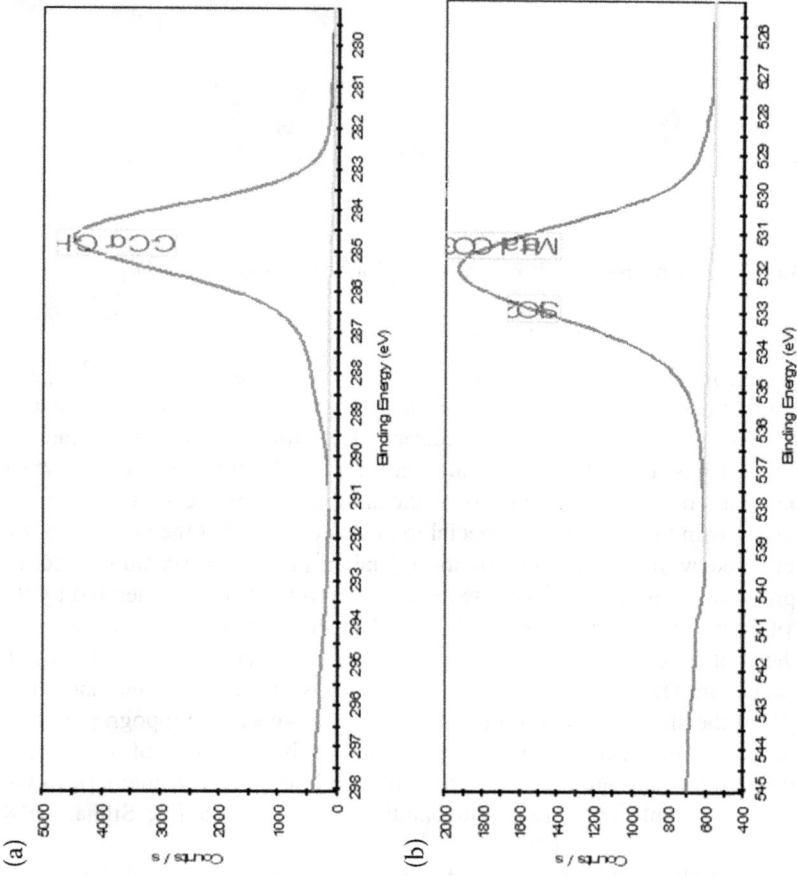

FIGURE 3.5 (a) XPS spectra showing the C1s envelope for conium maculatum fibers; (b) XPS spectra showing the O1s envelope for conium maculatum fibers (Kılınç et al., 2018).

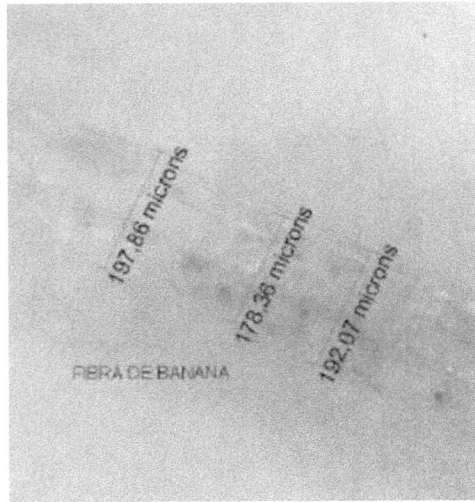

FIGURE 3.6   Optical micrograph of banana fiber (Vasconcelos et al., 2014).

3.8.2   SCANNING ELECTRON MICROSCOPY

Scanning electron microscopy (SEM) is under the category of an advanced type of microscopy that captures images of a specimen by scanning the entire surface with the application of a focused beam of electrons. The atoms of the sample interact with these electrons and different signals that contain the information of surface morphology and compositional analysis of the sample are produced. The scanning of an electron beam takes place in a special manner in a way that the position of the beam is coupled with the intensity of the signal to produce a scanning electron micrograph. In the common scenario, secondary electrons that are generated by the emission of atoms that has been excited by the electron beam are normally recorded with the help of a secondary type of electron detector that is called a Everhart-Thornley detector. The number of secondary electrons that can be seen, and so on the strength of the signal, depend on different factors like sample topography. Few scanning electron microscopes can attain resolutions in the range of 1 nm. In a traditional SEM, samples are normally observed in a high or low vacuum, or in wet conditions in a variable pressure or surrounding (G. L. Devnani & Sinha, 2018, 2019b; Rajeshkumar et al., 2017).

The scanning electron micrograph of the untreated and various alkali-treated Kans grass fibers are shown in Figure 3.7. It can be seen in Figure 3.7(a) of untreated Kans grass fiber that the different filaments of the fibers are joined due to the presence of waxy lignin, hemicellulose, and oily substances, and a comparatively smooth surface with impurities on the surface is visible. In Figure 3.7(b) at 3% alkali treatment, the process of fibrillation starts and roughness is increased all along the fiber surface and a comparatively cleaner surface with less impurities is observed, which is advantageous for the better adhesion of fiber with the polymer matrix and and exchange of tangential

(a)

(b)

(c)

(d)

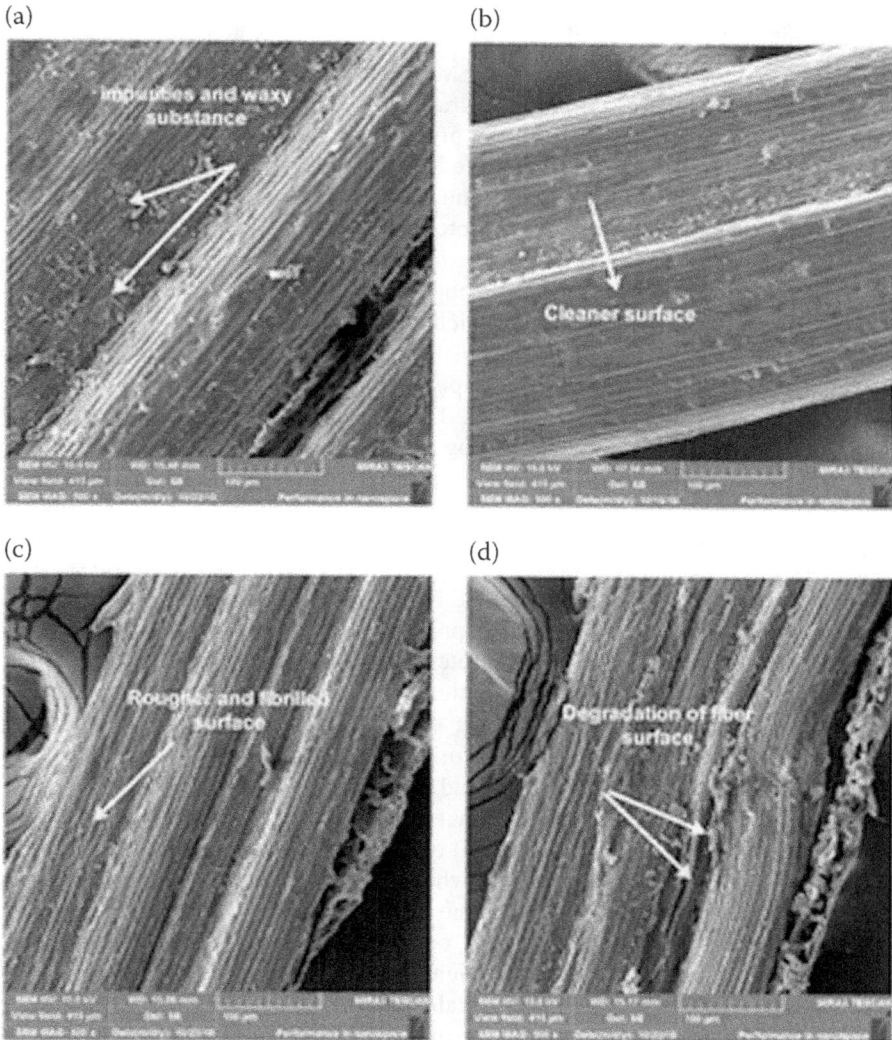

FIGURE 3.7   SEM images of (a) untreated; (b) 3% treated; (c) 5% treated; (d) 7% treated Kans fiber (G. L. Devnani & Sinha, 2019c).

stresses between reinforcement and the polymer matrix. Figure 3.7(c) shows a complete fibrilled and rougher surface in the case of 5% NaOH-treated Kans grass fiber that reflects the improvement in morphology by the alkalization process with an optimum concentration of alkali. Fibrillation is beneficial because it leads to increases in the available contact surface that will promote the fiber matrix interlocking. Excess NaOH concentration can cause damage on the fiber surface and can lead to degradation of the fiber, which is evident in Figure 3.7(d) of 7% NaOH-treated fiber.

3.8.3 Transmission Electron Microscopy

Transmission electron microscopy (TEM) is one of the advanced microscopy techniques in which a transmission of a ray of electrons occurs through a sample to get the micrograph. The thickness of the specimen is generally very thin, which is <100 nm thickness. The sample interacts with the electron transmission of the beam that takes place through the sample and the image is formed. The image is then magnified and focused onto an appropriate imaging device, like a fluorescent screen or a layer of photographic film, or a sensor such as a scintillator can be coupled with a charge-coupled accessory. Numerous applications are there in physical as well as biological and material sciences. Different operating modes can be exercised in the instrument, mainly conventional imaging, scanning TEM imaging, diffraction spectroscopy, and the different feasible combination of these different modes. TEM is regarded as an essential characterization technique for nanotechnology in both biological as well as material sciences (Reza et al., 2015).

3.8.4 Atomic Force Microscopy

Atomic force microscopy (AFM) is one of the most sophisticated and high-end and high-resolution types of scanning probe microscopy (SPM), with a resolution in the range of less than a nanometer and it is more than a thousand times better when we compare it with the optical diffraction limit. An AFM produces an image by scanning a tiny cantilever over the entire surface of a specimen. Figure 3.8(a) shows how a cantilever works. The sharp and sophisticated tip that can be an analyzed in the figure at the end of the cantilever touches the surface of the sample, and bends the cantilever that results in varying the amount of laser light reflected to photodiode. The height of the cantilever can be adjusted so that response signal produced is restored, which ultimately results in the measured cantilever height taking the trace of the surface of the given sample. Atomic force microscopy (AFM) is a highly beneficial and valuable morphological characterization method of modern science. It gives the three-dimensional topography of a sample surface and can calculate the relative and RMS roughness of the specimen used. It is useful in the three-dimensional (3D) characterization of the given specimen. AFM can be applied, usually in three different modes, depending upon which type of motion is taking place by the tip: (i) contact mode, which is also called a static mode (while the remaining two modes are considered as dynamic modes); (ii) tapping mode, which is also called intermittent contact; and (iii) AC mode, or vibrating mode. Different modes are applied according to different requirements of the experimentation and type of specimen we are taking for the morphological and topographical characterization. Figure 3.8 represents the tapping mode AFM micrograph of a kraft fiber, which is treated biologically by the *Trichoderma reesei* enzyme (Gholampour & Ozbakkaloglu, 2020).

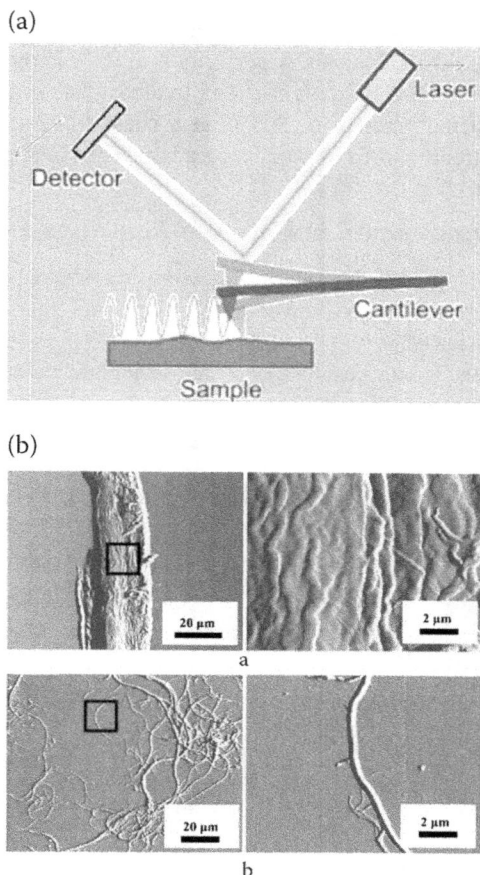

FIGURE 3.8 (a) AFM working; (b) AFM images of a kraft fiber treated by *Trichoderma reesei* enzyme and its cellulose aggregate fibrils (Gholampour & Ozbakkaloglu, 2020).

3.9 THERMOGRAVIMETRIC ANALYSIS

Thermal analysis of the natural fibers is as crucial as the mechanical and chemical analysis. Fiber type, fiber loading, surface modification, crystallinity, fiber size, and surface and bonding between the fiber and polymer matrix phase play a major role for the thermal properties of the composite material. These tests are very important for the quality control and application of these materials when exposed to higher temperatures. In TGA analysis, the thermal degradation of the natural fiber/composite material is determined by variation in the respective sample weight while the sample is heated at a constant heating rate of temperature or time under air or any inert gas atmosphere like nitrogen. This is a very useful quantitative analysis that depicts the thermal reaction accompanied by weight loss due to the evaporation, dehydration, and degradation. In a differential thermal gravimetric (DTG) study, the differential of temperature is plotted against time, or against temperature. This methodology gives an idea about the temperature at which a high rate of degradation takes place. TGA analysis is performed in a thermogravimetric analyzer under the flow of nitrogen. The untreated and treated fiber samples (8–10 mg) are heated from ambient temperature to 800°C at a constant heating rate. Figure 3.9 shows the effect of treatment on thermal degradation of Kans grass fiber.

3.9.1 THERMAL DEGRADATION KINETICS AND ACTIVATION ENERGY

Activation energy of thermal degradation of composites can be evaluated by different methods. There are many conversional procedures that can be used in the calculation of activation energy.

The kinetic models Table 3.6 work on a fundamental rate equation:

$$dx/dt = k.\ f(x)$$

where dx/dt is the conversion rate at constant temperature, k is rate constant, $f(x)$ is model of reaction, and conversion x is defined as $x = [(M_0 - M_t)/(M_0 - M_f)]$, where M_t, M_0, and M_f are weights at time t, initial and final weights of the sample, respectively. By introducing the heating rate of TGA analysis $\beta = dT/dt$ and rate constant $k = A \exp\left(-\frac{Ea}{RT}\right)$, where Ea is the thermal degradation activation energy in Equation (3.1), we get

$$dx/dT = \frac{A}{\beta} \exp\left(-\frac{Ea}{RT}\right) f(x)$$

Apart from this, there are so many other methods from which the calculation of activation energy of natural fibers and their reinforced composites is done (Table 3.7).

3.10 RECENT ADVANCEMENTS

3.10.1 ENERGY-DISPERSIVE X-RAY SPECTROSCOPY

EDX is known as energy-dispersive X-ray spectroscopy, which is a highly useful analytical technique in obtaining the elemental analysis of the given sample. The fundamental phenomena is based on interactions of X-rays produced from some X-ray generator and the given sample. It utilizes the basic principle that every element has unique and specific atomic structures that allow particular sets of peaks in its electromagnetic spectrum. Mosley's law is the common method for the prediction of peak positions. In this technique, the emission of characteristic X-rays from a given specimen of beam of concentrated and focused electrons into the specimen is to be examined. The incident ray of electrons has a tendency to excite the electron from an inner shell that results in the creation of an electron hole. The outer shell electrons that have higher energies fill the developed hole and the corresponding difference of energy values between a higher energy shell and a lower energy shell is emitted as X-rays. An

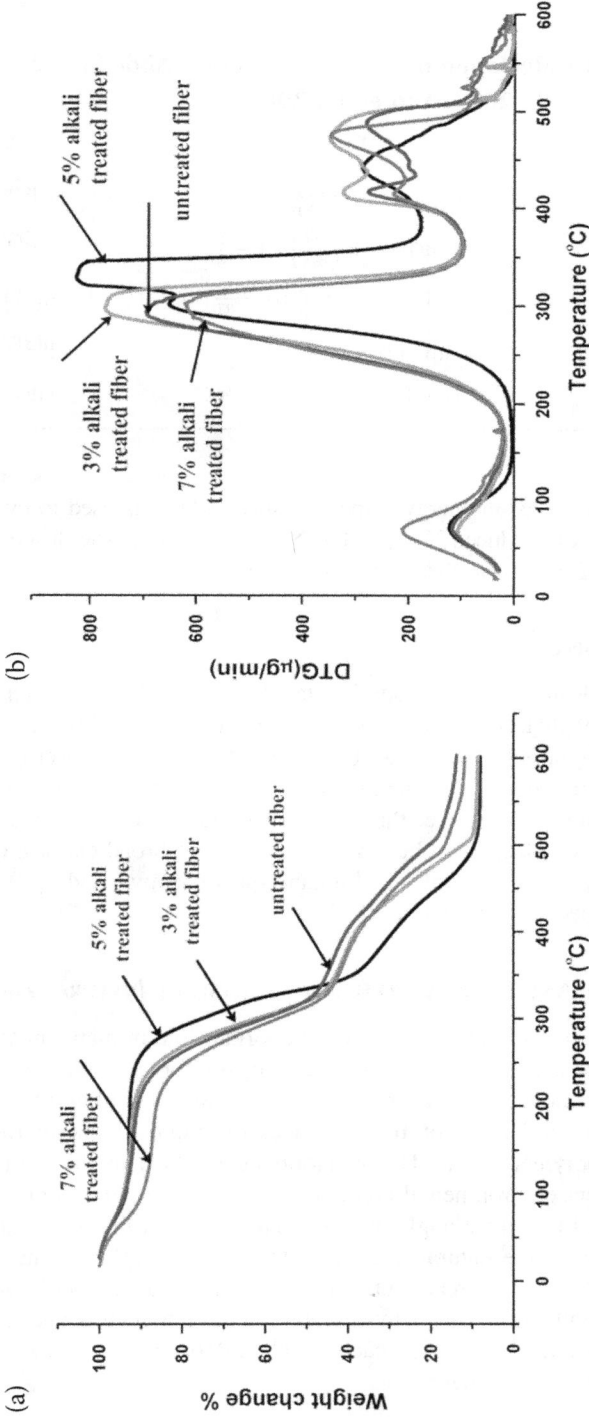

FIGURE 3.9 TGA and DTG thermograms of untreated and treated Kans fibers (G. L. Devnani & Sinha, 2019c).

TABLE 3.6

Kinetic Models for Calculation of Activation Energy (Abdullahi, 2018; Luiz et al., 2014; Oza et al., 2014; Yao et al., 2008)

Name of Method	Formula	Graphs
(FWO) (Integral)	$\ln \beta = c - 1.052\frac{E_a}{RT}$	$\ln \beta$ vs $1/T$
(KAS) (Integral)	$\ln(\beta/T^2) = \ln\left(\frac{AR}{E_a \times g(x)}\right) - \frac{E_a}{RT}$	$\ln(\beta/T^2)$ vs $1/T$
Friedman (Differential)	$\ln(dx/dt) = \ln[Af(x)] - \frac{E_a}{RT}$	$\ln\left(\frac{dx}{dt}\right)$ vs $1/T$
Kissinger (Integral)	$\ln(\beta/Tp^2) = \frac{E_a}{RTp} + \ln\left(\frac{AR}{E_a}\right)$	$\ln(\beta/Tp^2)$ vs $1/T$
Broido (Differential)	$\ln[\ln(1/Y)] = -(E/R)[(1/T) + K]$	$\ln[\ln(1/Y)]$ vs $1/T$

accessory that is known as an energy dispersive spectrometer is used to evaluate the energy and numbers of produced X-rays. Based on the characteristic difference, EDS gives composition of the given fiber sample (Mutalib et al., 2017).

3.10.2 RAMAN SPECTROSCOPY

The phenomena of Raman spectroscopy is related to interactions between light and matter and the light that is scattered inelastically. In the experimental system, a single wavelength light is focused onto the sample. Most often, a laser is used as a powerful monochromatic source. When interactions between photons of a laser and molecules of a sample take place, the spectrum is developed from the scattered photons. Raman spectroscopy can be categorized in three broad classes: (i) micro-Raman spectroscopy, (ii) resonance Raman spectroscopy, and (iii) surface-enhanced Raman spectroscopy (Cho, 2007).

3.10.3 CONTACT ANGLE AND WETTABILITY ANALYSIS OF NATURAL FIBERS

The wettability of natural fibers plays a very important role in many areas, such as textiles and material-selection process of the fiber-reinforced composite. Contact angles and measurements give information related to the wettability, which is considered in the modification of fiber surfaces or improvement in rheological properties of the polymer melts. The variation takes place in the contact angle measured in different environmental conditions. Temperature and humidity play a very important role for contact angle analysis. The variation in contact angle after surface modification of the natural fibers also tell about the effectiveness of physical, chemical, or biological treatment. Measurement of contact angle of a single fiber is a very difficult task, but many researchers have tried this and give novel solutions (Chen et al., 2012; Schellbach et al., 2016). Figure 3.10 shows the measurement of the contact angle.

TABLE 3.7

Methods for Calculation of Activation Energy

S. No.	Method
Non-Iterative Methods	
1.	Ozawa–Flynn–Wall (OFW) method
2.	Friedman method
3.	Kissinger
4.	Kissinger–Akahira–Sunose (KAS) method
5.	Augis and Bennett's method
6.	Coats and Redfern method
7.	Horowitz–Metzger method
8.	Li and Tang method
9.	Kofstad method
10.	Ingraham and Marrier
11.	Freeman and Carrolls method
12.	Malek and Mitsuhashi
13.	Tang and Starink Equation
14.	Desseyen method
15.	Broido method
16.	Flynn and Wall
17.	Modification of the Freeman and Carroll method
18.	Modification of the Achar, Brindley, and Sharp method
19.	Modification of the Coats and Redfern method
Iterative Methods	
20.	Model-free method
21.	Iterative equations of Kissinger–Akahira–Sunose
22.	Iterative equations of Ozawa–Flynn–Wall
23.	Modified Coats-Redfern method
24.	Iterative Linear Integral method
25.	Iterative Linear Integral Isoconversional method
Advanced Isoconversional Methods	
26.	Non-Linear Differential method
27.	Non-Linear Integral method
28.	NNL-Non-Linear method
29.	Modified Nonlinear method

CONCLUSION

Proper measurement and analysis of the characteristics of natural fibers is very essential so that proper selection can be made for the material. Depending upon this analysis, the application and utility can be determined. Traditional and advanced

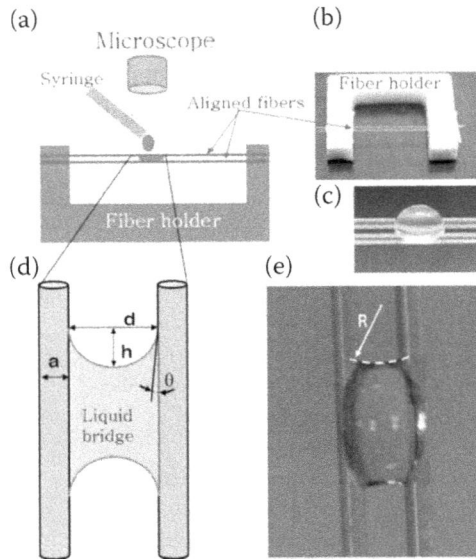

FIGURE 3.10 Measurement of contact angle.

techniques used by the scientific community for the characterization of natural fibers have been discussed, along with suitable examples.

REFERENCES

Abdullahi, M. (2018). Effect of visco-elastic parameters and activation energy of epoxy resin matrix reinforced with sugarcane bagasse powder (SCBP) using dynamic mechanical analyzer (DMA). *American Journal of Polymer Science and Technology*, *4*(3), 53. 10.11648/j.ajpst.20180403.11

Ayeni, A. O., Adeeyo, O. A., Oresegun, O. M., & Oladimeji, T. E. (2015). Compositional analysis of lignocellulosic materials: Evaluation of an economically viable method suitable for woody and non-woody biomass. *American Journal of Engineering Research*, *44*, 2320–2847. www.ajer.org

Chen, H., Fei, B., Wang, G., & Cheng, H. (2012). *Contact Angles of Single Fibers Measured in Different Temperature and Related Humidity*. 1–6. Proceedings of the 55th International Convention of Society of Wood Science and Technology August 27–31, 2012 – Beijing, CHINA

Cho, L. (2007). Identification of textil fiber by Raman microspectroscopy. *Forensic Science Journal*, *1*(1), 55–62. http://fsjournal.cpu.edu.tw/content/vol6.no.1/6(1)-4.pdf

Devnani, G. L. (2021). Recent trends in the surface modification of natural fibers for the preparation of green biocomposite. In B. P. Thomas, S. (Ed.), *Green Composites* (pp. 273–293). Springer Singapore. 10.1007/978-981-15-9643-8_10

Devnani, G. L., Mittal, V., & Sinha, S. (2018). Mathematical modelling of water absorption behavior of bagasse fiber reinforced epoxy composite material. *Materials Today: Proceedings*, *5*(9), 16912–16918. 10.1016/j.matpr.2018.04.094

Devnani, G. L., & Sinha, S. (2018). African teff straw as a potential reinforcement in polymer composites for light-weight applications: Mechanical, thermal, physical, and chemical characterization before and after alkali treatment. *Journal of Natural Fibers*, 1–15. 10.1080/15440478.2018.1546640

Devnani, G. L., & Sinha, S. (2019a). Effect of nanofillers on the properties of natural fiber reinforced polymer composites. *Materials Today: Proceedings*, *18*, 647–654. 10.1016/j.matpr.2019.06.460

Devnani, G. L., & Sinha, S. (2019b). Epoxy-based composites reinforced with African teff straw (*Eragrostis tef*) for lightweight applications. *Polymers and Polymer Composites*, *27*(4), 189–200. 10.1177/0967391118822269

Devnani, G. L., & Sinha, S. (2019c). Extraction, characterization and thermal degradation kinetics with activation energy of untreated and alkali treated Saccharum spontaneum (Kans grass) fiber. *Composites Part B: Engineering*, *166*, 436–445. 10.1016/j.compositesb.2019.02.042

Devnani, G. L., & Sinha, S. (2021). Utilization of natural cellulosic African teff straw fiber for development of epoxy composites: Thermal characterization with activation energy analysis. *Journal of Natural Fibers*. 10.1080/15440478.2021.1929646

Djafari Petroudy, S. R. (2017). Physical and mechanical properties of natural fibers. In *Advanced High Strength Natural Fibre Composites in Construction*. Elsevier Ltd. 10.1016/B978-0-08-100411-1.00003-0

Faruk, O., Bledzki, A. K., Fink, H. P., & Sain, M. (2012). Biocomposites reinforced with natural fibers: 2000-2010. *Progress in Polymer Science*, *37*(11), 1552–1596. 10.1016/j.progpolymsci.2012.04.003

Fuentes, C. A., Brughmans, G., Tran, L. Q. N., Dupont-Gillain, C., Verpoest, I., & Van Vuure, A. W. (2015). Mechanical behaviour and practical adhesion at a bamboo composite interface: Physical adhesion and mechanical interlocking. *Composites Science and Technology*, *109*, 40–47. 10.1016/j.compscitech.2015.01.013

Fuentes, C. A., Tran, L. Q. N., Van Hellemont, M., Janssens, V., Dupont-Gillain, C., Van Vuure, A. W., & Verpoest, I. (2013). Effect of physical adhesion on mechanical behaviour of bamboo fibre reinforced thermoplastic composites. *Colloids and Surfaces A: Physicochemical and Engineering Aspects*, *418*, 7–15. 10.1016/j.colsurfa.2012.11.018

Gholampour, A., & Ozbakkaloglu, T. (2020). A review of natural fiber composites: Properties, modification and processing techniques, characterization, applications. *Journal of Materials Science*, *55*(3), 829–892. Springer US. 10.1007/s10853-019-03990-y

Ioelovich, M. (2015). Methods for determination of chemical composition of plant biomass. *Journal SITA*, *17*(4), 208–214.

Kale, R. D., Getachew Alemayehu, T., & Gorade, V. G. (2018). Extraction and characterization of lignocellulosic fibers from Girardinia Bullosa (Steudel) Wedd. (Ethiopian Kusha plant). *Journal of Natural Fibers*, 1–15. 10.1080/15440478.2018.1539940

Kaushik, A., Singh, M., & Verma, G. (2010). Green nanocomposites based on thermoplastic starch and steam exploded cellulose nanofibrils from wheat straw. *Carbohydrate Polymers*, *82*(2), 337–345. 10.1016/j.carbpol.2010.04.063

Kılınç, A. Ç., Köktaş, S., Seki, Y., Atagür, M., Dalmış, R., Erdoğan, Ü. H., Göktaş, A. A., & Seydibeyoğlu, M. Ö. (2018). Extraction and investigation of lightweight and porous natural fiber from Conium maculatum as a potential reinforcement for composite materials in transportation. *Composites Part B: Engineering*, *140*, 1–8. 10.1016/j.compositesb.2017.11.059

Koronis, G., Silva, A., & Fontul, M. (2013). Green composites: A review of adequate materials for automotive applications. *Composites Part B: Engineering*, *44*(1), 120–127. 10.1016/j.compositesb.2012.07.004

La Mantia, F. P., & Morreale, M. (2011). Green composites: A brief review. *Composites Part A: Applied Science and Manufacturing*, *42*(6), 579–588. 10.1016/j.compositesa.2011.01.017

Luiz, H., Jr., O., Poletto, M., & Jose, A. (2014). Correlation of the thermal stability and the decomposition kinetics of six different vegetal fibers. *Cellulose, 21*, 177–188. 10.1007/s10570-013-0094-1

Madhu, P., Sanjay, M. R., Senthamaraikannan, P., Pradeep, S., Saravanakumar, S. S., & Yogesha, B. (2018). A review on synthesis and characterization of commercially available natural fibers: Part-I. *Journal of Natural Fibers*, 1–13. 10.1080/15440478.2 018.1453433

Madhu, P., Sanjay, M. R., Senthamaraikannan, P., Pradeep, S., Saravanakumar, S. S., & Yogesha, B. (2019). A review on synthesis and characterization of commercially available natural fibers: Part II. *Journal of Natural Fibers, 16*(1), 25–36. 10.1080/1544 0478.2017.1379045

Mishra, K. (2020). Development and assessment of Moringa oleifera (Sahajana) leaves filler/ epoxy composites: Characterization, barrier properties and in situ determination of activation energy. *Polymer Composites, 41*(12), 5016–5029. 10.1002/pc.25771

Mittal, V., Saini, R., & Sinha, S. (2016). Natural fiber-mediated epoxy composites – A review. *Composites Part B: Engineering, 99*, 425–435. 10.1016/j.compositesb.2016.06.051

Mutalib, M. A., Rahman, M. A., Othman, M. H. D., Ismail, A. F., & Jaafar, J. (2017). Scanning electron microscopy (SEM) and energy-dispersive X-ray (EDX) spectroscopy. In N. Hilal, A. F. Ismail, T. Matsuura, & D. Oatley-Radcliffe (Eds.), *Membrane Characterization* (pp. 161–171). Elsevier B. V. 10.1016/B978-0-444-63 776-5.00009-7

Nascimento, L. F. C., Monteiro, S. N., Louro, L. H. L., Luz, F. S. Da, Santos, J. L. Dos, Braga, F. D. O., & Marçal, R. L. S. B. (2018). Charpy impact test of epoxy composites reinforced with untreated and mercerized mallow fibers. *Journal of Materials Research and Technology, 7*(4), 520–527. 10.1016/j.jmrt.2018.03.008

Oza, S., Ning, H., Ferguson, I., & Lu, N. (2014). Effect of surface treatment on thermal stability of the hemp-PLA composites: Correlation of activation energy with thermal degradation. *Composites Part B: Engineering, 67*, 227–232. 10.1016/j.compositesb.2014.06.033

Pickering, K. L., Efendy, M. G. A., & Le, T. M. (2016). A review of recent developments in natural fibre composites and their mechanical performance. *Composites Part A: Applied Science and Manufacturing, 83*, 98–112. 10.1016/j.compositesa.2015.08.038

Rajeshkumar, G., Hariharan, V., & Scalici, T. (2017). Effect of NaOH Treatment on Properties of Phoenix Sp. Fiber. *Journal of Natural Fibers, 13*(6), 702–713. 10. 1080/15440478.2015.1130005

Reza, M., Kontturi, E., Jääskeläinen, A. S., Vuorinen, T., & Ruokolainen, J. (2015). Transmission electron microscopy for wood and fiber analysis – A review. *BioResources, 10*(3), 6230–6261. 10.15376/biores.10.3.reza

Sanjay, M. R., Madhu, P., Jawaid, M., Senthamaraikannan, P., Senthil, S., & Pradeep, S. (2018). Characterization and properties of natural fiber polymer composites: A comprehensive review. *Journal of Cleaner Production, 172*, 566–581. 10.1016/j.jclepro.2017.10.101

Sanjay, M. R., Siengchin, S., Parameswaranpillai, J., Jawaid, M., Pruncu, C. I., & Khan, A. (2019). A comprehensive review of techniques for natural fibers as reinforcement in composites: Preparation, processing and characterization. *Carbohydrate Polymers, 207*, 108–121. 10.1016/j.carbpol.2018.11.083

Schellbach, S. L., Monteiro, S. N., & Drelich, J. W. (2016). A novel method for contact angle measurements on natural fi bers. *Materials Letters, 164*, 599–604. 10.1016/j.matlet.2 015.11.039

Sluiter, J. B., Ruiz, R. O., Scarlata, C. J., Sluiter, A. D., & Templeton, D. W. (2010). Compositional analysis of lignocellulosic feedstocks. 1. Review and description of methods. *Journal of Agricultural and Food Chemistry, 58*(16), 9043–9053. 10.1021/ jf1008023

Sluiter, A., Sluiter, J., Wolfrum, E. J., Sluiter, A., Sluiter, J., & Wolfrum, E. J. (2013). Methods for biomass compositional analysis. In M. Behrens & A. K. Datye (Eds.), *Catalysis for the Conversion of Biomass and Its Derivatives* (pp. 213–254). Max Planck Research Library for the History and Development of Knowledge.

Sydow, Z., & Bieńczak, K. (2019). The overview on the use of natural fibers reinforced composites for food packaging. *Journal of Natural Fibers*, *16*(8), 1189–1200. 10.1080/15440478.2018.1455621

Truong, M., Zhong, W., Boyko, S., & Alcock, M. (2009). A comparative study on natural fibre density measurement. *Journal of the Textile Institute*, *100*(6), 525–529. 10.1080/00405000801997595

Varma, A. K., & Mondal, P. (2016). Physicochemical characterization and pyrolysis kinetic study of sugarcane bagasse using thermogravimetric analysis. *Journal of Energy Resources Technology*, *138*(5), 052205. 10.1115/1.4032729

Vasconcelos, G., Camões, A., Martins, A., Jesus, C., & Luciana Silva. (2014). Experimental characterization of gypsum-cork composite material reinforced with textile fibers. In *XIV Portuguese Conference on Fracture*, Régua, Portugal. February 2014.

Vinayaka, D. L., Guna, V., D., M., M., A., & Reddy, N. (2017). Ricinus communis plant residues as a source for natural cellulose fibers potentially exploitable in polymer composites. *Industrial Crops and Products*, *100*, 126–131. 10.1016/j.indcrop.2017.02.019

Wang, F., & Shao, J. (2014). Modified Weibull distribution for analyzing the tensile strength of bamboo fibers. *Polymers*, *6*(12), 3005–3018. 10.3390/polym6123005

Yao, F., Wu, Q., Lei, Y., Guo, W., & Xu, Y. (2008). Thermal decomposition kinetics of natural fibers: Activation energy with dynamic thermogravimetric analysis. *Polymer Degradation and Stability*, *93*, 90–98. 10.1016/j.polymdegradstab.2007.10.012

4 Surface Treatment of Natural Fibers (Chemical Treatment)

Nidhi Shukla
Chemical Engineering, Harcourt Butler Technical University
Kanpur, India

4.1 FACTORS AFFECTING PROPERTIES OF NATURAL FIBERS AND THEIR COMPOSITES

Composites are made of two materials i.e., natural and synthetic (Biagiotti et al., 2004). In the present time, synthetic material is generally used, but due to certain disadvantages, mainly environmental issues (non-biodegradable) are important and that we have to move on to natural material from synthetic ones . The demand of natural fiber is increasing day by day. Composites made by natural fibers also have huge advantages nowadays. Natural fibers also have different advantages in our daily life. These are biodegradable, low in cost, and low in density. Natural fibers made of cellulose or plant matter can be obtained from almost every part of the plant such as the root, stem or shoot, leaf, fruit, and bark from many tree species (Siakeng et al., 2019). There are various types of natural fibers like leaf fiber, reed fiber, bast fiber, grass fiber, etc. Leaf fiber (hemp, sisal, phormium etc.) is mainly obtained from sword-shaped leaves that are thick, fleshy, and hard (Madhu et al., 2019). Fibers are obtained from plant fibers like hemp, pineapple, sisal, jute, coir, banana, reed, organic cotton, nettle, grass, etc. Animal fibers like wool, mohair, cashmere, angora, yak wool, alpaca wool, camel hair, etc. and insect fibers like silk (Djafari Petroudy, 2017). Bast fibers are defined as those that are obtained from outer shell layers of stems of plants like ramie, jute, and flax, etc. This fiber is also called a skin fiber. They are used in making yarn and weaving cloth. Seed fibers are collected from seeds or seed cases, for example, cotton and kapok, etc. Reed fibers are obtained from reed plants that can be used to create papyrus sheets. Napier grass fiber is one kind of grass fiber (Yashas Gowda et al., 2018).

Natural fibers have different applications in various modes. Natural fiber products have different qualities like common color, texture, and belonging to earth. Two bamboo baskets are different in color while they could have the same form. Banana fiber is used in weaving the tradition Japanese fiber cloth called *bashofu* (Balla et al., 2019). Ancient communities must have used natural fibers to build shelters and thatched roofs. In Europe, North America, and Alaska, mats are made of grass, rush, and sedge; baskets are made from split wood (Thakur and Thakur, 2014). The use of bamboo in Bangladesh, China, and Japan is very extensive and is integral to the culture of the East. West Bengal is another region abundant in natural fibers. A different variety of grasses like madhurkati, khudi, taal beti, and benakati are used for making products. Different products like curtains, wallets, pouches, mobile holders, etc. are made using the woven mat material . Fiber consist of two layers: primary and secondary walls. A primary wall consists of cellulose amorphous microfibrills that contain hemicellulose and lignin and the secondary wall consists of cellulose crystalline microfibrills (Dar et al., 2020). The secondary wall is used to determine mechanical properties.

Figure 4.1 represents different constituents of fiber like cellulose, hemicellulose, lignin, and wax, etc. Every natural fiber has a different percentage of these quantities (Siakeng et al., 2019). The fiber collected from the environment is untreated fiber. To improve the property of fiber or for preparation of the composite, it is necessary to treat fiber with a different surface treatment method. These methods are helpful to improve adhesion between the fibers and matrix. The effect of surface treatment can be seen in mechanical properties. Different chemicals are used for chemical treatment (Gupta, 2019).

Table 4.1 represents chemical constituents of fiber. Cellulose percent is higher in a banana and ramie while lower in the case of bamboo and rice straw. The percent of hemicellulose and lignin is different for different fibers and in some cases it is lower and higher. Removal of hemicellulose and lignin is done by a chemical modification method.

Table 4.2 represents mechanical properties of some common fibers. Tensile strength is maximum for bamboo and flax fiber. This strength can be increase by reinforcing the matrix. Bagasse and coir fibers have lower tensile strength.

4.2 DIFFERENT CHEMICAL TREATMENT METHODOLOGIES

Figure 4.2 represents different types of chemical treatment. There are different types of chemical treatment alkali, coupling agent (silane, acylation, and graft

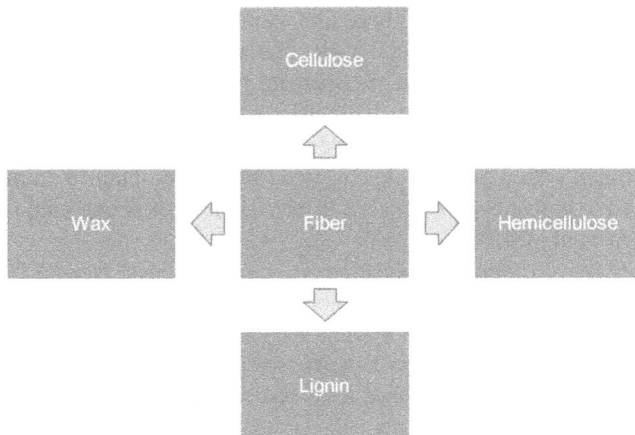

FIGURE 4.1 Constituent of fiber.

TABLE 4.1

Chemical Composition of Fibers (Jones et al., 2017)

Fiber	Cellulose %	Hemicellulose %	Lignin %	Wax %	Pectin %	Ash %
Abaca	53–63	20–25	7–9	3	–	–
Bamboo	26–43	30	1–31	10	–	–
Banana	63–80	19	63–83	11	–	–
Jute	61–72	18–22	12–13	0.5	0.2	0.5–2
Cotton	93	3	–	8–9	0.6	–
Coconut Coir	36–43	41–45	0.15–0.25	–	3–4	–
Ramie	69–91	5–15	0.4–0.7	–	1.9	–
Hemp	70–74	0.9	3.7–5.7	1.2–6.2	0.8	0.8
Flax	64–72	64–72	2–2.2	–	1.8–2.3	–
Kenaf	45–57	8–13	22	0.8	0.6	2–5
Wood	35–50	20–30	25–30	–	–	–
Sisal	78	10	8	2	–	1
Rice Straw	32.1	24	18	–	–	20

copolymerization), bleaching (reductive, oxidative), enzyme, peroxide, and per-mangate. These treatment have different advantages and disadvantages. Acetylation of flax fibers increases tensile and flexural strength. Xue Li et al., (2007) discussed various types of natural fibers and their effect on chemical treatment.

TABLE 4.2

Mechanical Properties of Some Common Natural Fibers (Djafari Petroudy, 2017)

Fiber	Tensile Strength (MPa)	Young's Modulus (GPa)	Elongation at Break (%)	Density (g/cm^3)
Bagasse	20–290	17	1–3	1.25
Flax	345–1,500	27.6	2.7–3.2	1.5
Wheat straw	273	4.76–6.58	2.7	1.15
Sisal	511–635	9.4–22	2.0–2.5	1.5
Hemp	690	70	1.6	1.48
Banana	529–914	29	3–10	1.35
Jute	393–800	13–26.5	1.16–1.5	1.3
Kenaf	930	53	1.6	1.2
Ramie	560	24.5	2.5	1.5
Oil palm	248	3.2	14	0.7–1.55
Pineapple	1020	71	1–3	0.8–1.6
Coir	140.5	6	30	1.2
Bamboo	391–1,000	11–30	2	0.6–1.1

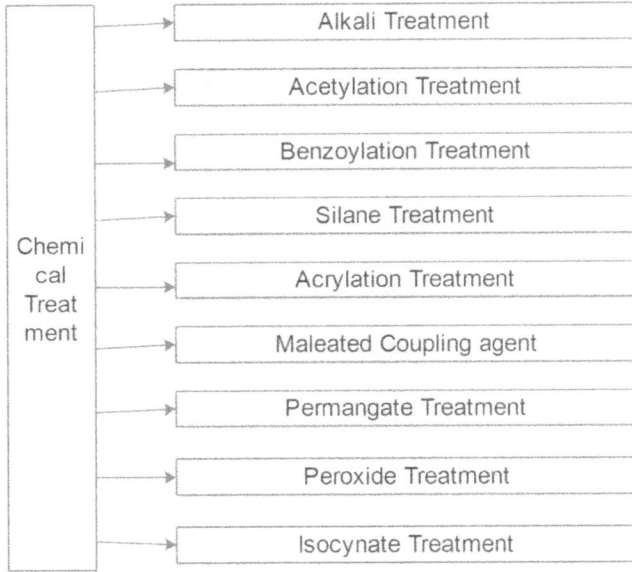

FIGURE 4.2 Types of chemical treatment.

Different areca fiber loading (40%, 50%, 60%, and 70%) is used for preparation of composites by a heat press machine. Composites of these fibers are used in lightweight applications. Zhu 2015 discussed untreated flax fiber has different properties from treated ones, so three different methods are used to improve the properties of flax fiber for the reinforcement of composites. Alkali, silane, and esterification methods are used for surface treatment. FTIR and thermal test are used for the comparison between untreated and treated ones and concluded that esterification gives better tensile strength. Chemical treatment increases adhesion between fiber and a matrix, resulting in an increase in mechanical properties. Surface treatment of *Acacia tortilis* fiber increases the mechanical properties. The application of this fiber acts as a green composites (Dawit et al., 2019).

The surface modification method has some advantages and also has disadvantages, which are discussed in Table 4.3. Alkali treatment roughens the surface and improves the adhesion between the fiber and matrix, but a high percentage of alkaline damages the fibers and a high soaking time also decreases tensile strength of the fiber. Silane-treated flax fiber improves hydrophobic and mechanical properties (Ali et al., 2016). In most of the cases, the alkali treatment is generally used, as it is very cost-effective method, while silane is a very costly treatment method. Silane-treated composite gives better strength. A permangate treatment is a very hazardous material and causes environmental pollution.

TABLE 4.3

Advantages and Disadvantages of Different Surface Treatment Methods

Alkali Treatment

Advantages	Disadvantages
Decreases water absorption of fiber	Mechanical properties depend on quantity of alkali solution is used as high percentage of alkali damage the fiber.
Removes lignin and hemicellulose content	Long time-taking procedure and take huge amount of water.
Improves adhesion between fiber and matrix	As soaking time increases, causes damage to the fiber.
Silane Treatment	
Improves thermal stability of composites	It is a coupling agent that depends on an alkali solution.
Increases toughness, water absorption reduction	Tensile strength of treated fiber decreases.
Acetylation Method	
Reduce water absorption, increase crystallinity	Mechanical properties decreases as an increase in the degree of acetylation.
Permanganate Treatment	
Improves better interlocking between the fiber matrix	Causes environmental pollution as it is a hazardous material.

4.3 EFFECT ON PROPERTIES

There are different types of chemical treatments. The following describes the various chemical treatment methodologies and their effects on properties of natural fibers and their reinforced polymer composites.

4.3.1 ALKALI TREATMENT

This is the most common and conventional method of treatment for natural fibers that improve the adhesion property with the matrix phase, increased surface roughness, and decreased moisture absorption after alkali treatment (Mittal et al., 2016). This is used as a pre-treatment method in other chemical treatments (Fuqua et al., 2017). The interfacial bonding between the fiber and the matrix is improved (Pickering et al., 2016). The alkali treatment is also called mercerization, which is named after the scientist John Mercer, who had proposed this method of treatment in 1850. In this method, the natural fiber is treated with a concentrated solution of NaOH in a range of 1% to 25% and the processing time depends on the nature of fibers. Fiber content, hemicellulose, wax, lignin, and oil, can be removed by this method. A high percentage of alkali can damage fiber and affect the mechanical properties. During this treatment, the hydroxyl group present in cellulose is converted into the -ONa group (Hashim et al., 2012).

$$Fiber - OH + NaOH = Fiber - O - Na + H_2O + surface\ impurities$$

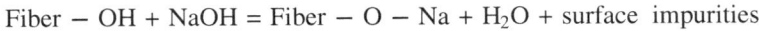

In this treatment, hydrogen is replaced by sodium ions and sodium ions have more ionic radius than hydrogen ions. Thus, the replacement by sodium ions has completely changed the crystalline structure of cellulose. This also affects the mechanical properties and surface modification (George et al., 2001). Untreated fibers have lower mechanical properties in comparison with treated ones. This method is affected by different parameters like operational temperature, temperature treatment time, strength of material, and different applied additives (Ahmad et al., 2019). Many researchers worked on an alkali treatment in a different field, which is explained in Table 4.4.

Figure 4.3 shows SEM images of bagasse fiber. It is clear from the figure that untreated bagasse fiber has impurities visible on the surface while alkali- and acrylic-treated fiber have rougher surfaces. One can observed fibrillation in the treated fiber surface (Devnani et al., 2018). A similar observation was found with increased surface roughness in the case of 3% alkali-treated Kans grass fiber, which cleaned the surface and improved adhesion between the fiber and matrix. Morphology and structure changes in the case of 5% alkali treatment, which is an optimum condition, as this will promote adhesion between them, but if the concentration of treatment is high, it will degrade the fiber surface and damage the fiber. Excess concentration of alkali also decreases mechanical properties (Devnani and Sinha, 2019).

4.3.2 SILANE TREATMENT

Silane treatment has a formula of SiH_4. This treatment reduces swelling properties as the fiber and matrix form covalent bonding (Jha et al., 2019).

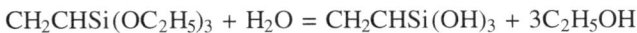

$$CH_2CHSi(OC_2H_5)_3 + H_2O = CH_2CHSi(OH)_3 + 3C_2H_5OH$$

$$CH_2CHSi(OH)_3 + Fiber{-}OH = CH_2CHSi(OH)_2O - Fiber + H_2O$$

Silane treatment is another effective method to increase the adhesion between the natural fiber and matrix phase (Xinxin Wang et al., 2010). A silane coupling agent can used as an adhesion promoter in both natural as well as inorganic filler for polymer composites (Y. Liu et al., 2019). A silane molecule has a bifunctional group that forms a bridge between the reinforcement phase and matrix phase (Xie et al., 2010). It is a multifunctional molecule that deposited on the fiber surface and makes a better linkage with the matrix through a siloxane bridge. A silane coupling agent has the general molecular formula $R_{(4-n)} -Si -(R'\ X)_n$, since the valency of silicon is 4. So n can be 1 or 2, R is the alkoxy group, R' is alkyl group (hydrocarbon part), and X represents the organo-functionality like amino, mercapto, glycidoxy, etc. The silane treatment amount for fillers is normally 0.5% to 2% of filler. Different types of silane are used like 3-Aminopropyl trimethoxysilane, 3-mercaptopropyl methyl dimethoxy

TABLE 4.4
Review Data Collection of Alkali Treated Fiber and Composites

Fiber	Matrix	Key Results	References
Ziziphus mauritiana	Epoxy	Extraction of fiber is done from novel Ziziphus Mauritiana plants. These fibers are suitable for lightweight structure. Composites are prepared with both treated and untreated fiber and compared both the physio-chemical and thermo-mechanical properties. Hand layup process is used for preparation of composites. Different characterization technique TGA, FTIR, and XRD is used for fiber testing. Results show that crystallinity index, thermal stability, and fiber strength improved by alkali treatment. Mechanical and absorption are also characterized. Ultimate tensile strength increased by 2.12 times. Flexural strength is increased by 1.38 times and sound absorption coefficient by 1.15 times.	(Vinod et al., 2020)
Kenaf fiber	Epoxy	This works deals with kenaf fiber–reinforced composites. Fiber is treated with 1%, 2%, 3% NaOH, and investigate mechanical properties. Tensile strength and surface friction coefficient are also calculated. After alkali treatment, surface roughness, tensile strength, and friction coefficient increases, but diameter decreases. Interfacial shear strength is maximum in 2% alkali-treated fiber-reinforced composites.	(Rawatan et al., 2019)
Kens grass	Epoxy	Fiber is treated with three different concentrations of alkali solution i.e., 3%, 5%, and 7%. 5% alkali concentration is the optimum condition to enhance mechanical properties. A different method is used to find activation energy by KAS, FWO, and Friedman. 5% alkali treatment shows maximum increase in activation energy. Crystallinity index and surface roughness improvement is also observed after 5% alkali treatment.	(Devnani and Sinha, 2019)
Jute fibers	Epoxy	Effect of hot-alkali treatment is discussed. 6 wt% alkali solution gives better results. Tensile strength, tensile modulus, flexural	(Xue Wang et al., 2019)

(Continued)

TABLE 4.4 (Continued)
Review Data Collection of Alkali Treated Fiber and Composites

Fiber	Matrix	Key Results	References
		modulus, and flexural strength are increased.	
ProsopisJuliflora	Epoxy	Different weight percentage of alkali is used for alkaline treatment. For preparation of composites, untreated natural fiber shows different disadvantages like higher moisture absorption, lack of bonding, etc. Alkali-treated fiber lowers water absorption and a different characterization technique is used for verification of these properties.	(P. V. Reddy et al., 2019)
Bamboo fiber	–	This work deals with thermal stability and wettability of bamboo fiber. Fiber is treated with 6, 8, 10, 15, and 25 wt% alkali solution. AFM technique is used for surface analysis. TGA analysis is also used for thermogravimetric analysis. At a lower concentration (6, 8, 10 wt%), thermal stability improves.	(Chen et al., 2018)
Cordia-cichotoma	Epoxy	5 wt% alkali solution is used in the treatment of cordia-dichotoma fiber-reinforced composites and the effect of this treatment on mechanical, water absorption, and chemical resistance properties. Hand layup method is used for preparation of composites. Water absorption decreases in the case of alkali treatment.	(B. M. Reddy et al., 2018)
Hemp	Polyamide	Elaborate the effect of alkali treatment on tribological and mechanical properties of hemp fiber–reinforced polyamide composites. Surface treatment with NaOH and NaClO2 gives better results. The combination NaClO2 and A-1160 is the most effective treatment for the improvement of the mechanical and tribological properties of HF/PA1010 biomass composites.	(Mukaida et al., 2017)
Ijuk	Epoxy	In this paper, NaOH, KOH, and NH4OH is used as alkaline treatment on Ijuk fiber. KOH solution is the strongest alkaline, followed by NaOH and NH4OH. Three different alkaline solutions are used for treatment. Results show that KOH gives maximum stiffness and tensile strength.	(You et al., 2017)

TABLE 4.4 (Continued)
Review Data Collection of Alkali Treated Fiber and Composites

Fiber	Matrix	Key Results	References
		Length of fiber is also an important part in tensile strength.	
Date palm fibers	Polyurethane	Different weight % of alkali is used in alkali treatment and investigate its effect on thermal, morphological, and mechanical properties. 5 wt% alkali solution is optimal condition.	(Oushabi et al., 2017)
Jute	–	Effect of alkali treatment is discussed on fatigue property of jute fiber–reinforced composites. 1 and 15 wt% of alkali solution is used for treatment. Fatigue strength increases in case of 1 wt% alkali treatment.	(Katogi et al., 2016)
Fan palm	Cement	Discussed the addition of fibers enhances the properties of composites. Fiber is treated with four different 1%, 2%, 4%, and 10% of alkali solution and it is concluded that 4% alkali solution gives better results i.e., increment in tensile strength. There is no major change in % elongation and modulus of elasticity.	(Elkordi, 2014)
Alfa fiber	Polyester	Fiber is treated with weight % of 1, 3, 5, and 7 of alkali solution. SEM analysis is used for morphology. Physico chemical changes are analysed by XRD analysis. Results show that tensile and flexural strength increases by 7 wt% alkali treatment method and this increment is about 30% to 50%.	(Benyahia et al., 2014)
Curua fiber	–	Curua fiber is used for alkali treatment with 5, 10, and 15 wt% of alkali solution. Tensile strength decreases compared to untreated fiber, while flexural strain increases.	(Gomes, n.d.)
Coir	Polyester	Coir fiber is treated with 5 wt% alkali solution with different soaking time. Soaking time of 72 hours will increase 31% in tensile strength for 96 hours. Results in 22% increment in flexural strength i.e., mechanical properties depend on soaking time and wt% of alkali solution used.	(Jayabal et al., 2012)

(Continued)

TABLE 4.4 (Continued)
Review Data Collection of Alkali Treated Fiber and Composites

Fiber	Matrix	Key Results	References
Kenaf	–	Discuss alkali treatment effect on mechanical and thermal properties of kenaf fiber–reinforced composites. Effect of alkali treatment is also discussed on physical properties of kenaf fiber. Fiber is treated with 2, 6, and 10 wt% alkali solution.There is a small change in fiber density of kenaf fiber after alkali treatment and reduction in diameter at high wt% of alkali solution.	(Hashim et al., 2012)
Kenaf fiber	–	Discussed heat and alkali effect on mechanical properties. At 140°C, tensile strength is maximum. Crystallinity decreases in case of alkali treatment due to removal of hemicellulose, lignin, and other impurities.	(Cao et al., n.d.)

silane, and 3-Glycidoxypropyl trimethoxysilane is used as a surface treatment method in preparation of different types of composites (Kavanagh, 2004). These silanes are compatible with different types of resin, either thermoplastic or thermosetting plastic.

Modification of the surface of kenaf fiber is done by using a silane coupling agent. The storage modulus of abaca fiber–reinforced polyester composites is improved by silane treatment (Faruk et al., 2012). To improve adhesion between curaua fiber and polyester, a silane treatment is used. Fibers were treated with 4 wt% solution of NaOH and then treated with 5 wt% solution of (3-aminopropyl) trimethoxysilane (AMPTS) or triethoxymethylsilane (TEMS) (Miguel et al., 2019). A pineapple leaf is treated with 2-methoxyetoxy silane and composite is prepared by a combination of these two. A forty percent increment in tensile strength is observed in mechanical properties. Sugar palm fiber is treated with alkali solution, silane solution, and a combination of alkali and silane solution. It is concluded from SEM images that untreated fiber have more impurities, while in the case of alkali with silane it is smooth. It is also investigated that alkali with silane treatment improves properties compared to untreated or single-treatment alkali or silane, which is described in Table 4.5 (Atiqah et al., 2017).

4.3.3 BENZOYL CHLORIDE TREATMENT

Inclusion of the benzyl group is called a benzoylation reaction. Benzoylation treatment of natural fiber decreases the hydrophilic nature in this treatment. First, natural fiber is pretreated with a diluted solution of NaOH for about half an hour to activate the hydroxyl group of lignin and cellulose in fiber. Then this preheated fiber

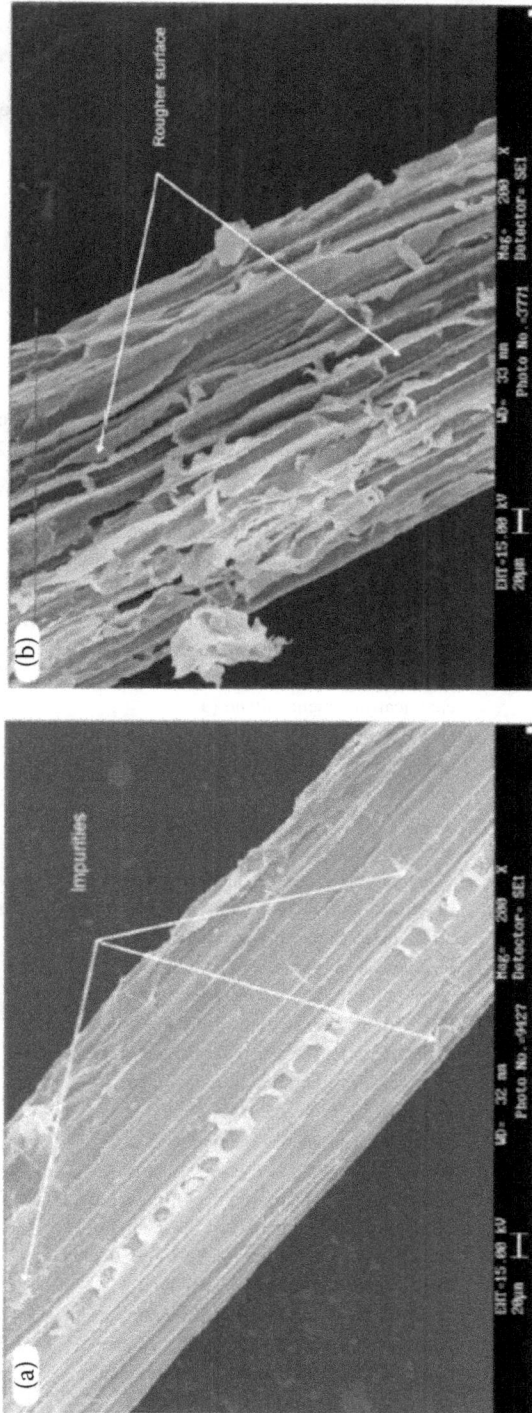

FIGURE 4.3 (a) SEM micrographs of untreated and (b) 1% NaOH followed by 1% acrylic acid–treated bagasse fiber (Devnani et al., 2018).

TABLE 4.5

Review Data Collection of Silane-Treated Fiber and Composites

Fiber	Matrix	Key Results	References
Rice husk	Polyethylene	Different chemical treatment is used for rice husk fiber and investigated its result on reinforced polymer composites. Alkaline, benzoylation, silane, permangate, isocyanate, and another method is used for the same and concluded that silane (3-aminopropyl triethoxy silane) gives a higher tensile strength than other treatment methods and ranges from 12–25 MPa.	(Halip et al., 2021)
Kenaf and pineapple fiber	Polypropylene	Discussed the effect of alkali treatment on mechanical properties of kenaf and pineapple-reinforced polymer composites. 5 wt% alkali and 3 wt% silane is an optimum value for treatment and the result concluded that pineapple leaf fiber have maximum mechanical properties.	(Feng et al., 2020)
Flax	Epoxy	To improve interfacial adhesion between fiber and matrix, fiber is needed to treat a specific surface treatment method, either physical or chemical treatment. Silane (3-aminopropyltrimethoxysilane) treatment is used in this work and improve adhesion between flax fiber and epoxy. This will result as an improvement in flexural properties and 20% reduction in water absorption.	(Fathi et al., 2019)
Rice husk flour	Natural rubber	Investigate the effect of silane treatment on rice husk flour–reinforced composites and its effect on dynamic, morphological, and mechanical properties. Tri-ethoxy-silyl-propyl tetrasulfide is used as a silane agent. Silane-treated fiber-reinforced composite has a maximum tensile strength than an untreated one. Tensile strength and tear strength increases by 12% and 10%.	(Srisuwan et al., 2018)
Sugar palm fiber	Polyurethane	Combination of alkali (6%) and triethoxy-(ethyl) silane (2%) is used for treatment purposes. SEM micrograph confirms removal of hemicellulose, lignin, and other impurities and concluded that silane and alkaline-treated fiber improves fiber-matrix bonding.	(Atiqah et al., 2017)
Hemp	Polyamide	Discussed the effect of three different types of silane coupling agents i.e., epoxy silane, amino silane, ureidosialne on tribological properties of hemp fiber–reinforced	(Nishitani, 2017)

TABLE 4.5 (Continued)
Review Data Collection of Silane-Treated Fiber and Composites

Fiber	Matrix	Key Results	References
		polyamide composites. Results concluded that ureidosialne act as a good coupling agent, comparing the other two silanes and give better mechanical and tribological properties.	
Sisal	Polyester	Investigated effect of combined treatment alkali, silane, and alkali with silane on mechanical properties of three grades of sisal fiber–reinforced composites and concluded that the alkali treatment increased the ultimate tensile strength while a silane decreases, it but a combination of alkali with silane gives the highest mechanical properties.	(Ibraheem et al., 2016)
Flax Yarn	Thermoplastic matrix	Three different chemical methods (alkali, acetylation, and silane) are used for single fiber treatment. FTIR, TGA, and DTG are used to see the effect of all treatments of fiber. Silane treatment gives better results from all other treatments. Moisture absorption is lower in a silane treatment.	(Rajan et al., 2012)
Borassus	Epoxy	Compare the effect of mechanical properties on both untreated and treated borassus fiber–reinforced epoxy-based composites. Fiber is treated with 3-aminopropyltriethoxysilane, followed by alkali treatment. SEM and FTIR analysis is done for surface and physical properties. It is concluded from results that silane-treated fiber shows the highest mechanical properties.	(K. O. Reddy et al., 2009)
Jute fiber	Polylactic acid	Combination of jute fiber with polylactic acid matrix-based reinforced composites have different advantages. This treatment improves the mechanical properties of these composites, resulting in a reduction in tensile and flexural creep than virgin composites at a temperature range of 40°C to 60°C.	(Takemura et al., n.d.)
Kenaf	polypropylene	Amino propyl triethoxy silane is used as a surface treatment for kenaf fiber. Tensile and flexural properties increased by this alkali-silane treatment. FTIR SEM analysis is used in this work.	(Asumani et al., 2012)

(Continued)

TABLE 4.5 (Continued)
Review Data Collection of Silane-Treated Fiber and Composites

Fiber	Matrix	Key Results	References
Sisal	Polyester	Alkali, benzoyl, and silane treatment is used for surface treatment of sisal fiber. Hand layup technique is used for preparation of untreated and treated sisal fiber–based reinforced composites. Surface treatment improves adhesion, resulting in increments in mechanical properties. Maximum tensile strength reported in silane treatment i.e., 27 MPa.	(Salisu et al., 2016)

is soaked in an alkaline solution of benzoyl chloride. The fiber is removed from the solution and soaked in ethanol for 1 hour to remove excess benzoyl chloride and washed with water and dried in an oven (Kalia et al., 2011). Table 4.6 represents different researchers' work on the benzoyl chloride treatment.

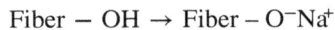

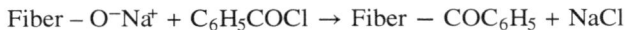

$$Fiber - OH \rightarrow Fiber - O^- Na^+$$

$$Fiber - O^- Na^+ + C_6H_5COCl \rightarrow Fiber - COC_6H_5 + NaCl$$

4.3.4 Isocyanate Treatment

Isocyanate is the organic compound with the general formula R–N=C=O, where R is the alkyl group. When isocyanate is treated with a hydroxyl group, a urethane linkage is formed. If di-isocyanate is treated with a compound containing two or more hydroxyl groups, polyurethane is formed (Xue Li et al., 2007). It acts as a coupling agent for surface modification of fiber. It also provides better bonding with the matrix to improve composite properties (Siakeng et al., 2019). The treatment of natural fiber with isocyanate reduces the hydrophilic nature of the fibers and enhances adhesion between fiber and the matrix phase (Xue Li et al., 2007). The reaction between the fiber and isocyanate coupling agent is shown below:

$$Fiber - OH + R - N = C = O \rightarrow Fiber - O - CO - NH - R$$

It works as promoter. It forms urea and coordination of the hydroxyl group with urea improves interfacial bonding between them. Polymethylene-polyphenyl-isocyanate is used in the treatment of pineapple fiber and also the silane treatment is used and it is comparable with the isocyanate treatment; it is better than the silane treatment described in Table 4.7 (Naveen and Naidu, n.d.).

TABLE 4.6

Review Data Collection of Benzoyl Chloride–Treated Fibers and Composites

Fiber	Matrix	Key Results	References
Cyperus pangorei fibers	–	Cyperus pangorei fiber is generally found in banks of rivers. Benzoyl chloride–treated fiber gives better improvement in results like an increase in physical, chemical, and mechanical properties than untreated ones.	(Babu, 2020)
Luffa fibers	Epoxy	Composite is prepared with both untreated and treated luffa fiber. Benzoyl chloride–treated fiber composites show increment in tensile strength, flexural strength, tensile modulus, and flexural modulus. Tensile strength and flexural strength increased by 27.21% and 41.84%.	(Chakrabarti et al., 2020)
Areca fibers	–	Alkaline, permangate, benzoyl, and acrylic treatment are done on areca fiber and investigate the density, water absorption properties, and mechanical properties. Benzoyl chloride treatment reduces water absorption and improves properties, but there is no change in density.	(Nayak and Mohanty, 2018)
Jute	Epoxy	Physicochemical and water absorption behavior is calculated and compare results with untreated ones.	(Tapas et al., 2017)
Kenaf	Epoxyrene	Benzoyl chloride treatment is used for kenaf fiber and prepared composites with different fiber loading and investigated tensile and morphological properties.	(Majid et al., 2016)
Jute fibers	Epoxy	Different methods of alkali, benzoyl, silane, and maleic anhydride are used for surface treatment. Surface-treated jute fiber shows improvement in storage modulus compared to untreated ones. In all four surface treatment methods, silane and benzoyl chloride–treated jute fiber–reinforced composites give better strength.	(Singhal and Tiwari, 2014)

4.3.5 Potassium Permanganate Treatment

Fiber is treated with an acetone solution by the potassium permanganate. The following reaction occurs in this treatment:

$$\text{Fiber} - \text{OH} + \text{KMnO}_4 \rightarrow \text{Fiber} - \text{O} - \text{H} - \text{O} - \text{Mn(OO)} - \text{OK}^*$$

TABLE 4.7

Review Data Collection of Isocyanate-Treated Fibers and Composites

Fibers	Matrix	Key Results	References
Pine fiber	Polypropylene	Addition of polyphenyl isocyanates improves strength by 11% and the modulus is improved by 77%. Maximum increments in tensile strength.	(Pickering et al., 2016)
Sisal fiber	Polyethylene	Sisal is treated with the isocyanate treatment method and prepared a composite by an injection molding method and reported increments in tensile strength.	(Pradesh, 2018)
Pineapple leaf fiber	Polyethylene	Fiber is treated with both isocyanate and silane treatments. Water absorption is reduced by a surface treatment and increases mechanical properties.	(Suwanruji et al., 2016)
*Eucalyptus*Kraft pulp fiber	–	N-octadecyl isocyanate is used for surface treatment. Treated fiber decreases water absorption. XPS is used to observe chemical changes.	(Henrique et al., 2013)
Pineapple leaf fiber	–	Polymethylene polyphenyl-isocyanate is used for the treatment for pineapple leaf fibers. Isocyanate treatment provides better mechanical properties than silane treatment.	(George et al., 2001)

Mn^{+++} is highly reactive and this will react with hydroxyl group 7; graft co-polymerization occurs. Chemical interlocking increases banana fiber treatment with this method and will increase flexural strength and modulus by 5% and 10% of polypropylene composites (Jha et al., 2019). In the potassium permanganate treatment, the alkaline-pretreated natural fiber is soaked in potassium permanganate – acetone solution of different concentrations for the time range of 1 to 3 minute. As a result of this treatment, the hydrophilic tendency and water absorption of fiber is reduced (Naidu, 2018). The hydrophilic tendency of natural fiber decreases with an increased concentration of KMNO4; however, it must not increase by 1% because beyond this range degradation of cellulose fiber occurs (W. Li et al., 2016). Jute fiber treatment with 0.03% of KMNO4 concentration shows better mechanical properties and this method is considerd one of the best methods to improve bonding between the fiber and matrix (Mondal et al., 2015). A combination of benzoyl chloride–treated fibers with a matrix increases mechanical and other properties and it is elaborated in Table 4.8.

TABLE 4.8
Review Data Collection of Potassium Permanganate–Treated Fibers and Composites

Fibers	Matrix	Key Results	References
Plantain	Epoxy	Discussed chemical and physical treatment that improves the bonding between matrix and fiber. Potassium permanganate–treated fiber improves the water absorption and mechanical properties of plantain fiber–reinforced epoxy composites. Water absorption characterized by Fickian diffusion. Thermal degradation of fiber is characterized by TGA analysis and rupture by SEM analysis.	(Imoisili and Jen, 2020)
Abaca	–	Alkali and permanganate treatment is used for surface modification. Increment in tensile strength can be observed in KMNO4 treatment and low moisture absorption. Higher concentration of treatment damaged the fiber.	(Batara et al., 2019)
Sansevieria	Epoxy	Sansevieria fiber is treated with 10%, 20%, 30%, and 40% of KMNO4 solution. Mechanical properties like tensile and flexural and increases. 30% KMNO4 solution gives better results than untreated fiber-reinforced composites.	(Hariharashayee et al., 2018)
Palmyra palm leaf fiber	–	Chemical and physical properties of untreated and treated palm leaf fiber is concluded. Alkali, benzoyl, and potassium permangate method is used for treatment. Diameter of fiber is reduced after alkali treatment and it is maximum in the case of $KMNO_4$ treatment i.e., 41%.	(Thiruchitrambalam et al., 2018)
Coconut fiber	Polyester	Coconut fiber is treated with a different alkali solution followed by a different concentration of KMNO4 and concluded the shear and tensile stress. This treatment provides better compatibility between the fiber and matrix.	(Arsyad and Soenoko, 2018)

(Continued)

TABLE 4.8 (Continued)

Review Data Collection of Potassium Permanganate–Treated Fibers and Composites

Fibers	Matrix	Key Results	References
Sugar palm	Polyurethane	Works on sugar palm fiber–reinforced polyurethane composites. Fiber is treated with 6% alkali solution followed by KMNO4 solution and investigated the properties. KMNO4 treated with three different concentrations of 0.033, 0.066, and 0.125. Composites are prepared by a hot press machine. Maximum tensile strength (8.986 Pa) is recorded at KMNO4 concentration of 0.125%.	(Abbas et al., 2017)
Ramie	Polypropylene	Ramie fiber–reinforced polypropylene composite is prepared. Fiber is treated with KMNO4 solution of 0.05% concentration and results in flexural strength, tensile strength, and impact strength increase compared to untreated ones.	(He et al., 2017)
Areca	Epoxy	Four different methods, alkali, benzoyl, potassium permangate, and acrylic acid, are used for surface treatment. Composite is prepared for untreated and treated fiber-reinforced epoxy matrix with four different fiber loading: 40%, 50%, 60%, and 70%. 60% fiber loading and acrylic acid chemically treat reinforced composites show higher flexural strength.	(Dhanalakshmi et al., 2015)
Bamboo fiber	Polyester	Bamboo fiber is treated with KMNO4 solution. Tensile properties of this fiber is better than untreated ones. Permanganate solution with 0.5% improves water absorption properties and increases tensile strength.	(Student, 2013)
Jute	Polypropylene	Jute fiber is treated with 0.1% w/v KMNO4 solution for preparation of composites. Thermal stability improves by treatment of jute fiber; mechanical properties also increases.	(Khan et al., 2012)

4.3.6 ACETYLATION TREATMENT

Fibers are first treated with acetic anhydride and this will remove the hydroxyl group with an acetyl group and become hydrophobic in nature. Process steps include that fiber is treated with acetic acid followed by acetic anhydride and this will make the surface rough. A catalyst also affects this reaction. If the reaction occurs in the presence of a catalyst, the reaction rate increases and without a catalyst it is lower (Bledzki et al., 2008). (Naidu, (2018) In this treatment, the alkali-treated natural fiber is soaked in acetic anhydride with one drop of concentrated sulphuric acid (which act as catalyst). The acetic anhydride substitutes the polymer hydroxyl group of the cell wall with an acetyl group that modifies the property of natural fiber, as shown in reaction. In this reaction, the acetyl group is introduced into an organic compound (Xue Li et al., 2007). To initiate a reaction, only a small amount of acetic acid is favorable as it will swell the wall of a cell. Reaction follows in the presence of a catalyst $(CH_3CO)_2CO$ with a concentration of H_2SO_4 (Kenned et al., 2020).

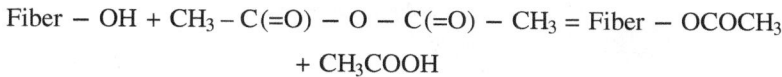

$$Fiber - OH + CH_3 - C(=O) - O - C(=O) - CH_3 = Fiber - OCOCH_3$$
$$+ CH_3COOH$$

Acetyl-treated flax fiber–reinforced composite, tensile, and flexural properties increase until 18%. Acetylation treatment of natural fiber reduces the hydroscopic nature, increases dimensional stability, and roughness. Acetylation-treated bamboo fiber changes properties. The matrix may be different for different fibers and it is discussed in Table 4.9. Acetic anhydride and acetic acid did not react with cellulosic fiber directly so first it is treated with acetic acid and then treated with an acetic anhydride solution. This will accelerate the reaction. This will provide roughness with less void content to promote better bonding between the fiber and matrix. Acetylated fiber shows increments in mechanical properties like acetylated-treated flax fiber increase tensile and flexural strength by 25% (Kabir et al., 2008).

4.3.7 PEROXIDE TREATMENT

[–O–O–] linkage is called peroxide linkage; peroxide easily decomposes into free radicals, and these free radicals remove hydrogen-free radicals from the matrix as well as natural fiber and increases adhesion between the matrix phase and reinforcement phase. For example, benzoyl peroxide is the most commonly used organic compound for this treatment that decomposes into free radicals at elevated temperatures, as shown by the following reaction (Xue Li et al., 2007):

Adhesion between the fiber and matrix improves by this method. It lowers moisture absorption.

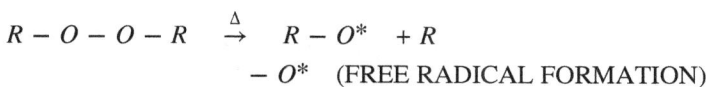

$$R - O - O - R \xrightarrow{\Delta} R - O^* + R$$
$$- O^* \quad \text{(FREE RADICAL FORMATION)}$$

TABLE 4.9

Review Data Collection of Acetylation-Treated Fibers and Composites

Fiber	Matrix	Key Results	References
Bamboo fiber	–	Discussed acetylation method with acetic anhydride. Bamboo fiber is treated with an acetylation method, resulting in a lower absorption rate compared to untreated ones. Algorithm design tool and experimental design is used to see changes occur in properties.	(Onyekwere et al., 2019)
Palm fiber	Polyvinyl chloride	Palm fiber is treated with acetyl and alkali methods. Loading rate is considered to be 10% and 30%. TGA analysis shows degradation of composites with a temperature range of 231°C for acetyl treated and it is lower for untreated and alkali treated. Water absorption rate is also lower in the case of acetyl treated.	(Boussehel et al., 2019)
Sugar palm fiber	Bisphenolic matrix	Bending strength of sugar palm fiber–reinforced composites is calculated. Hot press machine is used for preparation of composites. 40% fiber loading and 60% matrix combination is taken for preparation of composites. Bending strength depends on soaking time. As soaking time increases, bending strength and modulus increases. The fiber shows a maximum modulus and bending strength i.e., 3.3 GPa and 51.98 MPa.	(Diharjo et al., 2017)
Raffia palm fiber	Polyester resin	Alkali and acetylation are two different methods used for treatment of raffia palm fiber. Five different fiber loadings of 0%, 5%, 10%, 15%, and 20% are used. Acetylated fiber shows better ultimate tensile strength than alkali-treated fiber. 5% loading of acetylated fiber shows less absorption of water.	(Chukwudi et al., 2015)
banana	Polypropylene	Banana fiber is treated with acetylating solution (combination of toluene and acetic anhydride) and prepared reinforced composites with polypropylene matrix. SEM analysis is used to see adhesion between the fiber and matrix. Thermal stability is analyzed by TGA and DTG analysis and it is concluded that less water absorption in the case of treated	(Zaman et al., 2013)

TABLE 4.9 (Continued)
Review Data Collection of Acetylation-Treated Fibers and Composites

Fiber	Matrix	Key Results	References
		acetyling fiber compared to untreated one.	
Coconut fiber	polyethylene	Coconut fiber is treated with acetic acid and prepared a polyethylene-reinforced composite. Young's modulus, tensile strength, and elongation at break increases with this treatment. Increment in crystallinity is observed by DSC analysis. Thermal stability also increases for a treated fiber compared to untreated ones.	(Faisal et al., 2010)
Flax	Polypropylene	Flax fiber is treated with an acetylation surface modification method and studied moisture absorption, crystallinity, and surface morphology, etc. Both untreated and treated composites are prepared with 30% fiber loading and concluded that increments occur in tensile and flexural strength as increasing degree of acetylation but impact strength decreases.	(Bledzki et al., 2008)
Flax/hemp/ wood	–	Three different fibers are used for acetylation treatment. Hydrophilicity is decreased by this treatment, resulting in lower moisture absorption. SEM results in non-crystalline constituents disappears.	(Tserki et al., 2005)

where RO* is a free radical and reacts with the matrix.

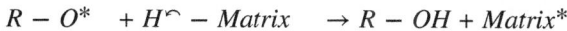

$$R - O^* \ + H^\frown - Matrix \ \rightarrow R - OH + Matrix^*$$

$$R - O^{\cdot} \ + H - -Fiber \ \rightarrow R - OH + Fiber^*$$

Fiber is first treated with alkali and then coated by dicumyl peroxide or benzoyl peroxide with approximately 6 wt% concentration (Kabir et al., 2012). This will improve thermal stability and reduction in moisture absorption. Table 4.10 described a peroxide treatment for different fibers.

TABLE 4.10

Review Data Collection of Peroxide-Treated Fibers and Composites

Fiber	Matrix	Key Results	References
Flax/cotton	Epoxy	Hand layup method is used for preparation of hydrogen peroxide–treated flax fiber–reinforced composite. Thermomechanical properties is studied. SEM analysis was used for structural change. Surface modification improves mechanical properties.	(Barczewski et al., 2020)
Kenaf	Polypropylene	Kenaf fiber is treated with two different chemical methods i.e., 5% hydrogen peroxide and 2% silane. SEM, FTIR, water absorption test is to be done and studied water absorption is lower in the case of hydrogen peroxide–treated reinforced composites compared to all other chemically treated fiber-reinforced composites and also enhances the mechanical properties.	(Sabri et al., 2020)
Pineapple leaf	Polyethylene	Injection molding is used for pineapple leaf–reinforced polyethylene composites. Different wt% of fiber is used as fiber loading. Tensile modulus, tensile strength, flexural modulus, flexural strength, and elongation at break is calculated for both untreated and treated ones. Untreated composites show lower mechanical properties than treated composites.	(Eze et al., 2017)
Peanut shell powder	Polypropylene	Compression molding is used for preparation of treated peanut shell powder–reinforced polypropylene composites with fiber loading 0 to 40 wt%. Lignin removal and surface roughness analyzed by SEM and FTIR analysis and concluded thermal and tensile properties enhanced in case of treated fiber.	(Zaaba et al., 2017)
Oil palm fiber	Poly butylene succinate	Oil palm fiber is treated with a hydrogen peroxide treatment method in an alkaline condition. The treatment removes hemicellulose, lignin, and wax content results roughen the surface and improve crystallinity. Fiber and matrix ratio is assumed to be 70:30 for	(Then et al., 2015)

TABLE 4.10 (Continued)
Review Data Collection of Peroxide-Treated Fibers and Composites

Fiber	Matrix	Key Results	References
		preparation of composites. Tensile strength, tensile modulus, and elongation at break is increased for treated fiber composites.	
Kenaf fiber	Poly lactic acid	Hydrogen peroxide is used for the surface treatment method for kenaf fiber and calculated mechanical properties of composites. Treatment improves adhesion between fiber and matrix resulting in an increase in mechanical properties.	(Razak et al., 2014)
Oil palm empty fruit bunch fiber	Poly lactic acid	Composite is prepared by the melt blending method. Fiber is treated with hydrogen peroxide and this bleach fiber–reinforced composites shows an improvement in mechanical properties than untreated ones.	(Marwah et al., 2014)
Silk	–	Silk fiber is treated hydrogen peroxide. Crystallinity, structure, and mechanical properties are observed of silk-treated fiber and compared with untreated silk. There is no change observed in elastic modulus; only a slight change is observed in crystallinity i.e., degree of crystallinity by some value.	(Xiang Li and Wang, 2013)
Jute fiber	Polyethylene	Jute fiber is treated with polyethylene composites. Composite is prepared by compression molding and fiber load ranges from 10% to 30%. Fiber is treated with (2-hydroxyethyl methacrylate) with benzoyl peroxide, resulting in improvement in mechanical properties.	(Miah et al., 2011)

4.3.8 Maleated Coupling Agent

Banana, bamboo, flax, and sisal, etc. are natural fibers treated with a coupling agent treatment. In most of the cases, maleic anhydride act as a polymer matrix. Surface roughness is improved by this treatment (Anbupalani et al., 2020). Maleic anhydride gives better mechanical properties i.e., twice that of silane treatment. Polypropylene and polylactide are a generally used matrix in this treatment. The reaction of fiber with maleic anhydride is described as (Kabir et al., 2012):

$$MAN + OH - Fiber - OH \rightarrow C - CH - C(=O) - O - Fiber - O$$
$$- C(=O) - CH_2$$

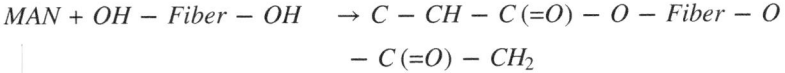

Maleic anhydride modifies surface adhesive bonding (Anbupalani et al., 2020). Table 4.11 shows the maleic anhydride treatment acts as a coupling agent.

TABLE 4.11

Review Data Collection of Maleated Coupling Agent–Treated Fibers and Composites

Fiber	Matrix	Key Results	References
Caranday palm fiber	Polypropylene	Fiber is treated with maleic anhydride treatment. Polypropylene is used as a reinforced matrix. Two-process method compression and injection molding is used for preparation of composites. Coupling agent change structure of fiber and increases mechanical properties of composites.	(Sammartino et al., 2016)
Jute fiber	Polypropylene	Injection molding was used to prepare a jute fiber–reinforced composite. Maleic anhydride was used as a coupling agent. Mechanical properties like tensile strength and tensile modulus are calculated and observed increment in tensile strength by 23.8% and modulus by 148% at 3% of maleic anhydride with polypropylene.	(Mirza et al., 2015)
Pine flou/raw bagasse/rice husk/rice straw	Polyethylene	Mechanical properties and crystallization were observed for four different natural fiber–reinforced composites. Fiber is reinforced with 2% maleated polyethylene. Tensile and flexural strength increases with 30% fiber loading and relativity impacted strength decreases. Maleated polyethylene–reinforced composites show lower crystallinity level than pure polyethylene.	(T. Liu et al., 2013)
Agro fiber (flax and jute)	Poly olefins	Agro fibers were used for preparation of composites. Flax and jute fiber are used as agro fiber. Fiber load with 30% and maleated poly olefins increases tensile and flexural strength.	(Keener et al., 2004)
Sawdust	Polystyrene	Discussed mechanical properties of maleated polystyrene–reinforced composites. Maleic anhydride coated fiber gives a better response.	(Maldas and Kokta, 1991)

4.3.9 ACRYLATION TREATMENT

.Acrylation is another surface treatment modification method in which fiber is first treated with an alkali treatment followed by acrylic acid with a 1% concentration. Fiber is then washed and dried at 70°C ($CH_2 = CHCOOH$), known as acrylic acid (Dhanalakshmi et al., 2015). Free radicals initiated this reaction. To modify fiber, acrylonitrile is used (Ravi et al., 2018). A small amount of work is done on this type of treatment, which is described in Table 4.12. The following reaction follows an acrylation reaction (Kenned et al., 2020):

$$Fiber - OH + CH_2 = CHCOOH \rightarrow Fiber - OOCC_2H_3$$

CONCLUSION

A different chemical modification method is used for the treatment of fibers. A combination of fiber with a matrix is different for different cases. These chemical

TABLE 4.12
Review Data Collection of Acrylation-Treated Fibers and Composites

Fiber	Matrix	Key Results	References
Alfa fiber	Acrylated epoxidized sunflower oil resin	Tensile and thermal properties is observed for both untreated and treated fiber. Mechanical and thermal properties increase in the case of acrylation-treated fiber.	(Kadem et al., 2018)
Abaca fiber	Epoxy	Different surface modification method is applied for surface treatment of fiber. Fiber is treated with different 20%, 30%, 40%, 50%, and 60% fiber loading. Tensile and flexural strength increases for treated fiber composites compared to untreated ones.	(Dhanalakshmi et al., 2015)
Hemp fiber	–	Hemp fiber is treated with acrylation, permangate, and peroxide treatment. Fiber is treated with acrylic acid and the peak of carboxylic acid confirms a peak of acrylic acid in FTIR analysis. Acrylated-treated fiber enhances thermal stability and crystallinity.	(Kalia and Kumar, 2013)
Jute	Epoxy	Fiber is treated with an alkali solution followed by 10 wt% acrylic acid treatment. Hand layup method is a preparation of composites. Increment in tensile and flexural strength was observed by 42.2% and 13.9% after this treatment.	(Patel and Parsania, 2010)

modifications improve the interlocking between a fiber and matrix. Surface roughness is improved by these methods. Generally, alkali treatment is used for surface treatment as it is a simple and cost-effective process. Different chemical methods have different advantages and disadvantages. Generally used chemical treatments are alkali, silane, and benzoyl chloride treatment. Mechanical properties, like tensile strength, flexural strength, impact strength, thermal stability, and other properties of fiber-reinforced composites improved after surface treatment. It may increase or decrease, depending on the types of fiber and matrix in use.

REFERENCES

Abbas, M. A., Bachtiar, D., and Hasany, S. F. (2017). Effect of potassium permanganate on tensile properties of sugar palm fibre reinforced thermoplastic polyurethane. February. *Indian Journal of Science and Technology*, *10*(7), 1–5. 10.17485/ijst/2017/v10i7/111453

Ahmad, R., Hamid, R., and Osman, S. A. (2019). Physical and chemical modifications of plant fibres for reinforcement in cementitious composites. *Advances in Civil Engineering*, *2019*.

Ali, A., Shaker, K., Nawab, Y., Jabbar, M., Hussain, T., Militky, J., and Baheti, V. (2016). Hydrophobic treatment of natural fibers and their composites—A review. *Journal of Industrial Textiles*, 1–31. 10.1177/1528083716654468

Anbupalani, M. S., Venkatachalam, C. D., and Rathanasamy, R. (2020). Influence of coupling agent on altering the reinforcing efficiency of natural fibre-incorporated polymers – A review. *Journal of Reinforced Plastics and Composites*, *39*(13–14), 520–544. 10.1177/0731684420918937

Arsyad, M., and Soenoko, R. (2018). The effects of sodium hydroxide and potassium permanganate treatment on roughness of coconut fiber surface. *MATEC Web of Conferences*, *204*(1), 05004.

Asumani, O. M. L., Reid, R. G., and Paskaramoorthy, R. (2012).The effects of alkali-silane treatment on the tensile and flexural properties of short fibre non-woven kenaf reinforced polypropylene composites. *Composites Part A-applied Science and Manufacturing*, *43*, 1431–1440.

Atiqah, A., Jawaid, M., Ishak, M. R., and Sapuan, S. M. (2017). Effect of alkali and silane treatments on mechanical and interfacial bonding strength of sugar palm fibers with thermoplastic polyurethane. *Journal of Natural Fibers*, 1–11. 10.1080/15440478.2017.1325427

Babu, L. G. (2020). Influence of benzoyl chloride treatment on the tribological characteristics of Cyperus pangorei fibers based non-asbestos brake friction composites. *Materials Research Express*, *7*(1), 015303.

Balla, V. K., Kate, K. H., Satyavolu, J., Singh, P., and Tadimeti, J. G. D. (2019). Additive manufacturing of natural fiber reinforced polymer composites: Processing and prospects. *Composites Part B: Engineering*, *174*(March), 106956. 10.1016/j.compositesb.2019.106956

Barczewski, M., Matykiewicz, D., and Szostak, M. (2020). The effect of two-step surface treatment by hydrogen peroxide and silanization of flax/cotton fabrics on epoxy-based laminates thermomechanical properties and structure. *Journal of Materials Research and Technology*, *9*(6), 13813–13824. 10.1016/j.jmrt.2020.09.120

Batara, A. G. N., Llanos, P. S. P., Yro, P. A. N. De, Sanglay, G. C. D., Eduardo, R., Batara, A. G. N., Llanos, P. S. P., and Yro, P. A. N. De. (2019 March). Surface modification of Abaca fibers by permanganate and alkaline treatment via factorial design. *AIP Conference Proceedings*, *2083*(1), 030007.

Benyahia, A., Merrouche, A., El, Z., Rahmouni, A., Rokbi, M., Serge, W., and Kouadri, Z. (2014). Study of the alkali treatment effect on the mechanical behavior of the composite unsaturated polyester-Alfa fibers. *Mechanics and Industry*, 15(1), 69–73. 10.1 051/meca/2013082

Biagiotti, J., Puglia, D., and Kenny, J. M. (2004). A review on natural fibre-based composites-part I. *Journal of Natural Fibers*, 1(2), 37–68. 10.1300/J395v01n02_04

Bledzki, A. K., Mamun, A. A., and Gutowski, V. S. (2008). The effects of acetylation on properties of flax fibre and its polypropylene composites. *Express Polymer Letters*, 2(6), 413–422. 10.3144/expresspolymlett.2008.50

Boussehel, H., Mazouzi, D., Belghar, N., Guerira, B., and Lachi, M. (2019). Effect of chemicals treatments on the morphological, mechanical, thermal and water uptake properties of polyvinyl chloride/palm fibers composites. *Revue des Composites et des Materiaux Avances*, 29(1), 1–8.

Cao, Y., Sakamoto, S., and Goda, K. (n.d.). Effects of heat and alkali treatments on mechanical properties of Kenaf fibers. 1–4. 16th International Conference on Composite Materials. https://iccm-central.org/Proceedings/ICCM16proceedings/contents/pdf/MonG/MoGM1-02ge_caoy223305p.pdf

Chakrabarti, D., Islam, S., Jubair, K., and Sarker, R. H. (2020). Effect of chemical treatment on the mechanical properties of Luffa fiber reinforced epoxy composite. *Journal of Engineering Advancements*, 1(02), 37–42.

Chen, H., Zhang, W., Wang, X., Wang, H., Wu, Y., Zhong, T., and Fei, B. (2018). Effect of alkali treatment on wettability and thermal stability of individual bamboo fibers. *Journal of Wood Science*, 64(4), 398–405. 10.1007/s10086-018-1713-0

Chukwudi, A. D., Uzoma, O. T., and Azuka, U. A. (2015). Comparison of acetylation and alkali treatments on the physical and morphological properties of raffia palm fibre reinforced composite. *Science Journal of Chemistry*, 3(4), 72–77. 10.11648/j.sjc.2015 0304.12

Dar, O. A., Malik, M. A., Talukdar, M. I. A., and Hashmi, A. A. (2020). Bionanocomposites in water treatment. In K. Mahmood Zia, F. Jabeen, M. Naveed Anjum, & S. Ikram (Eds.), *Bionanocomposites*, 505–518. Elsevier. 10.1016/b978-0-12-816751-9.00019-2

Dawit, J. B., Regassa, Y., and Lemu, H. G. (2019). Property characterization of Acacia Tortilis for natural fiber reinforced polymer composite. *Results in Materials*, 100054. 10.1016/j.rinma.2019.100054

Devnani, G. L., Mittal, V., and Sinha, S. (2018). Mathematical modelling of water absorption behavior of bagasse fiber reinforced epoxy composite material. *Materials Today: Proceedings*, 5(9), 16912–16918. 10.1016/j.matpr.2018.04.094

Devnani, G. L., and Sinha, S. (2019). Extraction, characterization and thermal degradation kinetics with activation energy of untreated and alkali treated Saccharum spontaneum (Kans grass) fiber. *Composites Part B: Engineering*, 166(October 2018), 436–445. 10.1016/j.compositesb.2019.02.042

Dhanalakshmi, S., Ramadevi, P., and Basavaraju, B. (2015). Influence of chemical treatments on flexural strength of Areca fiber reinforced epoxy composites.*Chemical Science Transactions*, 4(2), 409–418. 10.7598/cst2015.1033

Diharjo, K., Permana, A., Arsada, R., and Asmoro, G. (2017 January). Effect of acetylation treatment and soaking time to bending strength of sugar palm fiber composite. *AIP Conference Proceedings*, 1788(1), 3–7, 030049. 10.1063/1.4968302

Djafari Petroudy, S. R. (2017). Physical and mechanical properties of natural fibers. In M. Fan and F. Fu (Eds.), *Advanced High Strength Natural Fibre Composites in Construction*. Elsevier Ltd. 10.1016/B978-0-08-100411-1.00003-0

Elkordi, A. (2014). Alkali treatment of fan palm natural fibers for use in fiber reinforced concrete. *European Scientific Journal*, 10(12), 186–195.

Eze, I. O., Igwe, I. O., Ogbobe, O., Obasi, H. C., Ezeamaku, U. L., Nwanonenyi, S. C., Anyanwu, E. E., and Nwachukwu, I. (2017). Effects of compatibilization on mechanical properties of pineapple leaf powder filled high density polyethylene. *International Journal of Engineering and Technologies*, *10*(September), 22–28. 10.1 8052/www.scipress.com/ijet.10.22

Faisal, Z. H. T., Amri, F., Salmah, H., and Tahir, I. (2010). The effect of acetic acid on properties of coconut shell filled low density polyethylene composites. *Indonesian Journal of Chemistry*, *10*(3), 334–340.

Faruk, O., Bledzki, A. K., Fink, H. P., and Sain, M. (2012). Biocomposites reinforced with natural fibers: 2000–2010. *Progress in Polymer Science*, *37*(11), 1552–1596. 10.1016/ j.progpolymsci.2012.04.003

Fathi, B., Foruzanmehr, M. Elkoun, S. , and Robert, M. (2019). Novel approach for silane treatment of flax fiber to improve the interfacial adhesion in flax/bio epoxy composites. *Journal of Composite Materials*, *53*(16), 2229–2238. 10.1177/0021998318824643

Feng, N. L., Malingam, S. D., Razali, N., and Subramonian, S. (2020). Alkali and Silane treatments towards exemplary mechanical properties of kenaf and pineapple leaf fibre-reinforced composites. *Journal of Bionic Engineering*, *17*, 380–392.

Fuqua, M. A., Huo, S., and Ulven, C. A. (2017). Natural fiber reinforced composites. *Polymer Reviews*, *52*(3), 259–320. 10.1080/15583724.2012.705409

George, J., Sreekala, M. S., and Thomas, S. (2001). A review on interface modification and characterization of natural fiber reinforced plastic composites. *Polymer Engineering and Science*, *41*(9), 1471–1485. 10.1002/pen.10846

Gomes, A. (n.d.). *The Effect of Alkali Treatment on the Physical and Mechanical*. http:// www.escm.eu.org/docs/eccm/C059.pdf

Gupta, A. (2019). Improvement of physiochemical properties of short bamboo fiber-reinforced composites using ceramic fillers. *Journal of Natural Fibers*, 1–12. 10. 1080/15440478.2019.1584079

Halip, J. A., Selimin, A., Lee, S. H., Chuan, L. Te, and Saffian, H. A. (2021). A review: Chemical treatments of rice husk for polymer composites. *Biointerface Research in Applied Chemistry*, *11*(4), 12425–12433.

Hariharashayee, D., Hanif, M. A. A., Aravind, S., Yasar, S. M., and Azeez, M. A. (2018). Investigation of mechanical properties of KMnO4 treated Sansevieria cylindrical fiber reinforced polymer composite. *International Journal of Advance Research, Ideas and Innovations in Technology*, *3*(4), 294–298.

Hashim, M. Y., Roslan, M. N., Amin, A. M., Mujahid, A., and Zaidi, A. (2012). Mercerization treatment parameter effect on natural fiber reinforced polymer matrix composite: A brief review. *World Academy of Science, Engineering and Technology*, *6*(8), 1638–1644.

He, L., Li, W., Chen, D., Lu, G., Chen, L., Zhou, D., and Yuan, J. (2017). Investigation on the microscopic mechanism of potassium permanganate modification and the properties of ramie fiber/polypropylene composites. *Polymer Composites*, *39*(9). 10.1002/pc

Henrique, G., Tonoli, D., Mendes, R. F., Siqueira, G., and Bras, J. (2013). Isocyanate-treated cellulose pulp and its effect on the alkali resistance and performance of fiber cement composites. *Holzforschung*, *67*(8), 853–861. 10.1515/hf-2012-0195

Ibraheem, S. A., Sreenivasan, S. S., Abdan, K., Sulaiman, S. A., Ali, A., Laila, D., and Abdul, A. (2016). The effects of combined chemical treatments on the mechanical properties of three grades of sisal. *BioResources*, *11*(4), 8968–8980.

Imoisili, P. E., and Jen, T. (2020). Mechanical and water absorption behaviour of potassium permanganate (KMnO4) treated plantain (Musa Paradisiaca) fibre/epoxy bio-composites. *Integrative Medicine Research*, *9*(4), 8705–8713. 10.1016/j.jmrt.202 0.05.121

Jayabal, S., Sathiyamurthy, S., Loganathan, K. T., and Kalyanasundaram, S. (2012). Effect of soaking time and concentration of NaOH solution on mechanical properties of coir – Polyester composites. *Bulletin of Materials Science, 35*(4), 567–574.

Jha, K., Kataria, R., Verma, J., and Pradhan, S. (2019). Potential biodegradable matrices and fiber treatment for green composites: A review. *AIMS Materials Science, 6*(March), 119–138. 10.3934/matersci.2019.1.119

Jones, D., Ormondroyd, G. O., Curling, S. F., Popescu, C. M., and Popescu, M. C. (2017). Chemical compositions of natural fibres. In M. Fan & F. Fu (Eds.), *Advanced High Strength Natural Fibre Composites in Construction.* 10.1016/B978-0-08-100411-1.00002-9

Kabir, M. M., Wang, H., Cardona, F., and Aravinthan, T. (2008). *Effect of chemical treatment on the mechanical and thermal properties of hemp fibre reinforced thermoset sandwich composites.* In S. Fragomeni & S. Venkatensan (Eds.), Incorporating Sustainable Practice in Mechanics and Structures of Materials. CRC PRESS

Kabir, M. M., Wang, H., Lau, K. T., and Cardona, F. (2012). Composites: Part B Chemical treatments on plant-based natural fibre reinforced polymer composites: An overview. *Composites Part B, 43*(7), 2883–2892. 10.1016/j.compositesb.2012.04.053

Kadem, S., Irinislimane, R., and Belhaneche-Bensemra, N. (2018). Novel Biocomposites Based on Sunflower Oil and Alfa Fibers as Renewable Resources. *Journal of Polymers and the Environment, 26*(7), 3086–3096. 10.1007/s10924-018-1196-5

Kalia, S., Dufresne, A., Cherian, B. M., Kaith, B. S., Avérous, L., Njuguna, J., and Nassiopoulos, E. (2011). Cellulose-based bio- and nanocomposites: A review. *International Journal of Polymer Science, 2011.* 10.1155/2011/837875

Kalia, S., and Kumar, A. (2013). Surface Modification of Sunn Hemp Fibers Using Acrylation, Peroxide and Permanganate Treatments: A Study of Morphology, Thermal Stability and Crystallinity. *Polymer-Plastics Technology and Engineering, 52*(1), 24–29. 10.1080/03602559.2012.717335

Katogi, H., Takemura, K., and Sebori, R. (2016). *Fatigue property of natural fiber after alkali treatment. 166*(Hpsm). 10.2495/HPSM160321

Kavanagh, P. (2004). Where Open Source Is Successful. *Open Source Software,* 19–40. 10.1016/b978-155558320-0/50003-9

Keener, T. J., Stuart, R. K., and Brown, T. K. (2004). Maleated coupling agents for natural fibre composites. *Composites Part A: Applied Science and Manufacturing, 35*(3), 357–362. 10.1016/j.compositesa.2003.09.014

Kenned, J. J., Sankaranarayanasamy, K., and Kumar, C. S. (2020). Chemical, biological, and nanoclay treatments for natural plant fiber-reinforced polymer composites: A review. *Polymers and Polymer Composites.* 10.1177/0967391120942419

Khan, J. A., Khan, M. A., and Islam, R. (2012). *Effect of potassium permanganate on mechanical, thermal and degradation characteristics of jute fabric-reinforced polypropylene composite.* 10.1177/0731684412458716

Li, W., Meng, L., and Ma, R. (2016). Effect of surface treatment with potassium permanganate on ultra-high molecular weight polyethylene fiber reinforced natural rubber composites. *Polymer Testing, 55,* 10–16. 10.1016/j.polymertesting.2016.08.006

Li, Xue, Tabil, L. G., and Panigrahi, S. (2007). Chemical treatments of natural fiber for use in natural fiber-reinforced composites: A review. *Journal of Polymers and the Environment, 15*(1), 25–33. 10.1007/s10924-006-0042-3

Li, Xiang, and Wang, X. R. (2013). Effects of hydrogen peroxide treatment on structure and properties of natural color silk. *Advanced Materials Research, 821–822,* 90–93. 10.402 8/www.scientific.net/AMR.821-822.90

Liu, T., Lei, Y., Wang, Q., Lee, S., and Wu, Q. (2013). Effect of Fiber Type and Coupling Treatment on Properties of High-Density Polyethylene/Natural Fiber Composites. *BioResources, 8*(3), 4619–4632. 10.15376/biores.8.3.4619-4632

Liu, Y., Lv, X., Bao, J., Xie, J., Tang, X., Che, J., Ma, Y., and Tong, J. (2019). Characterization of silane treated and untreated natural cellulosic fibre from corn stalk waste as potential reinforcement in polymer composites. *Carbohydrate Polymers*, *218*(January), 179–187. 10.1016/j.carbpol.2019.04.088

Madhu, P., Sanjay, M. R., Senthamaraikannan, P., Pradeep, S., Saravanakumar, S. S., and Yogesha, B. (2019). A review on synthesis and characterization of commercially available natural fibers: Part II. *Journal of Natural Fibers*, *16*(1), 25–36. 10.1080/1544 0478.2017.1379045

Majid, R. A., Ismail, H., and Taib, R. M. (2016). Benzoyl Chloride Treatment of Kenaf Core Powder: The Effects on Mechanical and Morphological Properties of PVC / ENR / kenaf Core Powder Composites. *Procedia Chemistry*, *19*, 803–809. 10.1016/j.proche.2 016.03.105

Maldas, D., and Kokta, B. V. (1991). Influence of organic peroxide on the performance of maleic anhydride coated cellulose fiber-filled thermoplastic composites. *Polymer Journal*, *23*(10), 1163–1171. 10.1295/polymj.23.1163

Marwah, R., Ibrahim, N. A., Zainuddin, N., Saad, W. Z., Razak, N. I. A., and Chieng, B. W. (2014). The Effect of Fiber Bleaching Treatment on the Properties of Poly(lactic acid)/ Oil Palm Empty fruit Bunch Fiber Composites. *International Journal of Molecular Sciences*, *15*(8), 14728–14742. 10.3390/ijms150814728

Miah, M. J., Khan, M. A., and Khan, R. A. (2011). Fabrication and Characterization of Jute Fiber Reinforced Low Density Polyethylene Based Composites: Effects of Chemical Treatment. *Journal of Scientific Research*, *3*(2), 249–259. 10.3329/jsr.v3i2.6763

Miguel, A., Maria, S., Siva, I., Jappes, T. W., Amico, S. C. (2019). Effect of silane treatment on the Curaua fibre/polyester interface. *Plastics, Rubber and Composites*, *48*(4), 160–167. 10.1080/14658011.2019.1586373

Mirza, F. A., Rasel, S. M., Afsar, A. M., Kim, B. S., and Song, J. I. (2015). Injection molding and mechanical properties evaluation of short jute fiber polypropylene reinforced composites. *WIT Transactions on State-of-the-art in Science and Engineering*, *87*, 55–62. 10.2495/978-1-78466-147-2/006

Mittal, V., Saini, R., and Sinha, S. (2016). Natural fiber-mediated epoxy composites – A review. *Composites Part B: Engineering*, *99*, 425–435. 10.1016/j.compositesb.2016.06.051

Mondal, D., Ray, D. P., Ammayappan, L., Ghosh, R. K., Banerjee, P., and Chakraborty, D. (2015). Pre-treatment processes of jute fibre for preparation of biocomposites. *International Journal of Bioresource Science*, *2*(1), 7–14.

Mukaida, J., Nishitani, Y., and Yamanaka, T. (2017). Influence of types of alkali treatment on the mechanical properties of hemp fiber reinforced polyamide 1010 composites. *AIP Conference Proceedings*, *1779*(1), 1–6. 060005 (October 2016). 10.1063/1.4965526

Naidu, A. L. (2018). A study on different chemical treatments for natural fiber. *International Journal of Mechanical and Production Engineering Research and Development*, 8(5), 143–152.

Naveen, P., and Naidu, A. L. (2017). A study on chemical treatments of natural fibers. *International Journal for Research & Development in Technology*, 8(4), 2349–3585.

Nayak, S., and Mohanty, J. R. (2018). Influence of chemical treatment on tensile strength, water absorption, surface morphology, and thermal analysis of areca sheath fibers. *Journal of Natural Fibers*, 1–11. 10.1080/15440478.2018.1430650

Nishitani, Y. (2017). Effect of silane coupling agent on tribological properties of hemp fiber-reinforced plant-derived polyamide 1010 biomass composites. *Materials*, *10*(9), 1040. 10.3390/ma10091040

Onyekwere, O. S., Igboanugo, A. C., and Adeleke, T. B. (2019). Optimisation of acetylation parameters for reduced moisture absorption of bamboo fibre using taguchi experimental design and genetic algorithm optimisation tools. *Nigerian Journal of Technology*, *38*(1), 104–111.

Oushabi, A., Sair, S., Hassani, F. O., Abboud, Y., Tanane, O., and Bouari, A. El. (2017). The effect of alkali treatment on mechanical, morphological and thermal properties of date palm fibers (DPFs): Study of the interface of DPF–Polyurethane composite, *South African Journal of Chemical Engineering, 23*, 116–123. 10.1016/j.sajce.2017.04.005

Patel, V. A., and Parsania, P. H. (2010). Performance evaluation of alkali and acrylic acid treated-untreated jute composites of mixed epoxy-phenolic resins. *Journal of Reinforced Plastics and Composites, 29*(5), 725–730. 10.1177/0731684408100692

Pickering, K. L., Efendy, M. G. A., and Le, T. M. (2016). A review of recent developments in natural fibre composites and their mechanical performance. *Composites Part A: Applied Science and Manufacturing, 83*, 98–112. 10.1016/j.compositesa.2015.08.038

Pradesh, A. (2018). A review on mechanical behaviour of polymers/sisal fiber reinforced composites material. *International Journal of Scientific Development and Research, 3*(12), 190–199.

Rajan, R., Joseph, K., and Skrifvars, M. (2012). Evaluating the influence of chemical modification on flax yarn Conference: ECCM15 - 15th European Conference on Composite Materials, Venice, Italy, 24–28 June 2012.

Ravi, M., Dubey, R. R., Shome, A., Guha, S., and Anil Kumar, C. (2018). Effect of surface treatment on Natural fibers composite. *IOP Conference Series: Materials Science and Engineering, 376*(1). 10.1088/1757-899X/376/1/012053

Rawatan, K., Antara, S., Komposit, M., and Serabut, U. (2019). Effect of alkali treatment on interfacial and mechanical properties of kenaf fibre reinforced epoxy unidirectional composites. *Sains Malaysiana, 48*(1), 173–181.

Razak, N. I. A., Ibrahim, N. A., Zainuddin, N., Rayung, M., and Saad, W. Z. (2014). The influence of chemical surface modification of kenaf fiber using hydrogen peroxide on the mechanical properties of biodegradable kenaf fiber/poly (Lactic Acid) composites. *Molecules, 19*(3), 2957–2968. 10.3390/molecules19032957

Reddy, P. V., Krishnudu, D. M., and Prasad, P. R. (2019). A study on alkali treatment influence on prosopis juliflora fiber-reinforced epoxy composites . *Journal of Natural Fibers*, 1–13. 10.1080/15440478.2019.1687063

Reddy, K. O., Maheswari, C. U., Rajulu, A. V., and Guduri, B. R. (2009). Thermal degradation parameters and tensile properties of borassus flabellifer fruit fiber reinforcement. *Journal of Reinforced Plastics and Composites, 28*(18), 2297–2301. 10.1177/0731684408092380

Reddy, B. M., Reddy, Y. V. M., and Reddy, B. C. M. (2018). Effect of alkali treatment on mechanical, water absorption and chemical resistance properties of cordia-dichotoma fiber reinforced epoxy composites. *International Journal of Applied Engineering Research, 13*(6), 3709–3715.

Sabri, M. N. I. M., Bakar, M. B. A., Masri, M. N., Mohamed, M., Noriman, N. Z., Dahham, O. S., and Umar, M. U. (2020). Effect of chemical treatment on mechanical and physical properties of non-woven kenaf fiber mat reinforced polypropylene biocomposites. *AIP Conference Proceedings, 2213*. 10.1063/5.0000411

Salisu, A. A., Yakasai, M. Y. , and Aujara, K. M. (2016). Physico-mechanical properties of chemically modified Sisal fibre reinforced unsaturated polyester composites. *World Academy of Science, Engineering and Technology, 10*, 81–88.

Sammartino, E. K., Reboredo, M. M., and Aranguren, M. I. (2016). Natural fiber-polypropylene composites made from caranday palm. *Journal of Renewable Materials, 4*(2), 101–112. 10.7569/JRM.2014.634144

Siakeng, R., Jawaid, M., Ariffin, H., Sapuan, S. M., Asim, M., and Saba, N. (2019). Natural fiber reinforced polylactic acid composites: A review. *Polymer Composites, 40*(2), 446–463. 10.1002/pc.24747

Singhal, P., and Tiwari, S. K. (2014). Effect of various chemical treatments on the damping property of jute fibre reinforced composite. *International Journal of Advanced Mechanical Engineering*, 4(4), 413–424.

Srisuwan, L., Jarukumjorn, K., and Suppakarn, N. (2018). Effect of Silane treatment methods on physical properties of rice husk flour/natural rubber composites. *Advances in Materials Science and Engineering*, 2018, Article ID 4583974, 14 pages, 2018. https://doi.org/10.1155/2018/4583974

Student, M. T. (2013). Effect of chemical treatments on tensile and flexural properties of bamboo reinforced composite. *International Journal of Engineering Research & Technology*, 2(12), 2353–2365.

Suwanruji, P., Tuechart, T., Smitthipong, W., and Chollakup, R. (2016). Modification of pineapple leaf fiber surfaces with silane and isocyanate for reinforcing thermoplastic. *Journal of Thermoplastic Composite Materials*, 1–17. 10.1177/0892705716632860

Takemura, K., Takada, Y., and Katogi, H. (n.d.). Effect of treatment using silane coupling agent on creep properties of jute fiber reinforced composites. *WIT Transactions on The Built Environment*, 124, 417–424. 10.2495/HPSM120

Tapas, P., Swain, R., and Biswas, S. (2017). Influence of fiber surface treatments on physico-mechanical behaviour of jute/epoxy composites impregnated with aluminium oxide filler. *Journal of Composite Materials*, 51(28), 3909–3922. 10.1177/0021998317695420

Thakur, V. K., and Thakur, M. K. (2014). Processing and characterization of natural cellulose fibers/thermoset polymer composites. *Carbohydrate Polymers*, 109, 102–117. 10.1016/j.carbpol.2014.03.039

Then, Y. Y., Ibrahim, N. A., Zainuddin, N., Chieng, B. W., Ariffin, H., and Wan Yunus, W. M. Z. (2015). Influence of alkaline-peroxide treatment of fiber on the mechanical properties of oil palm mesocarp fiber/poly (butylene succinate) biocomposite. *BioResources*, 10(1), 1730–1746. 10.15376/biores.10.1.1730-1746

Thiruchitrambalam, M., Logesh, M., Shanmugam, D., and Muthukumar, S. (2018). The physical, chemical properties of untreated and chemically treated palmyra palm leaf fibres. *International Journal of Engineering and Technology*, 7, 582–585.

Tserki, V., Zafeiropoulos, N. E., Simon, F., and Panayiotou, C. (2005). A study of the effect of acetylation and propionylation surface treatments on natural fibres. *Composites Part A: Applied Science and Manufacturing*, 36(8), 1110–1118. 10.1016/j.compositesa.2005.01.004

Vinod, A., Vijay, R., Singaravelu, D. L., Khan, A., Sanjay, M. R., Siengchin, S., Verpoort, F., Alamry, K. A., and Asiri, A. M. (2020 July). Effect of alkali treatment on performance characterization of Ziziphus mauritiana fiber and its epoxy composites. *Journal of Industrial Textiles*, 1. 10.1177/1528083720942614

Wang, Xue, Chang, L., Shi, X., and Wang, L. (2019). Effect of hot-alkali treatment on the structure composition of jute fabrics and mechanical properties of laminated composites. *Materials*, 12(9), 1386.

Wang, Xinxin, Cui, Y., Xu, Q., Xie, B., and Li, W. (2010). Effects of alkali and silane treatment on the mechanical properties of jute-fiber-reinforced recycled polypropylene composites. *Journal of Vinyl and Additive Technology*, 16, 183–188. 10.1002/vnl.20230

Xie, Y., Hill, C. A. S., Xiao, Z., Militz, H., and Mai, C. (2010). Silane coupling agents used for natural fiber/polymer composites: A review. *Composites Part A: Applied Science and Manufacturing*, 41(7), 806–819. 10.1016/j.compositesa.2010.03.005

Yashas Gowda, T. G., Sanjay, M. R., Subrahmanya Bhat, K., Madhu, P., Senthamaraikannan, P., and Yogesha, B. (2018). Polymer matrix-natural fiber composites: An overview. *Cogent Engineering*, 5(1). 10.1080/23311916.2018.1446667

You, A., Be, M. A. Y., and In, I. (2017). Effects of alkaline treatment and fiber length towards the static and dynamic properties of ijuk fiber strengthened – epoxy composite. *030022*(October 2016). 10.1063/1.4965756

Zaaba, N. F., Jaafar, M., and Ismail, H. (2017). The effect of alkaline peroxide pre-treatment on properties of peanut shell powder filled recycled polypropylene composites. *Journal of Engineering Science, 13*, 75–87. 10.21315/jes2017.13.6

Zaman, H. U., Khan, M. A., and Khan, R. A. (2013). Banana fiber-reinforced polypropylene composites: A study of the physico-mechanical properties. *Fibers and Polymers, 14*(1), 121–126. 10.1007/s12221-013-0121-8

Zhu, J., Brington, J., Zhu, H., and Abhyankar, H. (2015). Effect of alkali, esterification and silane surface treatments on properties of flax fibres. *Journal of Scientific Research and Reports, 4*(1), 1–11. 10.9734/JSRR/2015/12347

5 Physical and Biological Treatment

G. L. Devnani and Himani Agrawal
Department of Chemical Engineering, Harcourt Butler
Technical University, Kanpur, U.P., India

Deepak Singh and Suresh Kumar Patel
Department of Chemical Engineering, Institute of
Engineering and Technology, Lucknow, India

INTRODUCTION

Nowadays, natural fibers are considered a promising substitute of synthetic fibers. Specific characteristics of natural fibers like sustainability, mechanical strength, low cost, and environmental friendliness attract scientific community. During the last decade, a lot of work has been done on the reinforcement of fiber in various polymer matrices (Koronis et al., 2013; Vinod et al., 2020). Some disadvantages are also there with the use of natural fiber that they have limited thermal stability and moreover due to presence of OH and other polar groups in their chemical structure, they have a high hydrophilic tendency. This hydrophilic tendency leads to degradation of properties of natural fibers and ultimately results in a poor-quality polymer composite. High water absorption also attracts degradation of fibers biologically. This also results in poor compatibility with the polymer matrix. Being lignocellulosic, natural fibers are hydrophilic in nature; on the other hand, these polymer matrices are hydrophobic, so compatibility between these two materials is very important for the quality of product commercialization. This issue can be remediated by different chemical and surface treatment methodologies. Different surface treatment methodologies like alkylation, silanization, acetylation, etc. have been suggested, experimented, and compiled by researchers to improve the compatibility and adhesion between natural fibers and polymer matrices. A seminal amount of work has been done on natural fiber surface treatment to make it suitable for reinforcement in polymer composites. Table 5.1 summarizes the review work done on the surface treatment of natural fibers.

Most of the treatment methodologies used by researchers are chemical in nature and the chemical used can create a harmful effect on the environment, so there is a need to work on environmentally friendly methods like physical and biological. Figure 5.1 represents the classification of environmentally friendly methods for the surface modifications of natural fibers.

DOI: 10.1201/9781003201724-5

157

TABLE 5.1
Review Work on Fiber Surface Treatment

S. No.	Areas Covered	Reference
1.	Different treatment methodologies for the treatment of natural fibers	(Mohanty et al., 2001)
2.	Physical as well as chemical treatment methodologies	(Belgacem & Gandini, 2005)
3.	Various chemical treatment methodologies for surface modification of natural fibers	(Li et al., 2007)
4.	Surface treatment methodologies to improve compatibility between natural fiber and polymer matrix	(John & Anandjiwala, 2008)
5.	Along with other treatment methodologies, plasma treatment was covered	(Kalia et al., 2009)
6.	Various physical treatment methodologies were also covered	(Mukhopadhyay & Fangueiro, 2009)
7.	Some new chemical treatment methodologies were discussed	(Kabir et al., 2012)
8.	Environmentally friendly methods like physical and biological treatment were discussed	(Kalia et al., 2013)
9.	Electric discharge and alkali treatment were emphasized, along with other treatment methodologies	(Adekunle, 2015)
10.	Fiber treatment method as per structure property relationship	(Zhou et al., 2016)
11.	Surface treatment methodologies for decreasing hydrophilic character of natural fibers	(Ali et al., 2018)
12.	Gamma radiation was also covered, along with other treatment techniques	(Le Moigne et al., 2017)
13.	Specific kenaf fiber was discussed to improve the properties by surface treatment	(Hamidon et al., 2019)
14.	Review on chemical treatment methods and their effect on fiber-matrix interfacial bonding mechanism	(Amiandamhen et al., 2020)
15.	Various physical, chemical, and biological treatment methods for the treatment of natural fibers	(Devnani, 2021)
16.	Vegetable fiber surface modification techniques	(Shukla et al., 2021)

The purpose of applying physical treatment for the natural fibers is to basically (a) separate the fiber bundles into filaments and (b) modification/improvement of the natural fiber surface so that it becomes more compatible with different polymer matrices (Belgacem & Gandini, 2005; Mukhopadhyay & Fangueiro, 2009). Figure 5.2 summarizes the various types of physical treatment methodologies.

5.1 PLASMA TREATMENT

Plasma methodology is the physical form of surface handling that is used to mitigate the fiber's surface in order for their application as reinforcing agents to

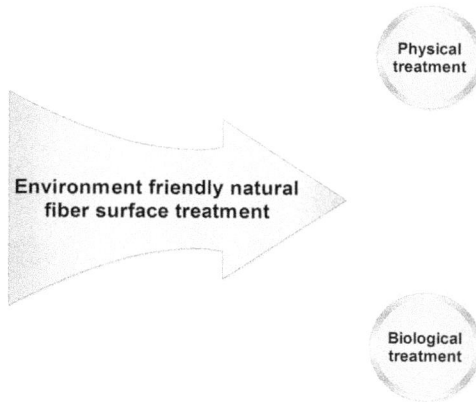

FIGURE 5.1   Classification of environmentally friendly treatment methods.

manufacture composites (Koohestani et al., 2019). Approaching the sample with plasma methodology can be done two ways: atmospheric or vacuum. Processing the probe with plasma is associated with the thermal phenomenon, employing the gas that is at a low degree of ionization (cold plasma), favoring the surface treatment of the fibers and the matrix by improving the strength of adhesion between them (Molina, 2016). Plasma is composed of gases like oxygen, nitrogen, air, and helium whose application on the specimen improves the wettability, lowers the resistance, and eliminates the weakly bounded layers from the surface, which in turn, magnify the morphology and chemical and mechanical properties of the treated fiber (Bhatia et al., 2016). The basic principle underlying the mechanism of this methodology is the engagement of the reactive species present in the plasma gas with the surface of the fiber present in the plasma reactor, resulting in the alteration of material and chemical characteristics of the specimen. The outcome of the modified surface depends upon the type of ionized gas used, the extent of microwave radiation, interval of exposure, and distance between the sample and plasma source.

Due to the simple and environmentally friendly nature, handling of vegetable fibers with plasma has caught the consideration of many scholars, due to their impeccable results. Peter et al. extensively researched the implications of the atmospheric plasma technique on the surface characteristics of the hemp fiber. They successfully developed an ecologically safe and cost-effective methodology by treating the hemp's surface at a low-temperature plasma at atmospheric pressure. The result presented shows the highly hydrophilic nature of the fiber, with no alternation in its natural color and flexural strength (Skundric, 2015). In an associated experiment, Gabriel et al. exercised the oxygen or air plasma treatment to competently eradicate the lignin supplemented amorphous layer on the coir surface. The findings were analyzed with the FTIR and concluded that the lignin-related FTIR peaks were decremented by tenfold for air plasma treated coir fiber, whereas by twentyfold for oxygen plasma treated fiber at 80 W power for 7.2 minutes. The aftermath also favors the physiochemical and adhesive properties of the oxygen

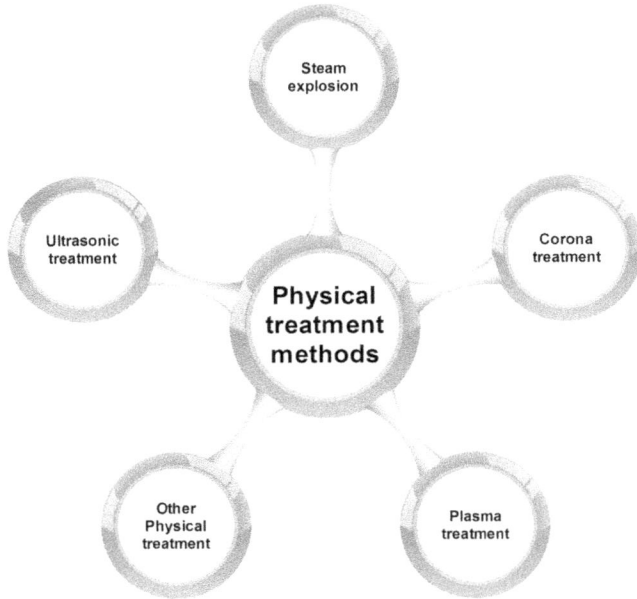

FIGURE 5.2   Different physical treatment methodologies.

plasma treated reinforced starch composites (Farias et al., 2017). Figure 5.3 shows the SEM analysis of untreated and plasma-treated coir fiber.

The treatment of fiber with different types of plasma gases produces the different outcomes. Camila et al. exploited the fiber surface with the cold plasma gas comprising of argon and carbon dioxide to investigate the adsorption kinetics and isotherms of sisal fiber over methylene blue. The repercussion involves the increment of the surface roughness and adsorption capacity of the fiber, along with the eradication of the hydroxyl group and alteration in the lignocellulosic components of the sisal fiber, leading to the better adhesion between the fiber and the matrix (Cristina et al., 2020). Wettability and dyeability are important traits required for the composite formed by the given fiber. Treatment of the bamboo fibers with the cold argon plasma leads to the enhancement of these properties, as studied by Xiangyu and his colleagues. They concluded the larger the duration of treatment, the more the roughness of the surface was enhanced and thus more wettability and dyeability (Xu et al., 2006). Several studies by the scholars show the enhancement in the interlocking characteristics of the plasma-treated fiber and the given polymeric matrix. Panikkassery et al. concluded better interfacial interaction between the plasma-treated coir fiber and the polypropylene matrix (Sasidharan et al., 2019), while the improvement in the hydrophobic nature of ramie fiber over a PP matrix was showed by Qian and his companion (Zhang et al., 2015). Adriana et al. analyzed the surface functionalization of the silane plasma–treated shredded sisal fiber along with its adhering strength with the high density polyethylene matrix (HDPE). The sample containing sisal and powdered HDPE is placed in the plasma glass

FIGURE 5.3 SEM of untreated and plasma-treated coir fibers from standard batches. (a & b) Untreated fibers; (c) fibers after air; and (d) oxygen plasma treatments in the standard batch (t = 7.2 minutes). Reprints with permission from Elsevier (Farias et al., 2017).

reactor, with rubber and steel rings installed on both the ends. The chamber is capable of gyrating at 13.56 MHz; the vapor and the gas flow into the reactor is maintained with the help of controlled valves. The cold plasma gas with the reactive species alters the fiber and the powdered matrix, whose results were abstracted with the help of electron spectroscopy for chemical analysis (ESCA). The findings from ESCA suggest the successful surface functionalization of silane-plasma treated sisal fiber reinforced with the HDPE matrix (Martin et al., 2002).

5.2   CORONA TREATMENT

To explore the eco-friendly path for the surface treatment of the fibers, scholars are increasing their interest over the physical treatment and one such methodology is the corona treatment of the vegetable fibers, which is ecologically safe and clean technology for the modification of the hydrophobic properties of fiber (Oudrhiri Hassani, 2020). Corona treatment comes under the category of low-power electric discharge, which functions at or near the barometric pressure, and is successful in transfiguring the passive polymeric surfaces like those of polyethylene, poly-propylene, and polystyrene (Cho et al., 2014). Corona discharge treatment (CDT) is inducted in the presence of air or in the blanket of non-reactive (inert) gases. The methodology is based on the principle of interposing the polar groups on the surface of the treating fiber, which in turn has a significant impact on the surface energy of the sample because of the creation of new functional groups, thus aimed to revamp the unevenness over the surface of the fiber in conjunction with adhesivity and wettability (Nabinejad & Debnath, 2019; Nemani et al., 2018). The electric dis-charge methodology imposed a significant impact on the mechanical properties of the cellulosic fibers due to the increment of the aldehyde groups on the surface of vegetable fiber by altering their superficial energy (Pizzi & Kumar, 2019). The treatment is procured by supplying high-magnitude voltage at the apex of the bundle of the sharp electrodes, which generate a shade of discharge known as corona, along with the high magnitude of electromagnetic fields. These radiations ionized the surface of the fibers, making the species excited and lead to the for-mation of molecular chains with the fulfillment of the aim of modifying their surface morphology (Mohit, 2018). CDT acts as the primary step for grafting the cellulosic fiber to a water-resistant, non-polar polymer matrix (Lu et al., 2015). In recent research reported by Ragoubi et al., they disclosed the impact of the corona treatment on miscanthus fiber reinforced with polypropylene or polylactic acid–based composites. Calorific and mechanical qualities of the composites are analyzed by numerous classical tensile tests and dynamic mechanical test. Fiber to be treated is placed in a glass reactor at a low frequency of 50 Hz and high voltage of 15 kV to generate a dielectric barrier for duration of 15 minutes. The composites are formed by molding techniques and the sample is tested in a laboratory for XPS, TGA, DMA, and tensile. The results so calculated prove the superiority of the corona-treated fiber composites over untreated fiber-based composites (M Ragoubi et al., 2012). In addition to fibers, the corona treatment improves the surface characteristics of matrices like HDPE, PP, and PET. The treatment improves the wettability and morphology, along with adhesivity and hydrophilicity of the matrix

(Guo et al., 2013; Louzi et al., 2019). Corona treatment finds wide applications in the processing and textile industries. Several research scholars exploited this methodology on the silk fabrics and obtained results that were in their favor. Mousa et al. in their research work, studied the enhancement of dyeability traits of the corona-treated silk fabric, pretreated with chitosan. The data when analyzed explained the increment from RY value of 1.84 to 4.58 and RB from 3.05 to 6.40. Although the ability to color fast was not affected, a potential and eco-friendly method of reactive dyeing of silk textile fabrics with CDT methodology was formulated (Sadeghi-kiakhani & Tayebi, 2016). In the similar report, Daives et al. did a parallel study between the traditional and corona surface modification techniques in order to increase the dyeability of the silk fabrics. High tensile strength along with great hydrophilicity are the prerequisites required for the fabrics to be dyed. The corona treatment displayed improvement in water-holding capacity and flexural strength of the corona-treated silk fabrics over the traditional method (Bergamasco & De Carvalho Campos, 2017). Apart from the textile industry, the packaging industry too employed the corona treatment for the manufacturing of antibacterial packaging sheets formulated by a corona-treated PLA/PBS blend, as reviewed by Nattakarn et al. The contact angle, adhesivity, and biodegradability of the PLA/PBS sheets were improved, along with the introduction of oxygen atoms on the surface of sheets that increased their elongation traits (Hongsriphan & Sanga, 2018).

5.3 ULTRASONIC TREATMENT

Sonication is basically applying the energy of sound to create agitation in particles present in the sample for different purposes, like in the extraction process of multiple compounds from plants, microalgae, and seaweeds. The frequencies used are in the range of the ultrasonic zone that is greater than 20 kHz and the process is commonly called ultrasonication. Although ultrasonication is not a very popular surface treatment method for natural fibers yet, it is an efficient method for the removal of different substances and impurities without using any kind of chemicals or surfactants from the fiber surface. It is a type of electromagnetic radiation with a frequency range from 10–400 nm. From the last decade, the ultrasonic method has been applied in many industries (Czaplicki & Ruszkowski, 2014). Renouard and his group showed that ultrasonic treatment can be an efficient method to improve lignocellulosic material composition. They took coir, hemp, and short flax fibers for their experimental work. After 1 day, they analyzed the optimal degradation and concluded with these findings that ultrasonication of these natural fibers only removed hemicelluloses present in the fibers (Renouard et al., 2014).

5.4 STEAM EXPLOSION

Mason developed the method of steam explosion for the pre-treatment methodology for biomass. In this method, saturated steam of water is used for heating natural fibers from 160°C–260°C for a particular duration ranging from seconds to minutes at a specific pressure and, after, this pressure is reduced to atmospheric pressure. Expansion of natural fiber takes place at high pressure and temperature conditions and gaps would be filled by steam when reduction of pressure takes place. Numerous hole formation takes place so that molecular substances can be released from the cell. This method is environmentally friendly, economically viable, and energy efficient and that's why it is gaining much popularity among the scientific community. A sharp change of pressure at high temperature values is responsible for affecting the plant cell. An outcome of this treatment splitting of lignocellulosic material takes place in its key components. An increase in cellulose percentage takes place in steam explosion and a crystallinity index of steam-exploded banana fiber (54.1), pineapple leaf (63.7), and jute fibre (52.9) was obtained (Abraham et al., 2011). Figure 5.4 represents the typical setup of the steam explosion process in which steam is generated in a steam generator and the sample is fed to the reactor, and the collection tank is there to collect the treated samples.

Apart from these physical methods, there are few other physical treatments like UV radiation also improves the performance of natural fibers (Devnani, 2021) and the improvement depends on various factors such the duration, intensity, and wavelength of UV radiation. Mahajan et al. did the experimental work related to effect of UV radiation on polymer composites (Mahajan, 2016). Another physical treatment technique is gamma radiation treatment that consists of very high energy that is capable to improve the properties of polymer surfaces. The optimum value of exposure of gamma radiation increases the mechanical properties of composite materials. The driving force of this upgrade in tensile properties was the high energy values gamma radiation is capable of making and crosslink among the natural fiber surface molecules. An appreciable amount of improvement due to exposure of surface in gamma radiations in the different composites like jute polyester, jute pp was examined by Kabir et al. (Shahriar Kabir et al., 2018).

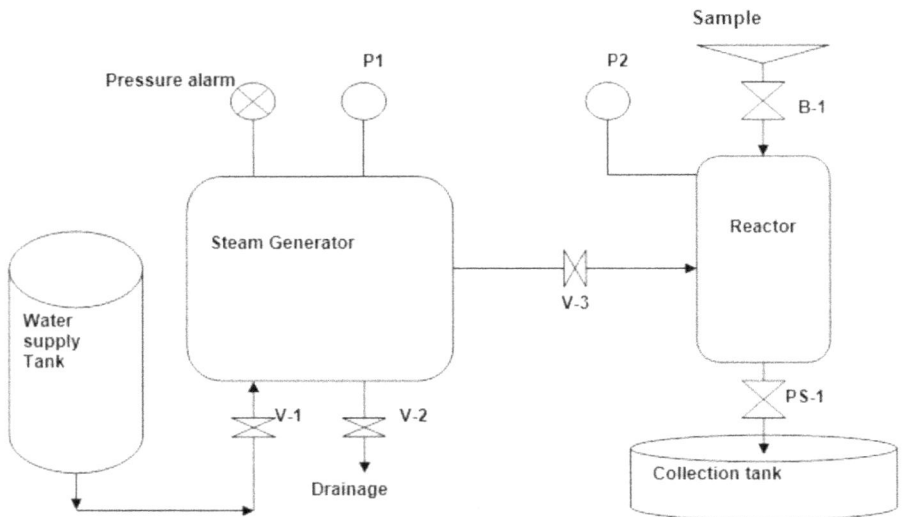

FIGURE 5.4 Process flow scheme of the steam explosion equipment (B: ball valve, PS: pressure relief valve, V: valve, P: digital pressure transducer) (Sheng et al., 2014).

5.5 BIOLOGICAL TREATMENT

Biological treatment is another environmentally friendly treatment technique for the treatment of natural fibers in which no toxic chemical is used and, moreover,it is energy efficient because the process parameters are easily achieved as most of the reactions occur at room temperature and maintaining high temperature and pressure requirements are less. The biological treatment of natural fibers consists of three categories: bacteria, fungi, and enzymes. Figure 5.5 represents these different categories (Kalia et al., 2013).

5.5.1 BACTERIAL CELLULASE

Bacterial cellulose refers to the group of enzymes that catalyze hydrolysis of cellulose. It consists of three enzymes: endoglucanase, cellobiohydrolase, and β-glucosidase. These have a tendency to decompose both crystalline and soluble cellulose derivatives.

5.5.2 FUNGI

It belongs to the eukaryotic class involving microorganisms including yeast, mushrooms, and molds. Basidiomycetous fungi degrades cellulose. White rot fungi from basidiomycetous are the fungi that decompose lignin. These fungi also affect the hydrophilicity of natural fibers.

5.5.3 ENZYMES

Modification of natural fibers with the help of enzymes is a novel environmentally friendly treatment method. Glucosidases, lipases, laccase, and peroxidase are the enzymes from different classes that are frequently used for surface modification of polymers. Milder reaction conditions and nondestructive transformations of the

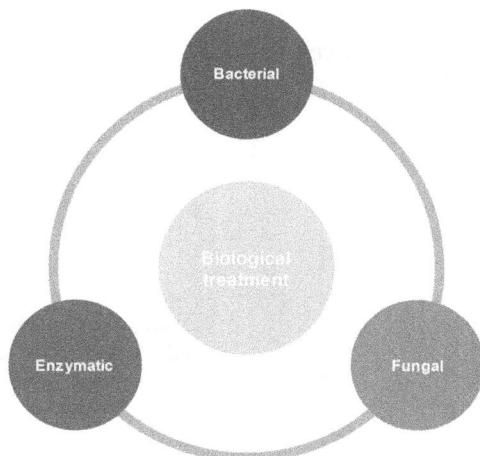

FIGURE 5.5 Various biological treatments for surface improvements of natural fibers.

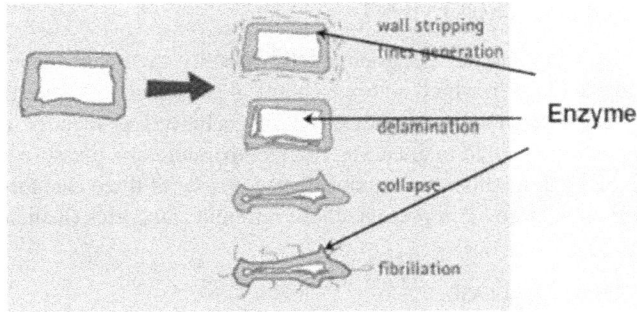

FIGURE 5.6 Enzyme activity on a natural fiber cell.

natural fiber surface are the key advantages of enzyme treatment methodologies. Figure 5.6 represents the mechanism on which an enzyme acts on the plant cell wall. Delamination, collapse, and fibrillation are the outcomes of enzyme activity on the cell of a plant fiber. It induces roughness also in the natural fiber that is desired for the interlocking of a fiber and different polymer matrix.

5.6 EFFECT ON PROPERTIES

These environmentally friendly methods have a positive effect on the properties of natural fibers. Roughness is induced on the fiber surface that make it more compatible with a polymer matrix. Hydrophilic nature of these natural fibers is also reduced due to these treatment methodologies. Treatment also improves cellulose percentage in the fiber, which is responsible for the tensile strength of fiber. Table 5.2 summarizes the improvement in properties as an effect of these environmentally friendly techniques.

TABLE 5.2
Environmentally Friendly Surface Treatment Methodologies and Their Positive Effect

S. No.	Fiber	Matrix	Treatment Technique	Key Highlights	Reference
1.	Cellulose	Thermoplastic Starch	Air Plasma	Surface of the cellulose fiber is enhanced, along with the great interfacial bond strength with the starch matrix. The XRD image of the composite shows a high degree of crystallinity while the SEM and FTIR show improved mechanical properties.	(Fazeli et al., 2018)

TABLE 5.2 (Continued)
Environmentally Friendly Surface Treatment Methodologies and Their
Positive Effect

S. No.	Fiber	Matrix	Treatment Technique	Key Highlights	Reference
2.	Coir	Epoxy	Liquid Plasma	H_2O and $NaHCO_3$ were used as a medium. The IFSS between the coir and epoxy matrix was studied with the help of a single fiber and pull-out test. The result shows the decrement in the tensile strength with enhancement in the interfacial strength due to the better interlocking between the fiber and matrix.	(Erwin et al., 2020)
3.	Jute	PLA	Plasma	Plasma-treated fibers were found to show superior properties than alkali-treated fibers. Interfacial adhesion between the jute and PLA shows enhancement.	(Gibeop et al., 2013)
4.	Hemp	PP	Corona	The surface properties of the corona-treated hemp-PP composites are studied like tensile strength and Young moduli. The treated composite showed an increment in these traits at 20% fiber loading. The fracture surface reveals the better adhesion between the treated fiber and matrix.	(M. Ragoubi, 2010)
5.	Cotton		Corona	Hardness, water absorption capacity, and recovery angle test were explored for the corona-treated cotton fiber. The surface morphology shows the roughness over the fiber surface with better adhesivity with nano-softeners, along with the increment of the –COOH groups on the surface of fiber.	(Nourbakhsh et al., 2018)

(Continued)

TABLE 5.2 (Continued)
Environmentally Friendly Surface Treatment Methodologies and Their Positive Effect

S. No.	Fiber	Matrix	Treatment Technique	Key Highlights	Reference
6.	Carbon fillers	Epoxy nanocomposites	Corona	Surface morphology reveals the enhancement in thermal and mechanical stability, with high adhesion strength with nano matrix, along with the high influence of the contact angle.	(Desai et al., 2018)
7.	Alpaca wool		Ultrasonic treatment	The use of ultrasound technology reduces water and detergent consumption and it shortens the time of scouring. The most important result is obtaining the scouring wool without damages and entanglements.	(Czaplicki & Ruszkowski, 2014)
8.	Coir, flax, and hemp		Ultrasonic treatment	It was found that ultrasonic treatment of coir, flax, and hemp fibers only decomposed hemicelluloses in the fibers.	(Renouard et al., 2014)
9.	Banana, jute, pineapple		Steam Explosion	Steam explosion method was applied on the alkali-treated fiber at a pressure of 137 Pa.	(Abraham et al., 2011)
10.	Date palm fiber	Polybutylene succinate	Enzymatic	Effect of different enzymatic treatments (xylanase, pectinase, and combination of xylanase +pectinase) on chemical properties, morphological natures, and mechanical properties of date palm fiber (DPF) and reinforced DPF composites were analyzed and it was concluded that enzymatically treated date palm fibers improve the rigidity of composite materials by about 42% and 29% for two different palm fibers.	(Chaari et al., 2020)

TABLE 5.2 (Continued)
Environmentally Friendly Surface Treatment Methodologies and Their Positive Effect

S. No.	Fiber	Matrix	Treatment Technique	Key Highlights	Reference
11.	Ramie	Polybutylene succinate	Enzymatic	In this study, ramie fibers were exposed with bacteria for the improvement of surface characteristics. The effect of various parameters like carbon and nitrogen sources along with pH, temperature, and incubation period on the hydrolysis of ramie cellulose were analyzed. Treated fiber showed improved compatibility with the polymer matrix.	(Thakur & Kalia, 2017)
12.	Flax	Epoxy	Enzymatic	All enzymes used in treatments resulted in finer fibers compared to untreated one and led to biocomposites with a reduced equilibrium moisture content and lower coefficient of diffusion.	(De Prez et al., 2019)
13.	Coir	Polylactic acid (PLA)	Enzymatic	Four different enzymes, namely lipase, lactase, pectinase, and cellulase, were used for the surface modifications of coir fiber. It was observed that enzymatic treatments enhanced the interfacial adhesion between coir fiber and PLA matrix by the waxes and fatty acid removal and/or the increment in surface roughness.	(Coskun et al., 2019)

CONCLUSION

Use of sustainable and biodegradable materials are gaining importance day by day. Natural fibers are renewable and cost-effective substitutes of synthetic fibers whose resources are depleting day by day. Green surface treatment methods, including physical and biological, for improving properties and compatibility of natural fibers with different polymer matrix is cost effective and will make the whole process environmentally friendly.

REFERENCES

Abraham, E., Deepa, B., Pothan, L. A., Jacob, M., Thomas, S., Cvelbar, U., & Anandjiwala, R. (2011). Extraction of nanocellulose fibrils from lignocellulosic fibres: A novel approach. *Carbohydrate Polymers*, *86*(4), 1468–1475. 10.1016/j.carbpol.2011.06.034

Adekunle, K. F. (2015). Surface treatments of natural fibres—A review: Part 1. *Open Journal of Polymer Chemistry*, *5*(03), 41–46. 10.4236/ojpchem.2015.53005

Ali, A., Shaker, K., Nawab, Y., Jabbar, M., Hussain, T., Militky, J., & Baheti, V. (2018). Hydrophobic treatment of natural fibers and their composites—A review. *Journal of Industrial Textiles*, *47*(8), 2153–2183. 10.1177/1528083716654468

Amiandamhen, S. O., Meincken, M., & Tyhoda, L. (2020). Natural fibre modification and its influence on fibre-matrix interfacial properties in biocomposite materials. *Fibers and Polymers*, *21*(4), 677–689. 10.1007/s12221-020-9362-5

Belgacem, M. N., & Gandini, A. (2005). The surface modification of cellulose fibres for use as reinforcing elements in composite materials. *Composite Interfaces*, *12*(1–2), 41–75. 10.1163/1568554053542188

Bergamasco, D. A., & De Carvalho Campos, J. S. (2017). Corona treatment applied in the processing of silk waste. study of properties such as hydrophilicity and tensile strength. *Procedia Engineering*, *200*, 18–25. 10.1016/j.proeng.2017.07.004

Bhatia, J. K., Kaith, B. S., & Kalia, S. (2016). Recent developments in surface modification of natural fibers for their use in biocomposites. In S. Kalia (Ed.), *Biodegradable Green Composites* (pp. 80–117). John Wiley.

Chaari, R., Khlif, M., Mallek, H., Bradai, C., Lacoste, C., Belguith, H., Tounsi, H., & Dony, P. (2020). Enzymatic treatments effect on the poly (butylene succinate)/date palm fibers properties for bio-composite applications. *Industrial Crops and Products*, *148*(July 2019), 112270. 10.1016/j.indcrop.2020.112270

Cho, D., Kim, H., & Drzal, L. T. (2014). Surface treatment and characterization of natural fibers: Effects on the properties of biocomposites. In *Polymer Composites* (pp. 133–177). Willey.

Coskun, K., Mutlu, A., Dogan, M., & Bozacı, E. (2019). Effect of various enzymatic treatments on the mechanical properties of coir fiber/poly (lactic acid) biocomposites. *Journal of Thermoplastic Composite Materials*, *34*(8), 1066–1079. 10.1177/0892705719864618

Cristina, C., Faria, A. De Aparecido, J., Dantas, S., Alves, M., Pasquini, D., Cipriano, E., Scarmínio, J., & Valentim, R. (2020). Influence of plasma treatment on the physical and chemical properties of sisal fibers and environmental application in adsorption of methylene blue. *Materials Today Communications*, *23*,101140. 10.1016/j.mtcomm.2020.101140

Czaplicki, Z., & Ruszkowski, K. (2014). Optimization of scouring alpaca wool by ultrasonic technique. *Journal of Natural Fibers*, *11*(2), 169–183. 10.1080/15440478.2013.864577

De Prez, J., Van Vuure, A. W., Ivens, J., Aerts, G., & Van de Voorde, I. (2019). Effect of enzymatic treatment of flax on fineness of fibers and mechanical performance of composites. *Composites Part A: Applied Science and Manufacturing, 123*(January), 190–199. 10.1016/j.compositesa.2019.05.007

Desai, B. M. A., Mishra, P., Vasa, N. J., Sarathi, R., & Imai, T. (2018). Understanding the performance of corona aged epoxy nano micro composites. *Micro and Nano Letters, 13*(9), 1280–1285. 10.1049/mnl.2018.0164

Devnani, G. L. (2021). Recent trends in the surface modification of natural fibers for the preparation of green biocomposite. In S. Thomas & B. Preetha (Eds.), *Green Composites* (pp. 273–293). Springer Singapore. 10.1007/978-981-15-9643-8_10

Erwin, A., Putra, E., Renreng, I., Arsyad, H., & Bakri, B. (2020). Investigating the effects of liquid-plasma treatment on tensile strength of coir fibers and interfacial fiber-matrix adhesion of composites. *Composites Part B, 183*(December 2019), 107722. 10.1016/j.compositesb.2019.107722

Farias, D., Cordeiro, R., Canabarro, B. R., Scholz, S., & Sim, R. A. (2017). Surface lignin removal on coir fibers by plasma treatment for improved adhesion in thermoplastic starch composites João Gabriel Guimarães de Farias a, Rafael Cordeiro Cavalcante a. *Carbohydrate Polymers, 165*, 429–436. 10.1016/j.carbpol.2017.02.042

Fazeli, M., Florez, J. P., & Simão, R. A. (2018). Improvement in adhesion of cellulose fibers to the thermoplastic starch matrix by plasma treatment modification. *Composites Part B, 163*, 207–216. 10.1016/j.compositesb.2018.11.048

Gibeop, N., Lee, D. W., Prasad, C. V., Toru, F., & Sun, B. (2013). Effect of plasma treatment on mechanical properties of jute fiber/poly (lactic acid) biodegradable composites. *Advanced Composite Materials, 22*(6), 389–399.

Guo, H., Geng, C., Qin, Z., & Chen, C. (2013). Hydrophilic modification of HDPE microfiltration membrane by corona-induced graft polymerization. *Desalination and Water Treatment, 51*(19–21), 3810–3813. 10.1080/19443994.2013.781729

Hamidon, M. H., Sultan, M. T. H., Ariffin, A. H., & Shah, A. U. M. (2019). Effects of fibre treatment on mechanical properties of kenaf fibre reinforced composites: A review. *Journal of Materials Research and Technology, 8*(3), 3327–3337. 10.1016/j.jmrt.2019.04.012

Hongsriphan, N., & Sanga, S. (2018). Antibacterial food packaging sheets prepared by coating chitosan on corona-treated extruded poly (lactic acid)/poly (butylene succinate) blends. *Journal of Plastic Film and Sheeting, 34*(2), 160–178. 10.1177/8756087917722585

John, M. J., & Anandjiwala, R. D. (2008). Recent developments in chemical modification and characterization of natural fiber-reinforced composites. *Polymer Composites, 29*(2), 187–207. 10.1002/pc.20461

Kabir, M. M., Wang, H., Lau, K. T., & Cardona, F. (2012). Chemical treatments on plant-based natural fibre reinforced polymer composites: An overview. *Composites Part B: Engineering, 43*(7), 2883–2892. 10.1016/j.compositesb.2012.04.053

Kalia, S., Kaith, B. S., & Kaur, I. (2009). Pretreatments of natural fibers and their application as reinforcing material in polymer composites—A review. *Polymer Engineering and Science, 49*(7), 1253–1272. 10.1002/pen

Kalia, S., Thakur, K., Celli, A., Kiechel, M. A., & Schauer, C. L. (2013). Surface modification of plant fibers using environment friendly methods for their application in polymer composites, textile industry and antimicrobial activities: A review. *Journal of Environmental Chemical Engineering, 1*(3), 97–112. 10.1016/j.jece.2013.04.009

Koohestani, B., Darban, A. K., Mokhtari, P., Yilmaz, E., & Darezereshki, E. (2019). Comparison of different natural fiber treatments: A literature review. *International Journal of Environmental Science and Technology, 16*(1), 629–642. Springer Berlin Heidelberg. 10.1007/s13762-018-1890-9

Koronis, G., Silva, A., & Fontul, M. (2013). Green composites: A review of adequate materials for automotive applications. *Composites Part B: Engineering, 44*(1), 120–127. 10.1016/j.compositesb.2012.07.004

Le Moigne, N., Sonnier, R., El Hage, R., & Rouif, S. (2017). Radiation-induced modifications in natural fibres and their biocomposites: Opportunities for controlled physicochemical modification pathways? *Industrial Crops and Products, 109*(August), 199–213. 10.1016/j.indcrop.2017.08.027

Li, X., Tabil, L. G., & Panigrahi, S. (2007). Chemical treatments of natural fiber for use in natural fiber-reinforced composites: A review. *Journal of Polymers and the Environment, 15*(1), 25–33. 10.1007/s10924-006-0042-3

Louzi, V. C., Sinézio, J., & Campos, D. C. (2019). Corona treatment applied to synthetic polymeric mono fi laments (PP, PET, and PA-6). *Surfaces and Interfaces, 14*(November 2016), 98–107. 10.1016/j.surfin.2018.12.005

Lu, N., Oza, S., & Tajabadi, M. G. (2015). Surface modification of natural fibers for reinforcement in polymeric composites. In V. K. Thakur & A. Singha (Eds.), *Surface Modification of Biopolymers* (pp. 224–237). Wiley.

Mahajan, S. (2016). Preface: International conference on recent trends in physics (ICRTP 2016). *Journal of Physics: Conference Series, 755*(1). 10.1088/1742-6596/755/1/011001

Martin, A. R., Manolache, S., Denes, F. S., & Mattoso, L. H. C. (2002). Functionalization of sisal fibers and high-density polyethylene by cold plasma treatment. *Journal of Applied Polymer Science, 85*, 2145–2154. 10.1002/app.10801

Mohanty, A. K., Misra, M., & Drzal, L. T. (2001). Surface modifications of natural fibers and performance of the resulting biocomposites: An overview. *Composite Interfaces, 8*(5), 313–343. 10.1163/156855401753255422

Mohit, H. (2018). A comprehensive review on surface modification, structure interface and bonding mechanism of plant cellulose fiber reinforced polymer based composites. *Composite Interfaces, 25*(5–7), 629–667, Taylor & Francis. 10.1080/09276440.2018.1444832

Molina, S. (2016). Modification of natural fibers using physical technologies and their applications for composites. In N. Belgacem & A. Pizzi (Eds.), *Lignocellulosic Fibers and Wood Handbook* (pp. 323–344). Scrivener Publishing.

Mukhopadhyay, S., & Fangueiro, R. (2009). Physical modification of natural fibers and thermoplastic films for composites – A review. *Journal of Thermoplastic Composite Materials, 22*(2), 135–162. 10.1177/0892705708091860

Nabinejad, O., & Debnath, S. (2016). Natural-fiber-reinforced polymer composites. In V. Mittal (Ed.), *Spherical and Fibrous Filler Composites* (pp. 101–125). John Wiley & Sons.

Nemani, S. K., Annavarapu, R. K., Mohammadian, B., Raiyan, A., Heil, J., Haque, M. A., Abdelaal, A., & Sojoudi, H. (2018). Surface modification of polymers: Methods and applications.*Advanced Materials Interfaces, 5*(24), 1–26. 10.1002/admi.201801247

Nourbakhsh, S., Parvinzadeh, M., & Jafari, S. (2018). Comparison between nano and micro silicon softener on corona discharge-treated cotton fabric. *Journal of Industrial Textiles, 47*(7), 1757–1768. 10.1177/1528083717708484

Oudrhiri Hassani, F. (2020). Effects of corona discharge treatment on surface and mechanical properties of Aloe Vera fibers. *Materials Today: Proceedings, 24*, 46–51. 10.1016/j.matpr.2019.07.527

Pizzi, A., & Kumar, R. N. (2019). Modification of natural fibers and polymeric matrices. In *Adhesives for Wood and Lignocellulosic Materials* (pp.367–388). Scrivener Publishing.

Ragoubi, M. (2010). Impact of corona treated hemp fibres onto mechanical properties of polypropylene composites made thereof. *Industrial Crops and Products, 31*(2), 344–349. 10.1016/j.indcrop.2009.12.004

Ragoubi, M., George, B., Molina, S., Bienaimé, D., Merlin, A., Hiver, J. M., & Dahoun, A. (2012). Effect of corona discharge treatment on mechanical and thermal properties of composites based on miscanthus fibres and polylactic acid or polypropylene matrix. *Composites Part A: Applied Science and Manufacturing*, *43*(4), 675–685. 10.1016/j.compositesa.2011.12.025

Renouard, S., Hano, C., Doussot, J., Blondeau, J. P., & Lainé, E. (2014). Characterization of ultrasonic impact on coir, flax and hemp fibers. *Materials Letters*, *129*, 137–141. 10.1 016/j.matlet.2014.05.018

Sadeghi-Kiakhani, M., & Tayebi, H. (2016). Eco-friendly reactive dyeing of modified silk fabrics using corona discharge and chitosan pre-treatment. *The Journal of The Textile Institute*, *108*(7), 1164–1172. 10.1080/00405000.2016.1222861

Sasidharan, P., Thomas, S., Spatenka, P., & Ghanam, Z. (2019). Effect of plasma modification of polyethylene on natural fibre composites prepared via rotational moulding. *Composites Part B*, *177*(March), 107344. 10.1016/j.compositesb.2019.107344

Shahriar Kabir, M., Hossain, M. S., Mia, M., Islam, M. N., Rahman, M. M., Hoque, M. B., & Chowdhury, A. M. S. (2018). Mechanical properties of gamma-irradiated natural fiber reinforced composites. *Nano Hybrids and Composites*, *23*(January 2019), 24–38. 10.4 028/www.scientific.net/nhc.23.24

Sheng, Z., Gao, J., Jin, Z., Dai, H., Zheng, L., & Wang, B. (2014). Effect of steam explosion on degumming efficiency and physicochemical characteristics of banana fiber. *Journal of Applied Polymer Science*, *131*(16), 1–9. 10.1002/app.40598

Shukla, N. et al. (2021). Natural composites: Vegetable fiber modification. In M. Jawaid & A. Khan (Eds.), *Vegetable Fiber Composites and Their Technological Applications* (pp. 303–325). Springer. 10.1007/978-981-16-1854-3_13

Skundric, P. (2015). Wetting properties of hemp fibres modified by plasma treatment. *Journal of Natural Fibers*, *January 2015*, 37–41. 10.1300/J395v04n01

Thakur, K., & Kalia, S. (2017). Enzymatic modification of ramie fibers and its influence on the performance of ramie-poly (butylene succinate) biocomposites. *International Journal of Plastics Technology*, *21*(1), 209–226. 10.1007/s12588-017-9178-3

Vinod, A., Sanjay, M. R., Suchart, S., & Jyotishkumar, P. (2020). Renewable and sustainable biobased materials: An assessment on biofibers, biofilms, biopolymers and biocomposites. *Journal of Cleaner Production*, *258*, 120978. 10.1016/j.jclepro.2020.120978

Xu, X., Wang, Y., Zhang, X., Jing, G., Yu, D., & Wang, S. (2006). Effects on surface properties of natural bamboo fibers treated with atmospheric pressure argon plasma. *Surface and Interface Analysis*, *38*(8), 1211–1217.

Zhang, Q., Jiang, Y., Yao, L., Jiang, Q., & Textile, N. (2015). Hydrophobic surface modification of ramie fibers by plasma-induced addition polymerization of propylene. *Journal of Adhesion Science and Technology*, *29*(8), 691–704. 10.1080/01694243.2014.997380

Zhou, Y., Fan, M., & Chen, L. (2016). Interface and bonding mechanisms of plant fibre composites: An overview. *Composites Part B: Engineering*, *101*, 31–45. 10.1016/j.compositesb.2016.06.055

6 Fabrication of Composites

Ishan Srivastava and G. L. Devnani
Department of Chemical Engineering, Harcourt Butler
Technical University, Kanpur, India

Shishir Sinha
Indian Institute of Technology Roorkee, India

INTRODUCTION

The never-ending war with environmentally hazardous materials has made the investigating society unturn every possible alternate. The last couple of decades foresaw the enormous scope in the field of reinforcements. As a result, a huge amount of implementation-based researches were carried out. The outcomes comprise terming the natural fibers as the most ideal reinforcements. The lignocellulosic fiber from the plants turned out to be a linchpin for the composites.

The cardinal properties of the natural fibers like tensile strength, mechanical strength, thermal strength, etc. make them the most reliable ones. The biodegradability, inexpensiveness, availability, and wide variety are the utmost characteristics of natural fibers. There are many other advantages of reinforcing fibers with the polymer matrix like light weight, shelf life, ease to build, enhanced energy recovery, ease of separation, reduced tool wear, acceptable mechanical strengths, free from health hazards, high toughness, etc. There are many sources of extracting natural fibers; Figure 6.1 gives a brief classification. The animal fibers mostly come from the wool, skin, and hair of the organisms, whereas the plant fibers almost constitute every part. The general examples include sisal and banana fibers as the leaf fiber, cotton fiber as the seed fiber, hemp and flax fibers as the bast fiber, rice and wheat fiber as the stalk fibers, etc.

The fibers mainly consist of cellulose, hemicellulose, lignin, pectin, and other small molecular substances. The cellulose is the most vital constituent of the fibers with 45%–70% occurrence in the cell wall, assisting them by providing a structure. Chemically these are the polysaccharides molecules tied through ß-1,4 glycosidic linkage. It has free hydroxyl groups that make the natural fibers hydrophilic. The greater the content of the cellulose, the greater the mechanical strength. The hemicellulose, lignin, and pectin are the non-cellulosic content that makes the fiber surface smooth. Hemicellulose is simpler in structure than cellulose; unlike cellulose, it is a cross-linked polysaccharide.

For a fiber to be implemented as a reinforcement material, it should have less content of the non-cellulosic substances. The hydrophilic fibers have very little

DOI: 10.1201/9781003201724-6

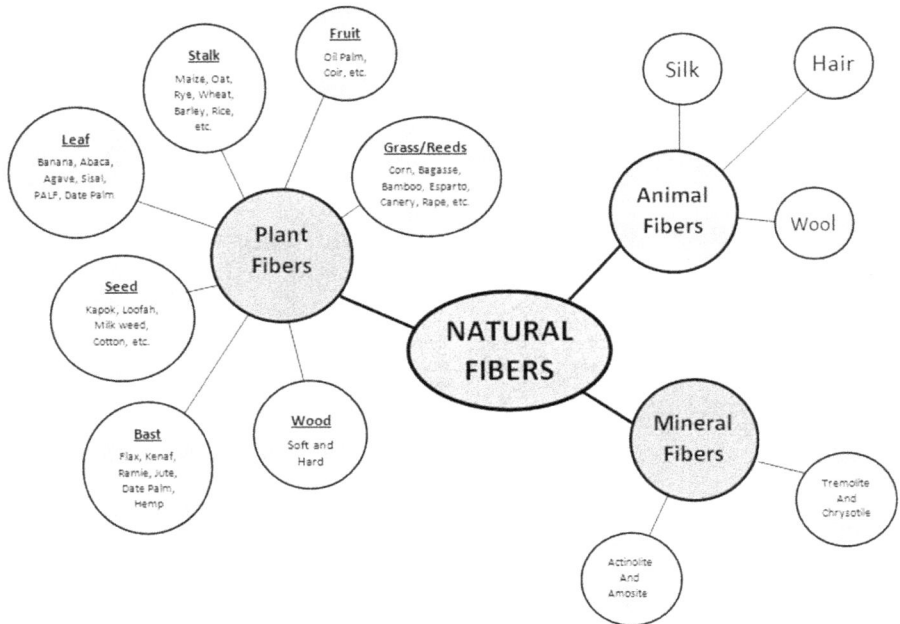

FIGURE 6.1 General classification of natural fibers.

adhesivity when those are blended with the hydrophobic polymer matrices. Nevertheless, implementing the natural fibers as the reinforcement have few additional problems. The natural fibers require a high degree of cautiousness while in storage, as they are easily degradable. Not only this, the moisture absorption capacity and tendency to form aggregations while processing also cause complications.

Therefore, implementing the raw fiber is practically of no use. Instead, while fabricating the composites, we have a complementary step of treatment of the natural fibers before contacting it with the polymer matrix. The fibers are exposed to treating agents like alkali, silane, benzoyl chloride, etc. These treatment methods largely affect the surface of the fibers, by partially discharging the non-cellulosic contents like hemicellulose, lignin, pectin, etc. The top waxy layer and oily coverings are also removed, thereby making the fibers much rougher than before. Out of all the treatment methods available, the mercerization (treatment with alkali) of fibers turns out to be the cornerstone. Hence, alkali treatment becomes the prerequisite for every other treatment technique. In silane treatment, trimethoxy silane acts as a coupling agent, which binds the fibers to the matrix. Results from the silane treatment are also more or less similar, but the alkali is cheaper than the silane. Mercerization is the most preferred treatment technique. The treated fibers, when tested, yield exceptional results. The properties like tensile strength, flexural strength, thermogravimetric properties, etc. get upgraded. The treated hydrophobic fibers are now capable of escaping themselves from absorbing the moisture. The water absorption capacity also gets reduced. The treated fibers are stored in the following form shown in Figure 6.2 before their reinforcement with the polymer resin.

The fabrication of composites is the most dynamic step in composite production. There are several methods available on the type of composite that is in demand. The most common methods include compression molding, injection molding, hot press/cold press, extrusion molding, hand layup, wet layup, vacuum bag, filament winding, etc. (Ambrosio et al., 2016). Each method has its importance and its different working procedure. Although the central working mechanism is the same, i.e., the treated fibers are made to contact the polymer matrix, in the presence of hardeners and catalysts. Maximum matrix-fiber wetting and minimum voids are the two key factors that are mandatory to be met at the time of producing the composites. The curing temperature and applied pressure also influence the composite fabrication to a large extent (Kuppusamy et al., 2020). Other guarding factors are listed in Table 6.1. Table 6.1 shows the characteristics of the raw material such as resin and the natural fibers that play a significant role in the shaping of the finished composite.

The resin undergoes several states. It transforms to gel from the liquid and later solidifies to a rigid solid. Apart from physical changes, the resin changes its chemical structure as well. The catalysts and other curing agents assist the resin to cross-link under a suitable temperature. The curing of resin is an exothermic process and ends up increasing its viscosity, signifying the hardness. Hence, the addition of heat results in the decrement of the curing time.

(a)   (b)   (c)

(d)   (e)

FIGURE 6.2   Different forms of natural fibers. (a) Short fibers; (b) yarn fiber; (c) randomly oriented fiber; (d) woven fiber; and (e) braided textile fiber. Reprints with permission from (Leong et al., 2013).

TABLE 6.1

Characteristics of Raw Materials That Affect the Fabrication of Composites

Reinforced Natural Fibers	Resin
Fiber permeability	Curing kinetics
Fiber volume fraction	Resin gelatine time
Fiber mat architecture	Viscosity
Mat porosity	Exothermy

Some methods are entirely mechanized while others employ manual labor. The mechanized processes are generally preferred at the manufacturing centers to cater to the large volumes of production, under a high efficiency. The design of the composite is majorly influenced by the content of fiber and the matrix. Thus, the selection of the composite preparation technique is based on production rate, shape, size, manufacturing cost, ease in making complex geometries, and high-performance parts. In this chapter, we will elaborate on the methods described in Figure 6.3.

6.1 COMPRESSION MOLDING

Compression molding is one of the most efficient methods to adopt while producing composites. Most often, the thermoplastic and thermosetting polymers are compression molded to form the composites (Asim et al., 2017). It is usually employed when dealing with large chunks of raw material. It can formulate high volumes by exerting high pressure. It is generally used to produce complex composites. The compression molding machine produces the finished composite in a rapid cycle time. This technique is fit for incorporating changes in the part designs. Compression molding is not only a cost-effective technique but also produces a good surface-finished composite. The low costs account for the automated process with very low labor costs. Hence, this is one of the ideal techniques that may be scaled up industrially.

A compression molding machine is described in Figure 6.4. The molds are generally mounted on the hydraulic press, which is typically made of alloy or cast metal (cast aluminum, cast iron, forged steel). Durable molds are usually steel and chromium-plated. The molds come in various configurations, such as single-cavity, multiple-cavity, etc. The molds are designed in such a way that it facilitates the easy inserts, side cores, etc.

The treated natural fibers were dispersed in the polymer matrix, and a lump was prepared that is commonly known as a charge (Gandhi et al., 2020). Meanwhile, the molds are preheated to 350°F–400°F. The charge is then put upon the hot molds, followed by imparting massive pressure. The bulk molding compound (BMC) and sheet molding compound (SMC) are the most common

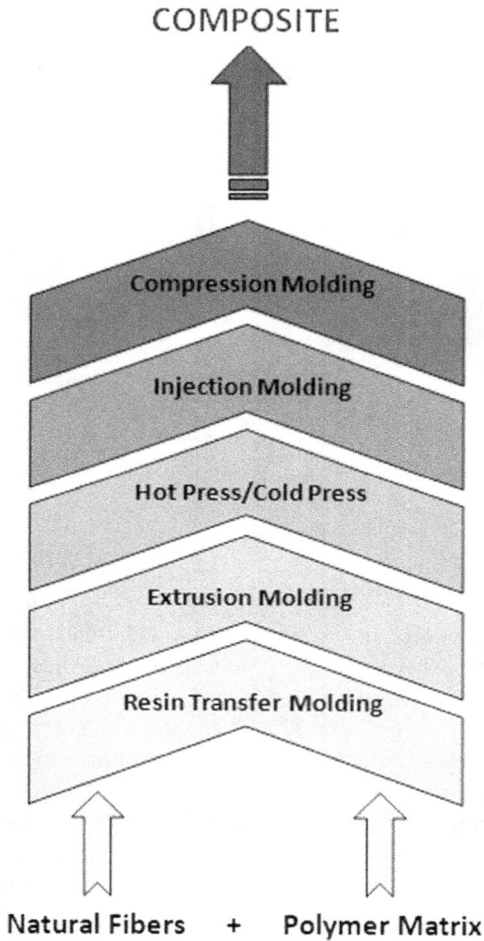

FIGURE 6.3 Different types of composite fabrication methods.

intermediate formed (Ambrosio et al., 2016). The curation period is entirely dependent on the shape, thickness, and size of the composite. This time usually ranges from 1 minute to 5 minutes (Summerscales & Grove, 2013). The molds are then opened and the composite is ready for implementation. The compression molding is generally used to manufacture brush and mirror handles, cookware knobs, fan blades, meter cases, electronic utensils, trays, circuit breakers, automotive parts, television cabinets, radio casing, pot handles, sockets, dinnerware, spoilers, water testing equipment buttons, aircraft terminal housing, and many more (Nair & Joseph, 2014). Also, the composites produced from the compression molding offer high mechanical properties in contrast to the aluminum (Kim & Chang, 2019).

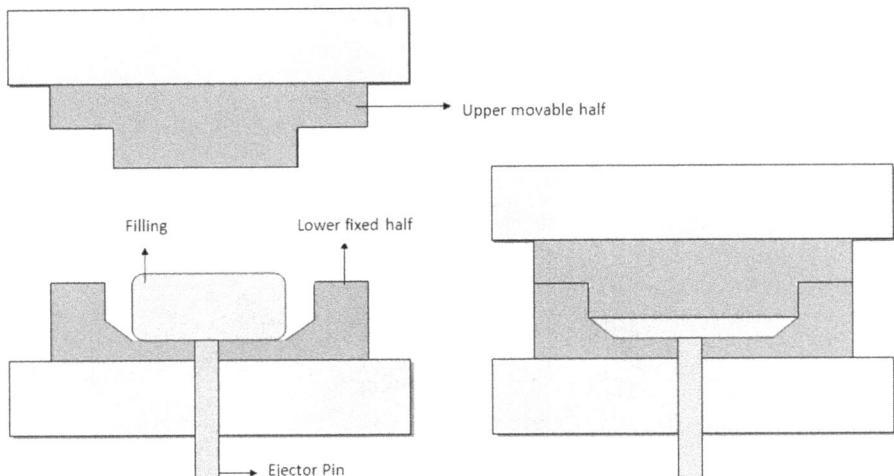

FIGURE 6.4 Compression molding machine.

6.2 INJECTION MOLDING

Injection molding is one of the most primitive and frequently used methods for composite fabrication. The composite produced prerequisites are very small-sized natural fibers, precisely in powder or particle form. A wide variety of design freedom, good quality composites, reduced assembly requirement, and consistency are a few of the standard parameters that this manufacturing process offers. The waste from the injection molding machine is nearly negligible. The injection molding is largely influenced by the configuration of the machine that is put to use. The factors such as volumetric flow rate and screw speed can be customized to meet the required characteristic of the composites. Apart from this, factors such as thermal conductivity, moisture, viscosity, thermal stability, temperature, crystallinity, and heat are a few of the significant considerations. Low values of viscosity are favorable, as it easifies transportation. The right amount of heat is needed for melting the pallets, alongside the suitable values of the thermal conductivity. The transference of heat can be controlled with the help of a control system. Crystallinity has the highest impact upon the fabrication of the composites as most often the polymers under consideration are semicrystalline, i.e., the increase in volume from its solid state to molten state. The recrystallinity index determines the cooling rate of the molds. The fiber should be free from moisture as moisture forms vapors under the application of heat. These vapors damage the composite structure producing voids and pores that are unacceptable (Ebnesajjad, 2003).

Nevertheless, the injection molding machine is highly expensive, due to its complex control system and its high-pressure requirement. Although it can process all kinds of molds within acceptable boundaries including complex geometry. Despite the generous initial investment, this method trades off well, with the other methods available. The cost per mold refines with the scale, and hence shrinks the overall operating cost. Figure 6.5 is showing the injection molding machine.

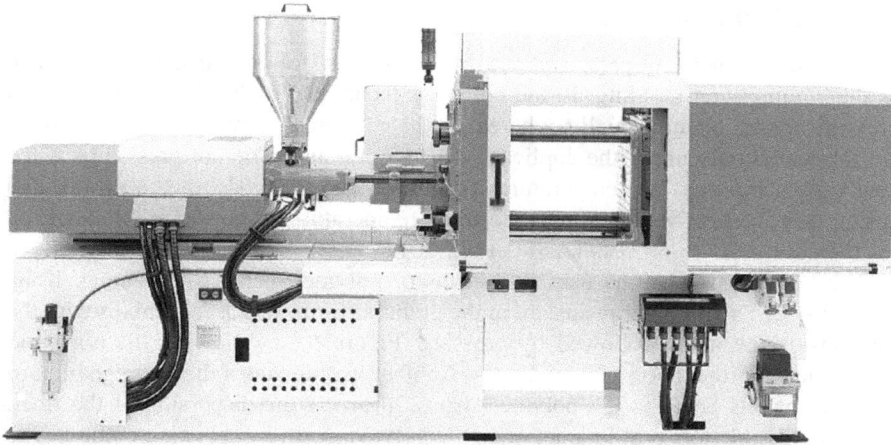

FIGURE 6.5   Injection molding machine. Reprints with permission from (Goodship, 2016).

The injection molding is broadly classified into reaction injection molding (RIM) and liquid injection molding (LIM). In the RIM, the pallets are melted and injected simultaneously into the molds, whereas in the LIM the pallets are melted in a tank at once, and later the whole of the molten mixture is discharged into the molds (Landrock, 1995). The LIM is usually capable of processing high-viscosity materials.

The injection molding machine can be classified into the following components: (i) injecting section (hopper), (ii) clamping assembly, (iii) molds and reciprocating screw, (iv) control system, and (v) tempering devices for the mold. The molds and reciprocating screws are an integral part of this system (Ebnesajjad, 2003). The process initiates from the granule formation by the intermixing of the chopped fibers with the polymer resin. The resultant pellets/granules are prepared in the twin-screw extruder and internal mixer. These granules are most commonly termed as the injection molding compound (IMC). The IMC is put into the hopper wherein it passes across a reciprocating screw and a heated barrel. While passing through this assembly, the pallets transit from the granules to their melt state, under the action of heat. The melted slurry is passed into the molds with the aid of the reciprocating screw that applies the shear force on the particles The curing starts as soon as the molds are occupied with the melted IMC. After the completion of its curation period, the composites are ejected from the machine (Ambrosio et al., 2016). The most common ejectors employed are pins and strippers, in some cases both.

The injection molding technique is one of the finest methods of producing rubber products. The intense automation of the injection molding process makes it the least labor-dependent method. The rubber composites not only cause less wear but are also less abrasive to the injection molding machine than the artificial polymer composites. Therefore, this method is fully suited to produce natural fiber–reinforced rubber composites as diaphragms, automotive parts, gaskets, etc. (Nair & Joseph, 2014).

6.3 HOT PRESS/COLD PRESS

The two techniques, hot press and cold press, broadly follow the same mechanism as the compression molding. In the hot press technique, both the temperature and pressure are maintained ideally, whereas in the cold press technique the composites are fabricated only under the application of the pressure. The hot press directs the heat (given through the temperature) to initiate and bring down the curing and curing time, respectively, whereas the room temperature is the sufficient condition for the cold press to process (Asim et al., 2017).

The hot press technique uses two forms of polymer resin (Arifuzzaman Khan et al., 2013). The polymer resins are in the granular form, which intermixes with the fiber inside the mold, followed by pressing. Figure 6.6 is showing the hot press technique for preparation of composites. Another way includes the impregnation of the fiber pallets between the polymer sheets, allowing the dispersion of the fiber. The two possible forms of interaction of the polymer resin and fiber are employed to be further treated by the application of heat and pressure (Shahinur & Hasan, 2020). The prepared molds are then heated and compressed to give the desired shape and size (Kumar et al., 2008). The cold press technique works in the same fashion apart from supplying heat. The cold press technique requires room temperature and compresses the molds with the application of pressure only.

6.4 EXTRUSION MOLDING

Extrusion molding is widely used for producing two-dimensional composite materials. Injection molding the reciprocating screw moves continuously in the extrusion molding. It is the closed-molding process (Kern & Gadow, 2009).

The extrusion molding process makes use of an extrusion molding machine, which is more or less similar to the injection molding machine. The entire assembly consists of an extrusion molder, cooling water jackets, and a pulling device. The extrusion molder is very much similar to the injection molding machine. The inlet feed is fed through a hopper, which directs the feed into the barrel. The barrel contained the screw that rotates continuously with the help of a motor that is electrically driven. The grooves of the screw impart the compressive and the forward push to the feed (Stevens & Covas, 1995). The region occupied by the reciprocating screw is broadly classified into three spaces: feed section, compression section, and the metering section. The name of these spaces attributes their functioning. The entire molding instrument is guarded by the heaters. The screw is followed by the breaker plate and a dieing outlet (Ambrosio et al., 2016; Rosato, 1998).

The inlet or the injection point for the reinforced fiber and the resin is done through a frustum-shaped hopper (Czerwinski, 2008). The palleted mixture of the fiber and resin is then acted upon the compressive force imposed by the reciprocating screw, which rotates continuously (Jiang & Zeng, 2019). The casing that covers the reciprocating screw is surmounted with the heaters over its surface, which supplies energy to liquidate the inside matter that is flowing across. The pallets are gradually crushed and transited across several states, ending up finally in the molten state. The molten matter is transported across several sections of the

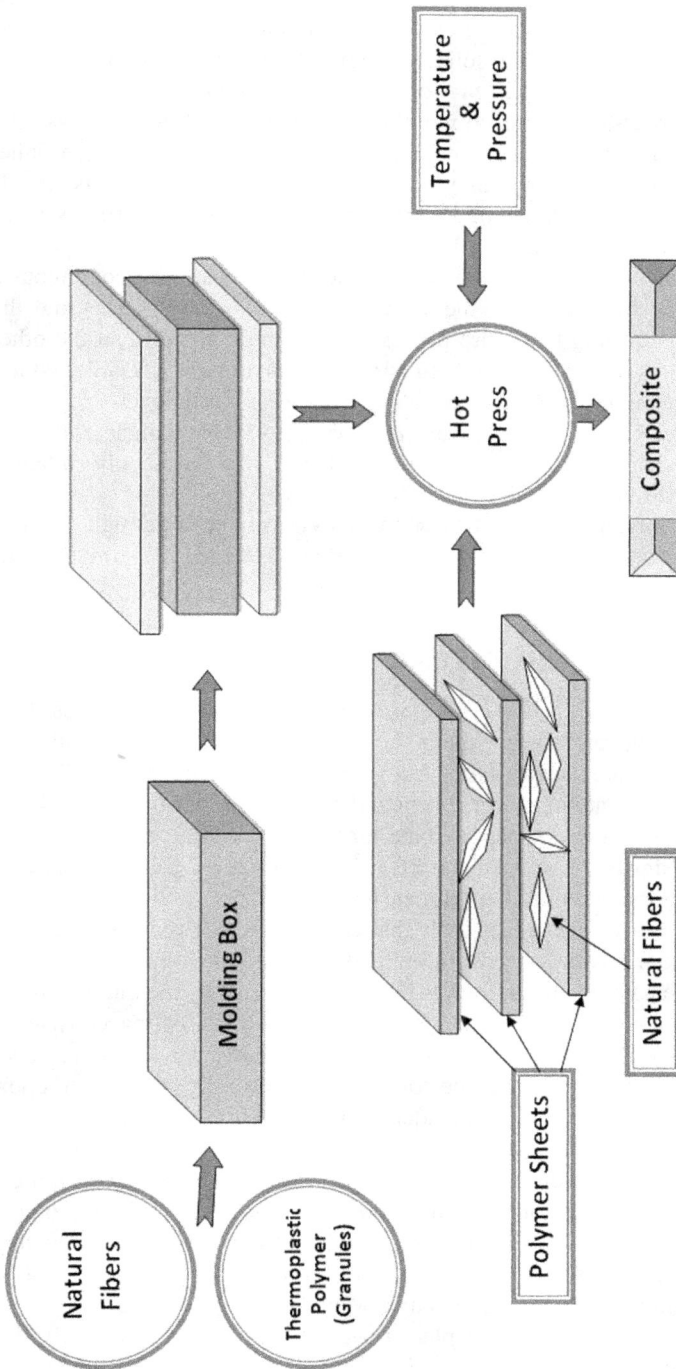

FIGURE 6.6 Preparation of composite through hot press technique.

barrel. The melt is segregated with the help of a breaker plate, which decreases the thickness of the molten stream. The reduced streams are passed over the cooling water through a die outlet, to solidify them. The extrudate is pulled across the pulling device, which results in the long strips as the end product.

Like other processes, extrusion molding is also exercised in two ways. The cold extrusion happens under room temperature. In the cold extrusion, the pallets are pre-heated before feeding them into the hopper. The other is hot extrusion, which is most often used on the industrial scale. The above-described process is the hot extrusion molding (Czerwinski, 2008).

The extrusion molding is ideal when it comes to producing the continuous length of products. The installation of the extrusion molding setup is easy and the cost involved in establishing the entire process is also much lower than the other processes. It requires less space. The extrusion process alone is versatile, which caps the lowest operating cost, out of all the process techniques (Processing & Handbook, n.d.; Rosato, 1998; Stevens & Covas, 1995). Nevertheless, the outcome from the extrusion molding assembly is unstable in size and is only suitable when long strip-shaped goods are desired.

Most commonly, the wood-plastic composites are prepared through the extrusion-compression process. The products of large formats and irregular structures could be created (Dai & Fan, 2013; Jiang & Zeng, 2019).

6.5   RESIN TRANSFER MOLDING (RTM)

Resin transfer molding (RTM), is a closed-molding fabrication process. It is high standard composite preparation method that works on low temperature and low pressure. The pressure usually varies from 3.5–7 bar. The dimensions of the molded composites are accurate to a large extent and provide a good surface finish. RTM is capable of dealing with an intermediate amount of volume and produces a two-ended surface finished composite. It offers a wide variety of composites that includes two-dimensional and three-dimensional materials.

RTM has the widest range of tools that can be used to produce the composites, which range from a very low cost to high cost. The tools of the RTM machine are classified as the hard and the soft tools that are influenced by the duration of run and types of resin and reinforced material in use. The hard tooling consists of cast metals and alloys, such as machined steel molds, machined aluminum, and electroformed nickel shells, whereas the soft tools are generally made up of epoxy and polyester matrices. The RTM is considered to be a robust fabrication technique that has a durable mechanism and toolings.

The resin transfer molding equipment consists of an assembly that is more or less similar to the compression molding machine. Although, unlike compression molding, the inlet of the feed has an entirely different mechanism in the resin transfer molding. The molds are flushed with nitrogen for the removal of oxygen and other moisture content (Wakeman & Månson, 2005). The reinforcement material (or the long fiber strands) is placed over the lower mold (or the fixed mold), pre-coated with a gel (anionic polymerizing agent), that is being compressed by the upper mold (or the movable mold). The upper mold is fitted with an injector that transports the resin accompanying the catalysts, fillers, and other curing agents (Erden & Ho, 2017). The pre-mixing of resin, catalysts, additives, and fillers takes place beforehand in a mixer, which is located above the mold. The thermoset, which is of lower viscosity, is most commonly used in the RTM. Different RTM machines may have different locations of the mixer. The mixer is electrically driven, which fetches the additives and resins from different containers. Later, as the molds are contacted, it is followed by clamping under the application of pressure and temperature. The clamping is of several types that include press clamping, perimeter clamping, etc. The curing of the composites starts both ways, at room temperature as well under heating. The heating of mold results in the shrinkage of the curation cycle. Nevertheless, the product consistency also gets increased with the addition of heat.

RTM nowadays has surpassed hundreds of transitions, which makes the process entirely automated and can leverage a wide variety of composites. The dimensions of the composite are determined by the size and shape of the mold cavity. It includes the use of a vacuum to upgrade the resin flows into the mold (vacuum assisted resin transfer molding–VARTM), producing low-cost composites, etc. (Erden & Ho, 2017; Marques, 2011). The VARTM involves fusion in the absence of positive pressure. The resultant decreased pressure, making the RTM produce larger molds, whereby bigger composites are produced. The molds of the RTM could themselves be lightweight, thereby making the overall process cheap (Brooman, 1996; Plummer et al., 2016).

The addition of natural fiber as reinforcement brings additional factors that are to be considered while producing the composites. The factors include fiber concentration, fiber washing, and edge flow (Dai & Fan, 2013). The production of wood fiber–reinforced composites through RTM has recorded enhanced values of strength and stiffness. The RTM is used in the construction sector, transportation, and wind power (Erden & Ho, 2017). It is also used in the fabrication of big containers, complex structures, bathtubs, hollow shapes, etc. (Devaraju & Alagar, 2019).

6.6   SHEET MOLDING COMPOUND (SMC)

Sheet molding compound or sheet molding composite is a fiber-reinforced thermoset material. It is a compression-molded compound, which is implemented in places where higher mechanical strength materials are needed. SMC consists of reinforced fiber, polymer resin, catalysts, inert fillers, stabilizers, and other additives. It has hefty dielectric properties that make it one of the best electrical insulators. The SMC can easily be molded into complex geometries. Therefore, it has a very wide scope of implementation. SMC is made up of glass fiber, liquid thermosetting polyester resin, mineral fillers such as slate (or calcium carbonate), inhibitors, catalysts, release agents and film;, the thickening agent is processed into malleable sheets. Figure 6.7 shows the preparation of a sheet molding compound.

The production of SMCs starts with the inter-mixing of all the constituent materials apart from the reinforced glass fiber. The reinforcement is typical of the order 25%–65% by weight. The approximate fiber length is 25 mm. The long glass fiber

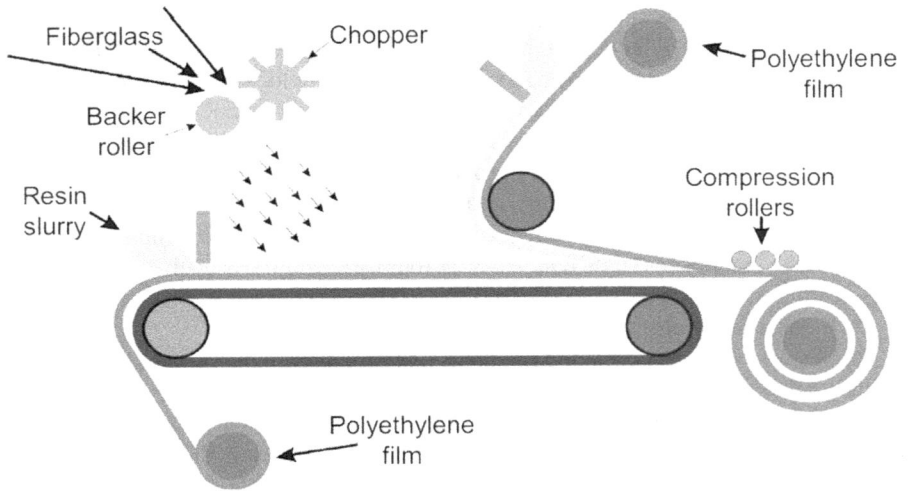

FIGURE 6.7 Preparation of sheet molding compound. Reprints with permission from (Yalcin, 2015).

strands are chopped before adding them to the polymer slurry. The chopping is done through a chopper, which converts the long fibers into small pieces. The conveyer belt rolling over the rollers is pre-coated with a carrier film for giving a good finish to the compound (Perez et al., 2020). The polymer slurry or paste is then rolled on this film that continuously moves forward. The polymer paste is then capped with the chopped fiber that comes directly from the chopper. The dispersion of the glass fiber is mainly done in two ways. Either the fibers are randomly oriented discontinuous fibers that are commonly termed SMC (or SMC-R) and the other in which the fibers are unidirectional and continuously arranged, which is termed SMC-CR. The conveyer belt is continuously moving, which exposes the topped fibers to get sandwiched between the layers of polymer paste, i.e., another layer of polymer paste is made to fall on the fibers from the top. This second layer of polymer paste is accompanied by the carrier film, which is used to give a good finish to the other side of the compound. The newly formed composite is then made to pass through the compacting rolls, followed by a gamma gauge. The final composite sheet is stiff and the carrier film is peeled off for the curation purpose. The sheets are then sent to a compression molding machine, where they dry at fast cycle rates. The cycle rate ranges from 3 to 10 minutes, depending upon the dimension of the composite.

It is ready-to-mold continuously produced compounds offer a wide range of advantages such as lowering the cost by employing cheap tools and integrating the parts, light in weight, manufactured in intermediate to large volumes It not only provides a good surface finish but also possesses the best weathering properties. It is recorded to bear the paint baking temperature of around 180°C. Nevertheless, the percentage of reinforcement largely affects the degree of complexity the composite can bear. The increase in reinforcement is inversely proportional to the level of complexity. The sheets are stored in an environmental chamber until they are not implemented somewhere (Mallick, 2012).

It is regarded as a flow-molding material that is heavily implemented in the automotive industry (Laribi et al., 2020; Yalcin, 2015). Its main implementation is in the automotive sector, where it is largely used in making exterior body parts like that of rear quarter panels, boot lids, tailgates, side extension panels, wings, etc. (Brooks et al., 2017; Brooman, 1996; Springer, 1984). Since it is dielectric, it cannot be used for making conductive panels; therefore, it is coated with a conductive primer allowing its extended use in the electrical industry (Mehta et al., 2005; Revellino et al., 2000).

6.7   DESIGN OF NATURAL FIBER COMPOSITES FOR DIVERSE APPLICATIONS AND STRUCTURAL ENGINEERING

The polymer composites have proven to be of great use. They are heavily implemented in almost every industrial field. The advancements to the composite brought the natural fibers as their reinforcements that twofold the benefits that the polymer composite alone offered. Being biodegradable, lightweight, and cheap made the composites reach new milestones (Książek, 2015; Neagoe et al., 2015). Researchers then started customizing the percentages of fibers and resins to rebuild and upgrade the mechanical strength. The composites now offered enhanced strength, stiffness, electrical conductivity, thermal conductivity, corrosiveness, shelf life, moisture absorption, etc. These characteristics of the composites guide their implementation in the industries like automotive, infrastructure, marine, aerospace, sporting goods, etc. (Sankaran et al., 2016; Wang et al., 2011; Yan & Chouw, 2015). The surveys and researchers on the implementation of green composite (natural fiber–reinforced polymer composite) inferred that it was valued at USD $4.46 billion in 2016, and is expected to rise at a cumulative annual growth rate of 11.8% till 2024. The Grand View Research report of 2018 stated that the present reinforcement materials usually comprise kenaf, wood, cotton, flax, and hemp. The wood fiber capitalized a 59.3% share of the total natural fiber revenue in 2015.

The aerospace industries employ carbon/epoxy composites for making the exterior parts of the aircraft. The multinational company Boeing used the composite material for the construction of vertical and horizontal tailplanes on the Boeing 777. The implementation of composites is not just confined to building the parts but also the full airframes on account of the light weight and increased fuel performance. The composites are also used in satellite structures and launch vehicles. The choppers and helicopters have employed the glass-reinforced composite blades for many decades due to their fatigue resistance. The use of composite is continuously expanding and is predicted to take away the entire aerospace industry Njuguna et al., 2011; Singh et al., 2011).

The enforcement of composite in the automotive industry is no less than the aerospace industry (Das & Das, 2021; Kishawy, 2012;). The robust mechanical strength and light weight make them the ideal raw material to be used. Apart from this, the smooth surface finish helps in the construction of stylish and glossy surfaces. Nevertheless, the fact that the composites are cost-effective can never be neglected; moreover, this is one of the integral elements while opting for it. The predictions made over the automotive industries include the savings of around 250 million barrels of crude oil, on account of the 25% reduction in the car weight. A 10%–30% reduction in the weight of the automotive is recorded when its part is built from natural fiber–reinforced polymer composites. The use of polypropylene (PP) and polyethylene (PE) as the polymer matrices in the reinforced composites dominates their use in the automotive industry. Other thermoplastics include polyamides and polystyrene. The parts such as door panels, side panels, brake, and clutch pads are usually built from banana fiber, areca fiber, and kenaf fiber–reinforced polymer composites.

The electrical and electronics industry utilizes composites to produce laptop casing, mobile casing, electrical switches, electrical pipes, building frames, etc. The end-of-life performance and light weight are the key properties for their vast implementation. The electronic applications include electrical circuitry, optoelectronic devices, thermoelectric devices, transistors, diodes, piezoelectric devices, micromachines, robotics, electrical interconnections, cables, etc. The optical devices include optical fibers, light sources, lasers, reflectors, optical lenses, optical data storage, etc. The magnetic devices include magnetic recorders, magnetic field sensors, magnetic particle inspection, magnetostriction, etc., (Chung, 2003) (Table 6.2) is showing the applications of these composite materials.

TABLE 6.2
Applications of the Natural Fiber–Reinforced Polymer Composites

S.No.	Application	Industry	Reference
1.	Automotive companies like Volkswagen, BMW, Audi Group, Mercedes, Ford, Daimler Chrysler, and Opel have started implementing the biocomposites. Parts such as door panels, dashboards, inner door panels, etc. were largely made from the reinforcement of sisal, flax, jute, hemp fibers. Polyurethane and polypropylene were the most commonly used matrices in the composites.	Automotive Industry	(Njuguna et al., 2011)
2.	Implementation of the barkcloth-reinforced green epoxy biocomposites in the manufacturing of the interior parts of the automobile. Parts such as interior paneling and dashboards can be constructed through this.	Automotive Industry	(Rwawiire et al., 2015)
3.	The biocomposites are used to construct seat backs, door panels, rear parcel shelf, boot lining, cargo area floor, seat padding, wheel box, trunk panel, sun visor, pillar cover panel, business table, glove box, floor trays, rear storage panel, instrumental panel, noise	Automotive Industry	(Faruk et al., 2014)

TABLE 6.2 (Continued)
Applications of the Natural Fiber–Reinforced Polymer Composites

S.No.	Application	Industry	Reference
	insulation panels, spare tire linings, etc. in the automotive companies like Saturn, Audi, Lotus, Volkswagen, Mitsubishi, Renault, Honda, Ford, Rover, Volvo, etc.		
4.	The literature infers the different methods available to choose the ideal bumper beam.	Automotive Industry	(Davoodi et al., 2011)
5.	The hemp fiber–reinforced epoxy composites can fully replace the glass-reinforced epoxy composites when it comes to implementation in the aeronautical industry. The Naca cowlings of ultra-light aircraft can easily be manufactured from biocomposites.	Aerospace Industry/ Aeronautical Industry	(Scarponi, 2015)
6.	The woven banana fabric–reinforced epoxy composite is used for the production of telephone stands, and can easily replace the conventional materials for building the telephone stands.	Furniture Industry	(Sapuan & Maleque, 2005)
7.	The oil-palm trunk biomass can be used to produce the binderless particleboard. The literature describes the optimum conditions for the manufacturing of the particleboard.	Furniture Industry	(Baskaran et al., 2015)
8.	The Elaeis palm can be used for the production of plywood. The literature gives detailed methods of improving the resin by several treatment methods.	Furniture Industry	(Hoong et al., 2015)
9.	The study revealed the fact that the energy consumption decreased by a large amount when the composites such as cellulose fiber–reinforced cement, reinforced concrete, etc are used.	Furniture Industry	(Lee et al., 2015)
10.	Flax fiber–reinforced polymer composites can be used for making small structural parts of the sailboats by resin infusion, hand layup, and vacuum bagging.	Marine Industry	(Mancuso et al., 2015)

CONCLUSION

Being regarded as tomorrow's material, composites are predicted to have a blazing future. Discovered back in the Paleolithic age, composites have a very long history. The natural fiber–reinforced composites provide a new dimension to this field. The composites have turned cheaper, lighter, sustainable, and more compatible to be used in almost every known industrial sector. The fabrication of composites is also not that complex. There is a wide range of methods available that can be used to produce them. The most commonly used methods include compression molding, injection molding, resin transfer molding, etc. More or less, all these methods are guided by the same mechanism that is implicitly elementary. The pallets/charge is prepared by intermixing the fibers and the resins that are then placed on the molds and compressed under the application of temperature and pressure. The nature of fibers and the polymer resin in hand largely affect the selection of the method. The selection also gets influenced by the volume that we are dealing with, cycle-time, dimensions, and the after-use of the composites. The use of composites is rising exponentially. Nowadays, composites find wide applications in fields such as the furniture industry, automotive industry, aerospace industry, electrical industry, sports products, electronics, transportation, packaging, etc. (Chung, 2003).

REFERENCES

Ambrosio, L., Carotenuto, G., & Nicolais, L. (2016). Composite materials. In W. Murphy, J. Black, & G. Hastings (Eds.), *Handbook of Biomaterial Properties* (Vol. 72) (pp. 205–260). Elsevier Inc. 10.1007/978-1-4939-3305-1_18

Arifuzzaman Khan, G. M., Alam Shams, M. S., Kabir, M. R., Gafur, M. A., Terano, M., & Alam, M. S. (2013). Influence of chemical treatment on the properties of banana stem fiber and banana stem fiber/coir hybrid fiber reinforced maleic anhydride grafted polypropylene/low-density polyethylene composites. *Journal of Applied Polymer Science, 128*(2), 1020–1029. 10.1002/app.38197

Asim, M., Jawaid, M., Saba, N., Ramengmawii, Nasir, M., & Sultan, M. T. H. (2017). Processing of hybrid polymer composites-a review. In *Hybrid Polymer Composite Materials: Processing* (pp. 1–22). Elsevier Ltd. 10.1016/B978-0-08-100789-1.00001-0

Baskaran, M., Hashim, R., Sulaiman, O., Hiziroglu, S., Sato, M., & Sugimoto, T. (2015). Optimization of press temperature and time for binderless particleboard manufactured from oil palm trunk biomass at different thickness levels. *Materials Today Communications, 3*, 87–95. 10.1016/j.mtcomm.2015.04.005

Brooks, R., Shanmuga Ramanan, S. M., & Arun, S. (2017). Composites in automotive applications: Design. *Reference Module in Materials Science and Materials Engineering*, 1–22. Elsevier. ISBN 9780128035818. 10.1016/b978-0-12-803581-8.03961-8

Brooman, E. W. (1996). Clean manufacturing technologies. *Advanced Materials and Processes, 149*(4). 10.1016/B978-0-08-101034-1.00003-7

Chung, D. D. L. (2003). Applications of composite materials. In *Composite Materials* (pp. 1–13). Springer. 10.1007/978-1-4471-3732-0_1

Czerwinski, F. (2008). Semisolid extrusion molding. In *Magnesium Injection Molding* (pp. 469–485). Springer. 10.1007/978-0-387-72528-4_11

Dai, D., & Fan, M. (2013). Wood fibres as reinforcements in natural fibre composites: Structure, properties, processing and applications. In A. Hodzic & R. Shanks (Eds.), *Natural Fibre Composites: Materials, Processes and Applications* (pp. 3–65). Woodhead Publishing Limited. 10.1533/9780857099228.1.3

Das, S., & Das, S. (2021). Properties for polymer, metal and ceramic based composite materials. In D. Brabazon (Ed.), *Reference Module in Materials Science and Materials Engineering*. Elsevier Ltd. 10.1016/b978-0-12-803581-8.11897-1

Davoodi, M. M., Sapuan, S. M., Ahmad, D., Aidy, A., Khalina, A., & Jonoobi, M. (2011). Concept selection of car bumper beam with developed hybrid bio-composite material. *Materials and Design, 32*(10), 4857–4865. 10.1016/j.matdes.2011.06.011

Devaraju, S., & Alagar, M. (2019). Unsaturated polyester-macrocomposites. In S. Thomas , M. Hosur, & C. J. Chirayil (Eds.), *Unsaturated Polyester Resins: Fundamentals, Design, Fabrication, and Applications* (pp. 43–66). Elsevier Inc. 10.1016/B978-0-12-816129-6.00002-8

Ebnesajjad, S. (2003). Injection molding. In *Melt Processible Fluoroplastics* (pp. 151–193). William Andrew Publishing. 10.1016/B978-188420796-9.50010-2

Erden, S., & Ho, K. (2017). Fiber reinforced composites. In M. Özgür Seydibeyoğlu, A. K. Mohanty & M. Misra (Eds .), *Fiber Technology for Fiber-Reinforced Composites* (pp. 51–79). Woodhead Publishing. 10.1016/B978-0-08-101871-2.00003-5

Faruk, O., Bledzki, A. K., Fink, H. P., & Sain, M. (2014). Progress report on natural fiber reinforced composites. *Macromolecular Materials and Engineering, 299*(1), 9–26. 10.1002/mame.201300008

Gandhi, U. N., Goris, S., Osswald, T. A., & Song, Y.-Y. (Eds). (2020). Special topic: Compression molding of discontinuous fiber material. *Discontinuous Fiber-Reinforced Composites* (pp. 371–432). Hanser. 10.3139/9781569906958.009

Goodship, V. (2016). Design and manufacture of plastic components for multifunctionality. In V. Goodship, B. Middleton, & R. Cherrington (Eds.), *Design and Manufacture of Plastic Components for Multifunctionality* (pp. 103–170). Elsevier. 10.1016/c2014-0-00223-7

Hoong, Y. B., Pizzi, A., Chuah, L. A., & Harun, J. (2015). Phenol-urea-formaldehyde resin co-polymer synthesis and its influence on Elaeis palm trunk plywood mechanical performance evaluated by 13C NMR and MALDI-TOF mass spectrometry. *International Journal of Adhesion and Adhesives, 63*, 117–123. 10.1016/j.ijadhadh.2 015.09.002

Jiang, T., & Zeng, G. (2019). An online extrusion-compression molding method to produce wood plastic composite packaging boxes. *Fibers and Polymers, 20*(4), 804–810. 10.1 007/s12221-019-1053-8

Kern, F., & Gadow, R. (2009). Extrusion and injection molding of ceramic micro and na-nocomposites. *International Journal of Material Forming, 2*(SUPPL. 1), 609–612. 10.1007/s12289-009-0487-8

Kim, H. S., & Chang, S. H. (2019). Simulation of compression moulding process for long-fibre reinforced thermoset composites considering fibre bending. *Composite Structures, 230*(July), 111514. 10.1016/j.compstruct.2019.111514

Kishawy, H. A. (2012). Turning processes for metal matrix composites. In H. Hocheng (Ed.), *Machining Technology for Composite Materials* (pp. 3–16). Woodhead Publishing. 10.1533/9780857095145.1.3

Książek, M. (2015). Use in the building cement composites impregnated with special polymerized sulfur. *Journal of Building Engineering, 4*, 255–267. 10.1016/j.jobe.2015. 09.007

Kumar, R., Choudhary, V., Mishra, S., & Varma, I. (2008). Banana fiber-reinforced biode-gradable soy protein composites. *Frontiers of Chemistry in China, 3*(3), 243–250. 10.1 007/s11458-008-0069-1

Kuppusamy, R. R. P., Rout, S., & Kumar, K. (2020). Advanced manufacturing techniques for composite structures used in aerospace industries. In K. Kumar & J. Paulo Davim (Eds.), *Modern Manufacturing Processes* (pp. 3–16). Woodhead Publishing. 10.1016/ b978-0-12-819496-6.00001-4

Landrock, A. H. (1995). Methods of manufacture. In A.H. Landrock (Ed.), *Handbook of Plastic Foams* (pp. 316–331). William Andrew. 10.1016/b978-081551357-5.50010-3

Laribi, M. A., TieBi, R., Tamboura, S., Shirinbayan, M., Tcharkhtchi, A., Dali, H. Ben, & Fitoussi, J. (2020). Sheet molding compound automotive component reliability using a micromechanical damage approach. *Applied Composite Materials, 27*(5), 693–715. 10.1007/s10443-020-09831-5

Lee, S., Kim, S., & Na, Y. (2015). Comparative analysis of energy related performance and construction cost of the external walls in high-rise residential buildings. *Energy and Buildings*, *99*, 67–74. 10.1016/j.enbuild.2015.03.058

Leong, Y. W., Thitithanasarn, S., Yamada, K., & Hamada, H. (2013). Compression and injection molding techniques for natural fiber composites. In A. Hodzic & R. Shanks (Eds.), *Natural Fibre Composites: Materials, Processes and Applications* (pp. 216–232). Woodhead Publishing Limited. 10.1533/9780857099228.2.216

Mallick, P. K. (2012). Failure of polymer matrix composites (PMCs) in automotive and transportation applications. In P. Robinson, E. Greenhalgh, & S. Pinho (Eds.), *Failure Mechanisms in Polymer Matrix Composites* (pp. 368–392). Woodhead Publishing Limited. 10.1533/9780857095329.2.368

Mancuso, A., Pitarresi, G., & Tumino, D. (2015). Mechanical behaviour of a green sandwich made of flax reinforced polymer facings and cork core. *Procedia Engineering*, *109*, 144–153. 10.1016/j.proeng.2015.06.225

Marques, A. T. (2011). Fibrous materials reinforced composites production techniques. In R. Fangueiro (Ed.), *Fibrous and Composite Materials for Civil Engineering Applications*. Woodhead Publishing Limited. 10.1533/9780857095583.3.191

Mehta, G., Mohanty, A. K., Thayer, K., Misra, M., & Drzal, L. T. (2005). Novel bio-composites sheet molding compounds for low cost housing panel applications. *Journal of Polymers and the Environment*, *13*(2), 169–175. 10.1007/s10924-005-3211-x

Nair, A. B., & Joseph, R. (2014). Eco-friendly bio-composites using natural rubber (NR) matrices and natural fiber reinforcements. In S. Kohjiya & Y. Ikeda (Eds.), *Chemistry, Manufacture and Applications of Natural Rubber* (pp. 249–283). Woodhead Publishing. 10.1533/9780857096913.2.249

Neagoe, C. A., Gil, L., & Pérez, M. A. (2015). Experimental study of GFRP-concrete hybrid beams with low degree of shear connection. *Construction and Building Materials*, *101*, 141–151. 10.1016/j.conbuildmat.2015.10.024

Njuguna, J., Wambua, P., & Pielichowski, K. (2011). Natural Fibre-Reinforced Polymer Composites and Nanocomposites for Automotive Applications. In S. Kalia, B. Kaith, & I. Kaur (Eds.), *Cellulose Fibers: Bio- and Nano-Polymer Composites*. Springer. https://doi.org/10.1007/978-3-642-17370-7_23

Perez, M., Prono, D., Ghnatios, C., Abisset, E., Duval, J. L., & Chinesta, F. (2020). Advanced modeling and simulation of sheet moulding compound (SMC) processes. *International Journal of Material Forming*, *13*(5), 675–685. 10.1007/s12289-019-01506-2

Plummer, C. J. G., Bourban, P.-E., & Månson, J.-A. (2016). Polymer matrix composites: Matrices and processing. *Reference Module in Materials Science and Materials Engineering*, *September 2015* (pp. 1–9). Elsevier. 10.1016/b978-0-12-803581-8.023 86-9

Revellino, M., Saggese, L., & Gaiero, E. (2000). Compression molding of SMCs. In A. Kelly & C. Zweben (Eds.), *Comprehensive Composite Materials* (pp. 763–805). Pergamon. 10.1016/b0-08-042993-9/00170-4

Rosato, D. V. (1998). The complete extrusion process. In *Extruding Plastics* (pp. 1–53). Springer. 10.1007/978-1-4615-5793-7

Rwawiire, S., Tomkova, B., Militky, J., Jabbar, A., & Kale, B. M. (2015). Development of a biocomposite based on green epoxy polymer and natural cellulose fabric (bark cloth) for automotive instrument panel applications. *Composites Part B: Engineering*, *81*, 149–157. 10.1016/j.compositesb.2015.06.021

Sankaran, S., Ravishankar, B. N., Ravi Sekhar, K., Dasgupta, S., & Jagdish Kumar, M. N. (2016). Syntactic foams for multifunctional applications. In K. K. Kar (Ed.), *Composite Materials: Processing, Applications, Characterizations* (pp. 281–314). Springer. 10.1007/978-3-662-49514-8_9

Sapuan, S. M., & Maleque, M. A. (2005). Design and fabrication of natural woven fabric reinforced epoxy composite for household telephone stand. *Materials and Design, 26*(1), 65–71. 10.1016/j.matdes.2004.03.015

Scarponi, C. (2015). Hemp fiber composites for the design of a Naca cowling for ultra-light aviation. *Composites Part B: Engineering, 81*, 53–63. 10.1016/j.compositesb.2015.06.001

Shahinur, S., & Hasan, M. (2020). Jute/coir/banana fiber reinforced bio-composites: Critical Review of Design, Fabrication, Properties and Applications. In S. Hashmi & I. A. Choudhury (Eds.), *Encyclopedia of Renewable and Sustainable Materials* (pp. 751–756). Elsevier Ltd. 10.1016/b978-0-12-803581-8.10987-7

Singh, B., Gupta, M., Tarannum, H., & Randhawa, A. (2011). Cellulose fibers: Bio- and nano-polymer composites. In S. Kalia, B. S. Kaith, & I. Kaur (Eds.), *Cellulose Fibers: Bio- and Nano-Polymer Composites* (pp. 701–720). Springer. 10.1007/978-3-642-173 70-7

Springer, G. S. (1984). Properties of sheet molding compounds. In J. C. Hilliard & G. S. Springer (Eds.), *Fuel Economy in Road Vehicles Powered by Spark Ignition Engines* (pp. 309–334). Springer. 10.1007/978-1-4899-2277-9_9

Stevens, M. J., & Covas, J. A. (1995). Practical extrusion processes and their requirements. In M. J. Stevens & J. A. Covas (Eds.), *Extruder Principles and Operation* (pp. 4–26). Springer. 10.1007/978-94-011-0557-6

Summerscales, J., & Grove, S. (2013). Manufacturing methods for natural fibre composites. In A. Hodzic & R. Shanks (Eds.), *Natural Fibre Composites: Materials, Processes and Applications* (pp. 176–215). Woodhead Publishing Limited. 10.1533/978085 7099228.2.176

Wakeman, M. D., & Månson, J. A. E. (2005). Composites manufacturing - thermoplastics. In A. C. Long (Ed.), *Design and Manufacture of Textile Composites* (pp. 197–241). Woodland Publishing. 10.1533/9781845690823.197

Wang, R.-M., Zheng, S.-R., & Zheng, Y.-P. (Eds). (2011). Introduction to polymer matrix composites. *Polymer Matrix Composites and Technology* (pp. 1–548). Woodhead Publishing. 10.1533/9780857092229.1

Yalcin, B. (2015). Hollow glass microspheres in sheet molding compounds. In S. E. Amos & B. Yalcin (Eds.), *Hollow Glass Microspheres for Plastics, Elastomers, and Adhesives Compounds* (pp. 123–145). Elsevier Inc. 10.1016/B978-1-4557-7443-2.00005-0

Yan, L., & Chouw, N. (2015). Effect of water, seawater and alkaline solution ageing on mechanical properties of flax fabric/epoxy composites used for civil engineering applications. *Construction and Building Materials, 99*, 118–127. 10.1016/j.conbuildmat.2015.09.025

7 Traditional and Advanced Characterization Techniques for Reinforced Polymer Composites

Amit Pandey
Department of Chemical Engineering, Institute of
Engineering and Technology, Lucknow, India

G. L. Devnani
Department of Chemical Engineering, Harcourt Butler
Technical University, Kanpur, U.P., India

Dhanajay Singh
Department of Chemical Engineering, Institute of
Engineering and Technology, Lucknow, India

7.1 PROPERTIES AND EVALUATION OF NATURAL FIBER COMPOSITES

As discussed previously, natural fibers in recent times has emerged as a topic of great interest to scientists and researchers across the globe. The reason is not only the tremendous potential that these natural fibers exhibit as a good alternative in the form of reinforced composites, but also due to their positive impact like cost effectiveness and less pollution during production. They can prove to be an outstanding solution to the problem of the current ecosystem. Hence, they are playing a major role in developing biocomposites in recent times. Properties of various natural fibers are utilized to manufacture commercial composite products. Several natural fibers like jute, kenaf, hemp, and sisal have been reported as excellent reinforcements in different matrices (thermoset and thermoplastic) and are currently being used in different applications such as the auto industry, packaging industry, and civil constructions. The main source of natural fibers is agricultural plants. So

DOI: 10.1201/9781003201724-7

they are easily and cheaply available as reinforcing material in any of the polymer matrices used in making several household items too like ropes, carpets, hand bags, etc. (Venkatesan & Bhaskar, 2020).

7.1.1 CLASSIFICATION OF NATURAL FIBERS

Natural fibers can broadly be classified into two categories based on their origin. Animal fibers originate from animals and plant fibers make their existence from plants. Wool, feathers, and silk (spider silk, cocoon silk) are animal-based fibers, which have a huge impact in medical applications like implants. The reason is that these bioproducts are required to be biodegradable, which means that they should be able to break down and merge back again inside the body. These fibers are also biocompatible, which ensures no harmful effects to the human body. Plant-based fibers, such as hemp, kenaf, bamboo, coir, sisal, and banana make their presence in a lot of NFRP composites (along with suitable polymer matrix), which find an extensive application in various useful domestic products. Such fibers are classified as renewable sources because they can be extracted easily without doing any harm to the environment (Lau et al., 2018).

As per Figure 7.1, we can easily conclude that the classification of natural fibers is based on the origin of these fibers, such as plant, animal, or mineral.

7.1.1.1 Plant Fibers

Plant fibers are lignocellulosic materials and their major component is cellulose e.g., cotton, pineapple, bagasse jute, banana flax, ramie, sisal, coir, and hemp, etc. These fibers can further be classified as seed fiber (meaning they are developed from a seed or seed case); an example of this category is cotton. The next category is leaf fiber, which is basically collected from the leaves of plants and trees; for example, sisal. Another type is skin fiber, which are normally collected from the skin surrounding the stem of the plant. These fibers have a superior mechanical performance than other fibers of their class. That is why these fibers have applications in durable yarn and fabric manufacturing; some examples of

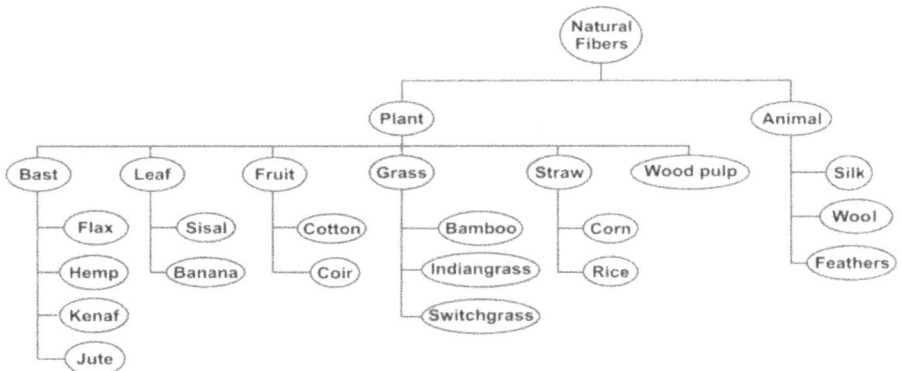

FIGURE 7.1   General classification of natural fibers.

this category are flax, hemp, banana, jute, and soybean. Next are fruit fibers, collected from the fruit of the plant; for example, coconut (coir) fiber. Another type is stalk fiber, which is actually the stalks of the plant. In this type, examples are wheat and rice straw and other common crops like grass and bamboo. Tree wood can also be considered such a fiber.

7.1.1.2 Animal Fiber

Animal fibers are rich proteins. They are extracted from different animal species or mammals that are hairy; for example, sheep wool, human hair, goat hair, silk alpaca hair, hair of horses, wool, etc. We can also get animal fibers from birds that are called avian fibers; e.g., feather fibers.

7.1.1.3 Mineral Fiber

All the fibers occurring naturally or fibers procured from minerals with slight modifications come under this category. These fibers can futher be sub-categorized into asbestos, ceramic fibers, and metal fibers. The only naturally occurring fiber is asbestos, with amphiboles, serpentine, and anthophyllite as different forms of it. The ceramic fiber category includes glass fibers (a combination of wood, quartz, and glass), Al_2O_3, SiC, and B_4C, while metal fibers include aluminium fibers. A significant amount of literature is available to prove that these NFs have a potential to substitute glass fiber in a variety of engineering uses as the mechanical properties of the NFRP composites are quiet encouraging when compared to that of glass fiber–reinforced polymer (GFRP) (Lau et al., 2018).

7.1.2 Properties of Natural Fibers Composites

Physical characteristics of NFCs are greatly related to the lignocellulosic composition of the single fiber (cellulose, lignin, hemicellulose, and water content), grooving (climate, soil feature, aging conditions), and processing method conditions. Chemical composition along with tensile characteristics of various natural fibers are tabulated in Table 7.1 and Table 7.2.

TABLE 7.1
Lignocellulosic Composition of Natural Fibers (Thiruchitrambalam et al., 2010)

Natural Fibers	Cellulose (in %)	Hemicellulose (in %)	Lignin (in %)
Sun hemp	74 ± 4	18.5 ± 0.5	4.5 ± 0.5
Jute fiber	62 ± 1	13	9 ± 4
Sisal fiber	63.5 ± 3.5	12.5 ± 2.5	10 ± 2
Banana fiber	62.5 ± 2.5	7 ± 1	7.5 ± 2.5
Wood fiber	47.5 ± 2.5	23	27
Flax fiber	71 ± 1	14	4.5 ± 0.5

TABLE 7.2

Mechanical Properties of Some Common Natural Fibers (Abilash & Sivapragash, 2013; Lau et al., 2018)

Fibers	Density (g/cm³)	Tensile Strength (Mpa)	Young Modulus (Gpa)	Faliure Strain (%)
Jute	1.4 ± 0.1	485 ± 285	37 ± 17	2.5 ± 0.5
Sisal	1.5	450 ± 350	15.5 ± 6.5	5 ± 2
Coir	1.2	180	5 ± 1	3.0
Flax	1.5	700 ± 350	49 ± 21	3 ± 1
Hemp	1.5	690	50 ± 20	2.7 ± 1.2
Bamboo	0.8 ± 0.2	185 ± 45	14 ± 3	5.5 ± 1.5

The other factors that may influence these mechanical properties of natural fibers are the length and diameter of the individual fibers, along with the experimental conditions that may vary in many cases.

7.1.3 EVALUATION OF MECHANICAL PERFORMANCE OF NATURAL FIBER COMPOSITE

Several factors are combined together to prepare a composite. So obviously there are a lot of optimizations involved among these factors for the preparation of any reinforced composite. After the preparation, the mechanical performance of these reinforced composites is tested for different mechanical properties, which will be discussed later in this chapter in detail. The factors that can affect the mechanical performance of NFCs are discussed in brief below.

7.1.3.1 Fiber Selection

As discussed previously, fiber type, which is commonly classified on the basis of its origin (plant, animal, and mineral), is an important aspect. The basic structural component in a plant fiber is cellulose, whereas for animal fiber, protein is considered to be the major structural component. Mineral-based natural fibers were once used extensively in composites, but they are being discarded now because of the presence of asbestos minerals, which is harmful and banned in many countries. On the other hand, silk, which has very high strength, is relatively expensive, shows less stiffness, and is also not easily available (Shah et al., 2014). Hence geography related to fiber availability holds a key factor in fiber selection (Mustafa et al., 2015). With higher cellulose content with cellulose microfibrils aligned more in the direction of fiber, higher structural requirements can be met and the performance can be enhanced. Thus, the properties of natural fibers vary significantly with their chemical composition and structure, which is related to the type of fiber along with its growing conditions, extraction method, treatment, harvesting time, and storage procedures.

7.1.3.2 Matrix Selection

The matrix plays a key role in fiber-reinforced composites. Apart from protecting the surface of the fibers from mechanical damage, it also transfers loads to the fibers and hence acts as a defending barrier against unfavorable environments. The most commonly used matrices in natural fiber composites are polymeric because of their light weight and low temperature processing. Both thermoplastic and thermoset polymers are being widely used as matrices with natural fibers (Holbery & Houston, 2006). Matrix selection is often limited by the fiber's degradation temperature. Natural fibers that are used as reinforcement usually exhibit unstable thermal behavior above 200°C, although under some circumstances it is made possible to process them at a higher temperature for a shorter span of time (Summerscales et al., 2010). Hence, only thermoplastics like polyethylene, polystyrene, PVC, polyolefin, PP, and thermosets are preferred as matrix material because they become soft below this temperature (Dos Santos et al., 2008). The thermosets used are epoxy resin, VE resins, polyester, and phenol formaldehyde. Thermoplastics are softened by applying heat and hardened by cooling and are capable of being easily recycled, which is most favored in recent commercial applications, and also better realization of the fiber properties is achieved while using thermosets. Petroleum-based matrices are replaced with bioderived matrices. In terms of mechanical property, PLA is used, which has shown higher strength and stiffness when used with natural fibers than polypropylene (Faruk et al., 2014).

7.1.3.3 Interface Strength

Interface bonding exists between a fiber and matrix. This plays an important role when it comes to determining the mechanical properties of any composite. The transfer of stress takes place between the matrix and fiber; hence, sufficient interface bonding is required to get optimum reinforcement. There is always a possibility to have a strong interface, enabling crack propagation, which in turn reduces toughness as well as strength. Plant-based fiber composite exhibits feeble interaction between the hydrophilic fibers and hydrophobic matrices, which leads to very limited interfacial bonding. Thus, the mechanical performance as well as low moisture resistance are encountered, which affects the long-term properties. Fiber and matrix are brought in intimate contact for bonding; wettability is considered to be an important precursor to interface bonding. Fiber wettability affects the strength (tensile and flexural) of composites. Proper chemical and physical treatment can enhance the wettability of the fiber, therefore improving the interfacial strength. Mechanisms such as electrostatic bonding, mechanical interlocking, inter-diffusion bonding, and chemical bonding are attributed to interfacial bonding. Mechanical interlocking occurs when the surface of the fiber is rough, which increases the interfacial shear strength but has less impact on the transverse tensile strength. Electrostatic bonding is only applicable on metallic surfaces. Chemical bonding occurs when there is a presence of chemical groups in the fiber surface and in the matrix, which leads to a chemical reaction to form bonds. Use of a coupling agent acting as a bridge between the fiber and matrix is the main reason behind chemical bonding. The interaction between atoms and molecules present in the fiber and matrix at the interface results in inter-diffusion bonding (Raja et al., 2017).

7.1.3.4   Fiber Dispersion

Fiber dispersion is a prime factor affecting the properties of short fiber composites and, hence, is a matter of prime concern for NFCs, which generally have hydrophobic matrices and hydrophilic fibers. Longer fibers can be used to increase their agglomerating tendency. Good fiber dispersion enhances excellent interfacial bonding and reduces voids by ensuring that the fibers are properly surrounded by the matrix material (Heidi et al., 2011). Dispersion can also be affected by various processing parameters like temperature and pressure; additives such as stearic acid to increase interfacial bonding increases the fiber matrix interaction. The use of an intensive mixing process like a twin screw extruder can be preferred over a single screw extruder for better fiber dispersion. But a high risk of damage to the fiber and reduction in length of the fibers is also involved during the process due to screw configuration and temperature (Beckermann & Pickering, 2008; Sanadi et al., 1997).

7.1.3.5   Fiber Orientation

The best of mechanical properties in a composite can be achieved when the fiber alignment is parallel to the direction where the load is applied. Continuous synthetic fibers are easier to align than natural fibers, although with injection molding some alignments are made possible to achieve, depending on viscosity of the matrix and the design of the mold (Joseph et al., 1999). Before impregnating the matrix, natural fibers are placed in sheets and carded to achieve a higher degree of fiber alignment. Conventional textile processing techniques like spinning can be used for producing a continuous yarn. Aligned fiber yarns are produced by wrap spinning, which is very common in the textile industry since 1970. The influence of fiber orientation on the mechanical performance of composites can be understood from the fact that with an increase in fiber orientation angle relative to the test direction, the strength and Young's modulus of the product get compromised.

7.1.3.6   Manufacturing

Common methods involved in the preparation of NFCs are injection molding, compression molding, and extrusion. RTM (resin transfer molding) and pultrusion are also other methods that are used with some thermoset matrix composites. Factors determining the properties are speed of the process, temperature, and pressure. It is possible for the fiber to degrade when it is subjected to high temperature; hence, the thermoplastic matrices cannot be used because their degradation temperature is much higher than the melting point. The extrusion process involves softening and mixing of thermoplastics in bead or pellet form with the fiber with the help of a single or two rotating screws. It is compressed and forced out at a constant rate through a die. The high speed of the screw can lead to excessive melting temperatures, entrapment of air, and breakage of fiber, whereas a low screw speed can lead to insufficient mixing and poor wetting of the fibers.

Better mechanical performance and fiber dispersion is achieved through the twin screw system rather than the single screw extruder (Malkapuram et al., 2009).

Injection molding can be performed with thermoset or thermoplastic matrices, in which thermoplastic matrices are most often used. Fiber orientation may significantly vary across the mold with shear flow along the walls due to friction resulting in alignment of the fiber along the wall of mold, while a higher stretch rate at the center can produce a transversely aligned fiber with respect to the direction of flow. The structure is known as a skin core structure. The strength of the composite can be compromised with the presence of residual stress in composites involving thermoplastics. This residual stress can be present due to any of the mentioned factors like polymer chain alignment, pressure gradients, non-uniform temperature profiles, or due to difference in the coefficients of thermal expansion of a fiber and matrix (Ho et al., 2012). Injection molding generally has a limitation to produce composites with fiber content no more than 40%, as per the viscosity constraints. The extrusion process involves reduction in the length of fibers. CM (compression molding) can be used for thermoplastic matrices with loose chopped fiber that is either short or long, oriented randomly or aligned or mats along with thermoset matrices. The fibers are stacked with alternate sheets of thermoplastic matrix before the curing temperature and pressure is applied. Proper control of matrix viscosity is required during the application of pressure and temperature, especially for thick samples to ensure proper impregnation of matrix material in between the fibers. Quality composites can only be produced by controlling certain key parameters like temperature, pressure, and holding time viscosity, depending upon the type of matrix and fiber. Film stacking involves a single temperature cycle and hence is recommended to limit any degradation. Temperature holds a key aspect as sometimes there is not much difference between a matrix processing temperature and the fiber degradation temperature. As reported by Jiang et al. in 1999, fiber strength weakens at a temperature range of 150°C to 200°C and the reduction in strength is reported as 10% in just 10 minutes. So, in order to achieve both the goals of good wetting and to avoid fiber degradation, an optimum temperature is very essential to define for any composite material. The optimum temperature for composite tensile properties is reported as 150°C (Jiang & Hinrichsen, 1999). In compression molding, sheet molding composites have been used as an alternative to film stacking. RTM involves injection of liquid thermoset resin into a mold containing a preformed fiber. The important variables to control in this process are injection pressure, temperature, mold configuration, and preform architecture. RTM have advantages over other process in the form of lower temperature and no involvement of thermomechanical degradation. The structure of natural fibers defines the compaction in this process. Because of a lower degree of fiber alignment, NFCs are much more compactable than glass fiber composites. Good compact strength is achieved in this process, which makes it suitable for low production runs.

7.1.3.7 Porosity

Mechanical properties of composites are largely influenced by porosity and much research and effort have been put in to reduce the porosity in synthetic fiber composites. It basically occurs due to several factors, including air inclusion while processing, limited or poor wettability of fibers, presence of some hollow features within the bundles of fiber, and low compaction ability of fibers (Madsen et al., 2009). Porosity in NFCs are reported as directly proportional to fiber content, once the geometrical compaction limit is exceeded, based on which flax/PP composites were reported with an increase of porosity from 56% to 72 % (Madsen & Lilholt, 2003). Its inclusion in models marked a significant improvement in its strength as well as stiffness.

7.2 MECHANICAL CHARACTERIZATION

The major applications of these materials are in structural and semi-structural. Mechanical properties like tensile strength and modulus and flexural strength and modulus play a very important role in determining the quality of composites.

7.2.1 TENSILE TEST

The tensile test determines the ability of the material to withstand the force that tends to pull apart the specimen and the extent to which the specimen stretches before breaking. Tensile modulus is a denotation of the relative stiffness of the material and can be obtained from the slope of the stress-strain plot.

Tensile strength and tensile modulus are determined numerically as follows:

$$\text{Tensile strength} = \text{Force/Cross section area} = F/A$$

$$\text{Tensile modulus} = \text{Tensile stress/Tensile strain} = \frac{F/A}{\Delta L/L}$$

where:
F = Force exerted on a specimen under tension,
A = cross-sectional area of specimen through which force is applied,
ΔL = the amount by which the length of the specimen changes,
L = original length of the specimen.

Tensile test is basically conducted to calculate the tensile behavior of the reinforced composite. This behavior may include certain properties like tensile strength, tensile modulus, tensile strain, Poisson's ratio, etc. In fibrous-based composites, the tensile test is performed as per ASTM D 3039 standard (ASTM, 2014). ASTM D3039 testing determines various tensile characteristics of PMC materials reinforced by high-modulus fibers. Different forms of composite material can be tested involving continuous or discontinuous fibers in a reinforced state whose laminate is balanced and symmetric with respect to the direction of test. This test technique is designed to know the tensile property data for various purposes like research and development, material specifications, quality evaluation, and structural design purposes. A minimum of five specimens are usually tested as per defined testing conditions or less if valid results are obtained in fewer specimens like in the case of any designed experiment. There are lot of configurations that can be tested well without using tabs; for example, fabric-based materials and multidirectional laminates and randomly reinforced sheet-molded compounds. Tabs are greatly recommended while testing unidirectional materials or highly unidirectional laminates to failure in the direction of fiber. Tabs are also recommended while testing unidirectional materials in the direction of a matrix to avoid any damage caused due to gripping. A continuous E-glass-reinforced polymer matrix has been the most commonly used tab material in a laminate configuration of [0/90]. The tab material is generally kept at an angle of 45° with the direction of applied force to ensure a soft interface. Composite adhesive bonding can be utilized at the end tabs to avoid any failure within the area of tab. The specimen to be tested is kept fixed in the tensile fixture within the grip and a standard crosshead velocity is provided by the user through a software. Transverse and longitudinal strains can also be measured as strain gauges are also provided for strain measurement.

The specimen is mounted first in the grips of the testing machine (Figure 7.2) to determine the elastic modulus of the specimen. A universal testing machine requires a constant rate of movement. The machine consists of one grip that is in stationary mode and a second grip with a movable mode. A load-indicating mechanism capable of indicating the tensile load is used with good accuracy. An extensometer is also used to find out the distance between the two points placed within the gauge length of the test sample when the sample is stretched.

The procedure for conducting a tensile test is as follows:

- The specimen is tightened vertically in the grips of the machine; it should not slip.
- The dimension of the specimen is given to the machine, testing speed is set, and the machine is started.

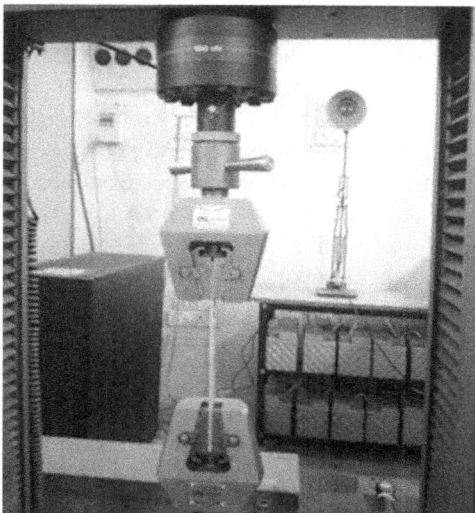

FIGURE 7.2   ASTM D-3039 testing system setup.

- Specimen starts to elongate, and resistance of the specimen is recovered by the load cell.
- Elongation of the specimen continues until the breakup occurs.
- Finally, the display of the machine shows a value of tensile strength, tensile modulus, stress vs strain curve, elongation, etc.

There is also a digital plotting of load deformation or load strain curves. Hence, a tensile test basically determines the following tensile properties of the composite materials:

 I. **Tensile Strength:** The maximum stress that can be applied on the specimen or before it actually breaks.
 II. **Ultimate Tensile Strain:** The strain recorded for the specimen when it actually breaks.
 III. **Tensile Modulus:** Maximum stretch or deformation encountered by the specimen in response to the applied stress during the test.
 IV. **Poisson's Ratio:** Change in transverse to longitudinal strain ratio between two longitudinal strain points. The points usually are the same as those used to determine the modulus i.e., 0.1% to 0.3%.
 V. **aTransition Strain:** Important aspect where a yield behavior of the material is displayed as a slope change in stress-strain response. The strain value where the slope change actually occurs is known as a transition strain.
 VI. **Failure Mode:** Examination of broken specimens are done and their failure type, area, and location are recorded using a three-character code.

7.2.2 IMPACT TESTING

The impact test is utilized to evaluate the shock-absorbing capacity of composite materials subjected to a load applied suddenly. The impact strength of the material is associated with the toughness of the material. It is the ability of the material to absorb the applied energy. It can be defined as the ability to resist the fracture under the stress, which is applied at high impact. Izod impact strength is determined using ASTM Standard D256. The test specimen has a dimension 63.5 mm in length and 12.7 mm in breadth of the material, which is cut from a composite sheet of the respective material. Figure 7.3 (a) shows the epoxy composite reinforced by untreated and alkali-treated mallow fiber and Figure 7.3(b) shows the impact strength of these composites (Nascimento et al., 2018).

The procedure to conduct an impact test is as follows:

- The specimen should be tightened vertically in the grips of the machine.
- The dimension of the specimen is given to the machine and the striking hammer is locked in its position.
- Now, press down the pendulum and release over lever so that the hammer strikes the test specimen and breaks it.
- Finally, the display of machine shows value of impact strength.

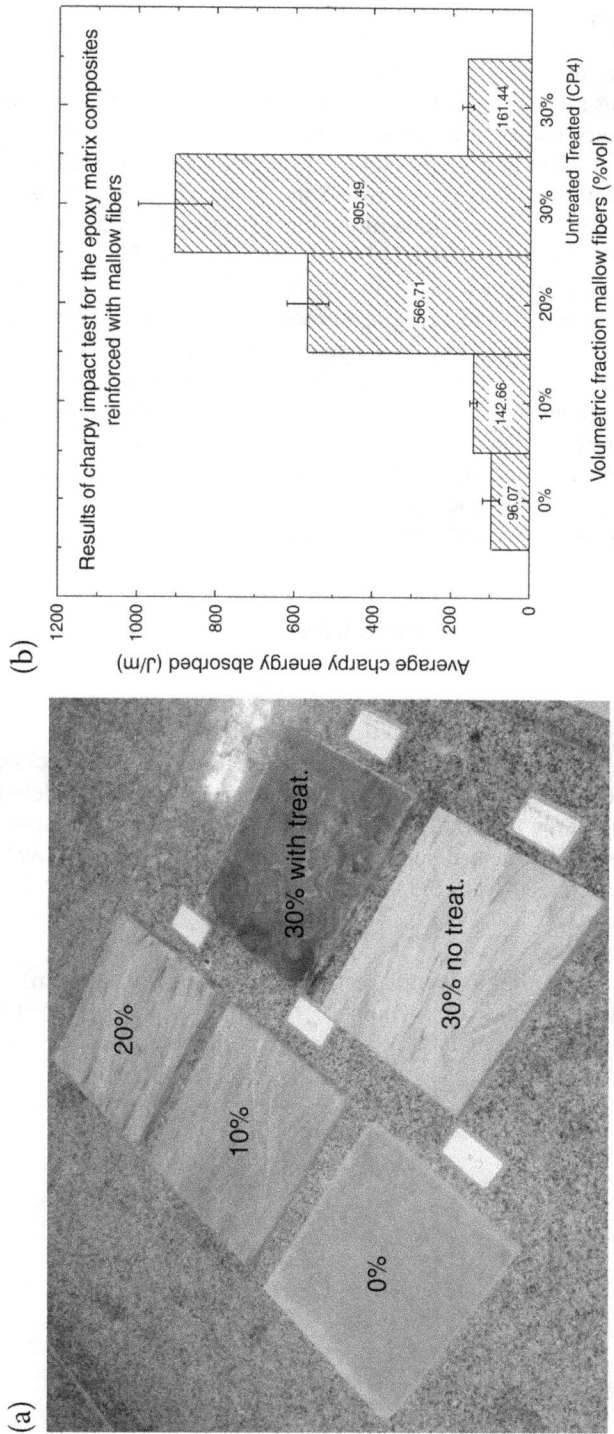

FIGURE 7.3 (a) Plate composite fiber percentage from 0 to 30 vol% without treatment and 30 vol% with treatment (mercerization – 5% NaOH) (Nascimento et al., 2018); (b) results of Charpy impact test for the epoxy matrix composites reinforced with mallow fibers (Nascimento et al., 2018).

7.2.3 FLEXURAL TEST

The flexural testing method is a common practice to test FRP composite materials. The proper loading rate along with a convenient loading fixture is used to test the specimen. It's a very common practice to determine the strength and modulus of FRP laminate composites. Stress distribution gets complex during flexural loading. Hence, it gets a little difficult to determine the mechanical properties using this method. This testing method for determination of flexural properties of PMC is standardized as per ASTM D7264/D7264M-15 (ASTM, 2015).

This test technique is widely used for testing and determining various flexural properties of PMC materials like strength, stiffness, and load/deflection behavior. The flexural test method was established for optimum use with continuous FRP matrix composites and is different in various aspects from other techniques that include the use of a standard span-to-thickness ratio of 32:1 versus the 16:1 ratio. Flexural properties can be utilized in different applications like quality control, specification, and designing applications apart from estimating the flexural properties of structures. The flexural test provides a graphical relation between the load (P) and displacement (h).

By applying equation 7.1 and 7.2 below stress, (σ) and strain (ε) can easily be calculated as

$$\sigma = 3Pl/2bd^2 \tag{7.1}$$

$$\varepsilon = 6hd/l^2 \tag{7.2}$$

where b is the width of specimen, d represents the thickness of specimen, and I determines the specimen span length. Mostly the material strength is determined as the stress taken up by the material. This value of stress becomes the maximum load that can be carried by the tested specimen after the test is over. The value of the modulus can be calculated from the slope of the initial portion of the stress-strain plot, which is linear.

Procedure for conducting a flexural test (three-point loading system)
- The specimen bar lies on two supports (Figure 7.4) and is loaded by the help of a loading nose, which is in the middle between two supports.

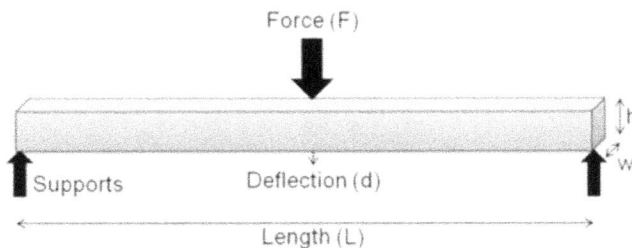

FIGURE 7.4 Diagrammatic representation of flexural strength measurement.

- The loading nose and the support must have the cylindrical surfaces to avoid the stress concentration.
- The dimension of the specimen is given to the respective machine and the load is applied to the specimen at the specified speed.
- The specimen start to deform; resistance of the specimen is recovered by the load cell.
- Bending of the specimen is continued until the breakup occurs.
- Finally, the display of the machine shows a value of flexural strength, flexural modulus, stress vs strain curve, etc.

7.2.4 INTERLAMINAR SHEAR STRESS OR SHORT BEAM SHEAR TEST

The three-point short beam is a shear test carried on a specimen for a minute time. It basically simulates failure through interlaminar shear (J. K. Kim & Mai, 1998). A beam with the bending load always has a shear stress. This shear stress is directly proportional to the amount of load applied, irrespective of length of span. Thus, the short beam shear (SBS) involves a short support span so that an excessive interlaminar shear failure can be created ahead of bending failure. This test is conducted as per ASTM D 2344 (Laminates, 2000).

7.2.5 CREEP TEST

To study the creep rupture of FRP composites ASTM developed and demonstrated ASTM D7337/D7337M-12 (ASTM International, 2019). The initial result of the test is the million-hour creep rupture size of the sample. This technique finds an application in testing those FRP matrix composite bars that are used as tensile components in pre-stressed, reinforced, or post-tensioned concrete and, hence, the study of tensile creep rupture of such composites becomes very essential. Creep behaviors of such structures are also important for design. To test the creep rupture of FRP bars, the main variables are size and type of FRP bars, force application period, and the magnitude of the force applied. Creep rupture of FRP bars can happen below their static tensile strength, unlike the bars using steel as a reinforcing agent that can sustain heavy stress. Therefore, it becomes critical to analyze the creep rupture strength of FRP bars used as reinforcement before resisting any significant load. Creep rupture strength may vary as per the type and size of the FRP bars. Creep rupture time of FRP bars can also be calculated using this method, provided that certain environmental situations and force ratios are controlled.

7.3 DYNAMIC MECHANICAL CHARACTERIZATION (DMA)

The DMA technique is considered to be very useful in the characterization of composite structures with damping as a function of temperature, time, frequency, stress, atmosphere, or multiple combinations of these different parameters taken (Romanzini et al., 2013). Composites that are considered multicomponent systems require theories of complex constitutive equations and micromechanics to evaluate their dynamic mechanical response. This response depends on the different sets of combinations that are physical or structural arrangements of phases like interfacial behavior, morphological property, and the nature of composite constituents (Jawaid & Khalil, 2011; Sreekala & Thomas, 2005). A lot of literature is available to state that the dynamic mechanical characteristics of a given composite material are governed by some important factors that include orientation of fibers and the mode of testing, fiber loading, treatment with compatibilizer, and additives and fillers (Jacob et al., 2006). The dynamic mechanical analysis (DMA) technique is used to evaluate the viscoelastic characteristics of polymers. It measures properties of material such as modulus or stiff nature and damping along with energy dissipation as they get deformed under the application of dynamic stress. Much quantitative information can be provided regarding the performance of materials by this measurement technique. This technique is currently being used for the evaluation of a variety of polymeric materials such as elastomers, composites, thermosets, fibers, and different films along with coatings and adhesives. Dynamic mechanical analysis is considered to be a very precious technique because it is assumed as the most precise thermal analysis for measuring the glass transition temperature zone, T_g. Secondary relaxation events can only be observed by DMA among all other thermal techniques. The storage modulus (E'), also known as the dynamic modulus, is closely related to the Young's modulus. It is considered a measurement of stiffness of a given specimen that determines if the sample is stiff or flimsy. E' can be defined as the tendency or nature of a material to develop storage of the applied energy for future purposes. Loss modulus (E"), or dynamic loss modulus, is defined as the tendency of the material to dissipate the applied energy. E" can also be defined as a viscous reaction of a material (Jawaid et al., 2015). This dynamic loss modulus is in close relation with ''internal friction''. The best way to understand this phenomenon is to allow a ball to bounce. It will result in dissipation of some energy and some energy will be stored for the future as well, as shown in Figure 7.5.

tan δ is a number without a dimension, defined as the ratio of loss modulus to storage modulus and is considered the mechanical damping factor. Mathematically, tan δ = (E")/(E') is shown in Figure 7.6. Loss modulus, storage modulus, and tan δ can be correlated within the DMA graph plotted along the temperature (refer to Figure 7.7). The resultant component that is derived from the given plot is known as

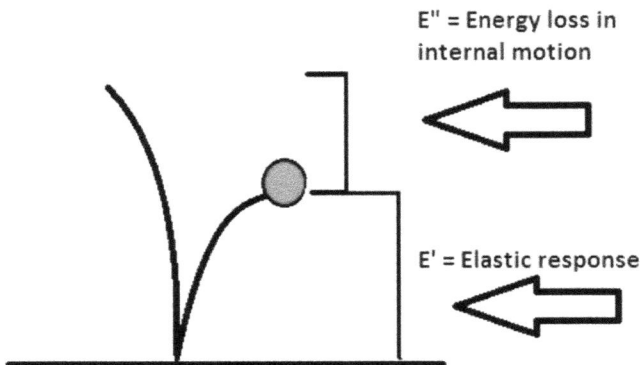

FIGURE 7.5   Illustration of the loss modulus and storage modulus.

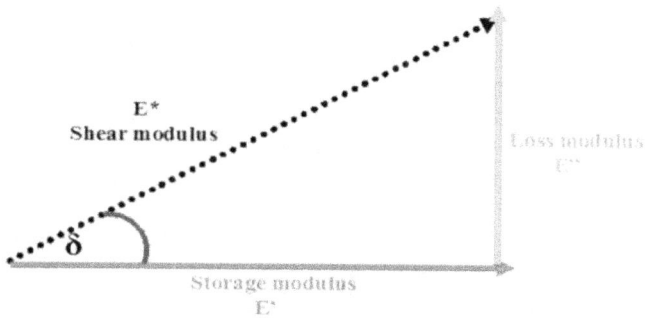

FIGURE 7.6 Relationship between E', E", and tan (δ).

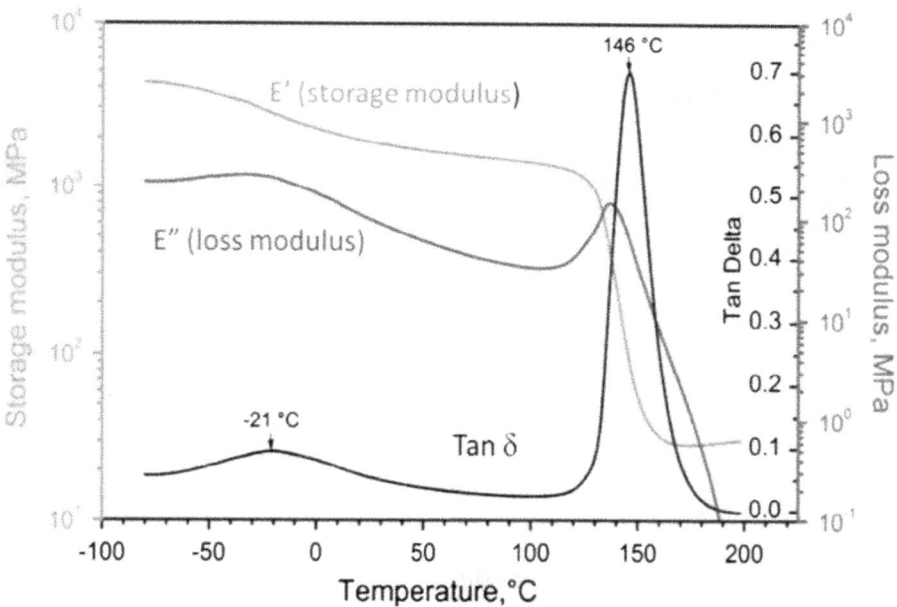

FIGURE 7.7 Relationship between E', E", and tan δ vs temperature in the DMA.

a complex modulus or shear modulus. It is represented by (E*). A higher value of tan δ means that the material has a high, nonelastic strain component, whereas a lower value defines material with high elasticity. If the fiber/matrix interface bonding is increased in FRP, it will result in a reduction of the damping factor because the movement of the different chains of the molecule at the interface of the fiber/matrix also get reduced with the increase in interfacial bonding between the fiber and matrix. Thus, the lower the loss of energy with respect to its storage capacity, the greater is the tan δ (E"/E') value in the system. The damping factor can be related to molecular movements and viscoelasticity apart from certain defects that contribute towards damping such as dislocations, phase boundaries, grain boundaries, and several interfaces (Zhang et al., 2012).

So, it can be concluded that the dynamic mechanical analysis (DMA) is a useful and valuable technique that goes in line with the observations provided by other discussed traditional thermal characterization techniques like differential scanning calorimetry (DSC) or thermogravimetric analysis (TGA). Several dynamic parameters discussed, such as storage modulus (E'), loss modulus (E"), and damping factor (tan δ), depend on temperature to provide enough useful information about interfacial bonding between the reinforced fiber and polymer matrix of any composite material (Saba et al., 2016). DMA can also predict the effects of time and temperature on polymer sealants' viscoelastic performance under different environments. Hence, DMA can be regarded as a measure of dynamic mechanical properties of natural fiber–reinforced polymer composites along with hybrid- and nano-composites.

7.4   WATER DIFFUSION ANALYSIS

Water absorption characteristics of untreated and various treated fiber-epoxy composites were examined according to ASTM D570. The equation used for analysis follows:

$$\% \ M = [(M_f - M_i)/(M_i)] * 100$$

where:
M_i = dry initial weight,
M_f = weight after immersion in water,
% M = water absorption.

Water absorption of natural fiber–based reinforced composites depends on two types of methods: Fickian diffusion and non-Fickian diffusion (Célino et al., 2013).

To analyze diffusion behavior, the following equation can be used:

$$F_s = (M_t/M_m) = kt^n$$

Where:
M_t = percentage of the water is absorbed in the sample at time t,
M_m = maximum percentage of the water absorbed,
K &n = kinetic parameter.

The diffusion coefficient (D) for the water absorption by a composite can be calculated by:

$$F_s = (M_t/M_m) = kt^n = (4/h)(Dt/\Pi)^{1/2}$$

where h = thickness of sample.

Table 7.3 shows the compilation of the work of different researchers in this field.

TABLE 7.3

Review Work on Water Absorption Analysis of Composites

Fibers	Matrix	Method	Key Findings	Reference
Flax	Polypropylene	Thermos-compression method	Discussed maximum moisture absorption capacity. Composites with coupling agent absorb less moisture than those without coupling agent.	(El Hachem et al., 2019)
Sisal	Epoxy	Hand layup	Mechanical properties of sisal-reinforced composites like tensile strength, impact strength, and flexural strength reduces due to water absorption.	(Gupta, 2018)
Wood	Polyolefin	Injection molding	Discussed water absorption in wood-olefin-based polymer composites. Experimental and numerical both approaches are taken into consideration.	(Mrad et al., 2018)
Wood	Polypropylene	Injection molding	Water immersion test was done to inquire the water absorption kinetics.	(Hosseinihashemi et al., 2016)
Hemp	Polyester	Hand layup	With the increment in moisture content, there was the decrement in flexural and tensile properties of HFRUPE samples. The water absorption ways of the composites were found to follow Fickian behavior at room temperature but non-Fickian at elevated temperature.	(Dhakal et al., 2007)
Flax	Bioepoxy resin	Injection molding	Mechanical properties do not get affected negatively due to water absorption.	(Muñoz & García-Manrique, 2015)
Oil palm fibre	Polyester	Open molding	The water absorption nature of composites of oil palm fiber and oil palm ash reinforced with polyester have been inquired.	(Oke et al., 2013)

(Continued)

TABLE 7.3 (Continued)
Review Work on Water Absorption Analysis of Composites

Fibers	Matrix	Method	Key Findings	Reference
Jute	Epoxy	Injection molding	The study was done on water absorption behavior, thickness, and volume swelling of bio-based composites made from woven jute fibers and epoxy.	(Masoodi & Pillai, 2016)
Flax fiber	Polyester	Hand layup	Discussed water absorption nature of chemically treated flax fiber–reinforced composites.	(Alix et al., 2011)
Jute	Polylectide	Hot pressed method	Moisture absorption characteristics, microstructure evolution, and tensile strength of short jute fiber reinforced with polylactide composite in hygrothermal environment.	(Hu et al., 2010)
Jute	Polyester	Thermoset pultrusion machine	In the study, the water immersion test of jute fiber reinforced with polyester composite to inquire about the effects of water absorption on mechanical property.	(Akil et al., 2009)
Jute	Polyester	Vacuum-assisted resin transfer molding	Relation between dielectric behavior, and water absorption behavior of jute reinforced with polyester composite material.	(Fraga et al., 2006)
Bagasse	Epoxy resin	Hand layup	The study of water absorption nature was done for both treated and untreated bagasse fiber reinforced with epoxy resin composite.	(Devnani et al., 2018)
Areca	Epoxy resin	Hand layup	Areca is turning out to be more promising alternative due to the effect of water absorption on mechanical property.	(Venkateshappa et al., 2011)

TABLE 7.3 (Continued)
Review Work on Water Absorption Analysis of Composites

Fibers	Matrix	Method	Key Findings	Reference
Sisal	Epoxy resin, vinyl ester resin	Resin transfer molding	The study describes the effect of drying and wetting cycles on the mechanical properties of sisal fiber reinforced with vinyl ester and epoxy composite.	(H. J. Kim & Seo, 2006)
Banana	Vinyl ester	Hand layup	The study was done on the effect of ultrasonic treatment on moisture absorption nature and mechanical properties on woven banana fiber–reinforced composite.	(Ghosh et al., 2014)

7.5 THERMO GRAVIMETRIC AND DSC ANALYSIS

Change in properties (physical or chemical) of Fiber reinforced plastic (FRP) with an increase in temperature or at constant temperature in a isothermal environment can be studied by conducting TGA. Any variation reported during the test is primarily because of structural transformation or thermal degradation of either the reinforcement or the matrix in FRP. TGA can provide useful information related to physical changes like phase transitions, absorption along with desorption, and chemical changes that include phenomena like chemisorption, thermal degradation, and reactions of solid–gas. The yesting atmosphere can also be controlled while testing argon, nitrogen, air, or vacuum atmosphere can be created while testing any FRP composite. A thermo-gravimetric analyzer consists of an accurate weighing balance provided with a pan inside a programmable temperature control heater or furnace (Figure 7.8). A small amount of specimen, approximately 1–150 mg, is kept in the pan (holder) before increasing the temperature at a constant rate to initiate a thermal reaction. A reference pan is also provided to compare the properties of the given specimen and the reference material. Mostly alumina is taken in the reference pan as it shows high thermal stability. A thermocouple is installed in each holder to receive DTA signal output. Thermal decomposition analysis for the prepared composites can easily be done with TGA. It can also help in understanding other processes like dehydration, oxidization, heat resistance mechanism, and kinetics analysis associated with the prepared composites. The mass of the sample to be tested along with the reference substance are measured separately. This difference value of mass is sent as a thermogravimetric signal. This mass measurement difference confirms that the beam expansion effect, the convection flow, and buoyant

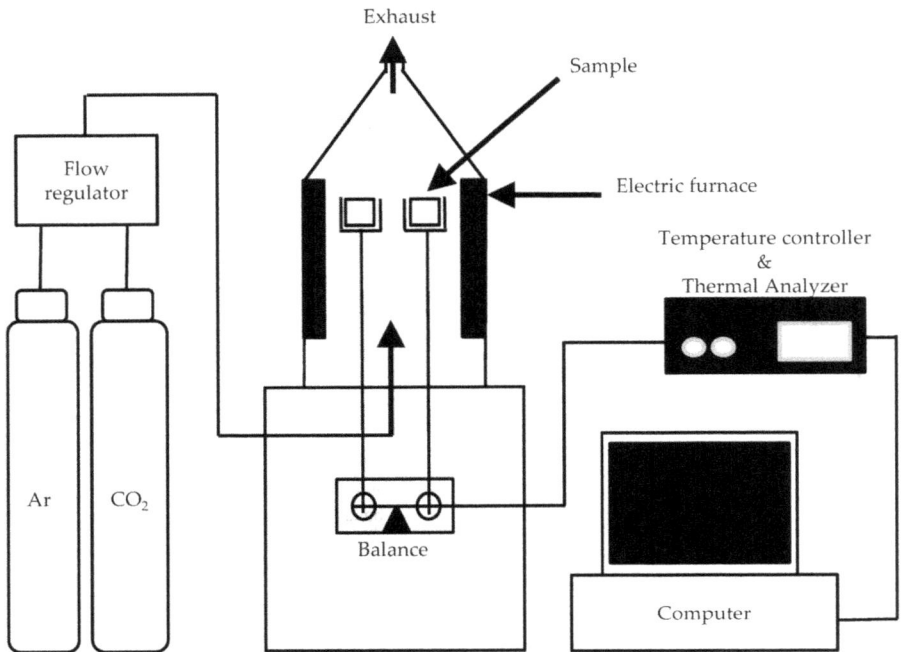

FIGURE 7.8   Diagram of thermogravimetric analyzer (TG-DTA) (Sarkar & Wang, 2020).

force are cancelled to achieve extreme sensitivity in measurement. The measurement of the mass of the specimen and the standard are done by the drive coils facilitating ease in adjustment of the thermogravimetric baseline drift. The variation in mass of the sample with respect to time or temperature gets converted into a graph/plot of mass or percentage of initial mass to temperature/time. The plot obtained is known as thermogravimetric analysis (TGA) curve. TGA, combined with other measurement methods, can provide a variety of valuable results can be achieved from just a single sample. The first derivative of this TGA curve is referred to as a derivative thermogravimetric (DTG) curve. This curve can also be used for in-depth analysis for the prepared composite (Lila et al., 2019). In thermogravimetry, the change in mass can be related to the extent of conversion (α) as:

$$\alpha = \frac{W0 - W(t)}{W0 - W\infty}$$

where
   $W(t)$ = the sample mass at time t,
   W_0, W_∞ = initial mass and residual mass, respectively.
   α lies between 0 and 1. Any recognizable changes in the thermogram are individual processes that can only be identified through certain primary variables such

as temperature (T), pressure (P), and time (t). Hence, the extent of conversion with respect to time can be given as:

$$\frac{d\alpha}{dt} = k(T)f(\alpha)h(P)$$

where:

  k = the simplified Arrhenius equation.

$$k(T) = k_O \exp\left[-\frac{Ea}{R}\left(\frac{1}{T} - \frac{1}{TO}\right)\right]$$

where:

  E_a = the activation energy of thermal decomposition,

  k_0 = the kinetic rate constant at reference temperature T_0.

Pressure also is a key parameter in reaction kinetics involving gaseous compounds. The extent of the reaction greatly depends on the partial pressure of gases introduced in the system with the purge gas.

The differential scanning calorimetry (DSC) technique came into existence as a commercial instrument during the early 1960s and was discovered as a convenient and useful method to measure certain important properties like (i) glass transition/ melting/crystallization temperatures of composites. It is also useful for the estimation of degree of cure of the product obtained; finally, the estimation heat of reaction can also be done using this technique. The basic advantage associated with DSC analysis is that a relatively smaller sample size (20 mg) is required to get quick and easy data on the overall rection kinetics (Barton, 1986).

7.6   CONTACT ANGLE AND WETTABILITY ANALYSIS

Wetting properties of fluids on a fiber surface plays a key role in polymer composite engineering. A significant amount of research has been done to study the interfacial properties of a fiber matrix system that includes the wetting property of a fiber surface along with the interface bonding strength between matrix and fibers. The toughening ofa composite material and the damage mechanisms involved during fiber reinforcement of composites are greatly affected by the interfacial properties of a fiber matrix system (Wu & Dzenis, 2006). Natural fibers are generally hydrophilic in nature and when they are made to interact with some hydrophobic thermoplastic matrix like polyethylene or polypropylene, they produce very low interfacial interactions. This leads to a poor interfacial strength and with that comes the need to study all the aspects related to the composite interface (Fuentes et al., 2014). The wetting property of the droplet defines the geometry of a droplet on a fiber. The wetting property of microdroplets on monofilaments has drawn a significant amount of attention in the past. The wetting shape of a microdroplet on a flat surface is defined as a partial sphere, but in case a fiber it does not exhibit the same geometry. Wettability studies basically revolve around the measurement of contact angles. Contact angles act as an important tool to describe the degree of wetting during a solid liquid interaction. Small contact angles (<<90°) signify high wettability; on the other hand, large contact angles (>>90°) can be associated with low wettability.

7.6.1 DEFINITION OF CONTACT ANGLE

If a liquid drop is made to rest on a flat, horizontal solid surface, then the contact angle can be given as the angle formed by the intersection of the liquid-solid interface and the liquid-vapor interface. This can be obtained geometrically by drawing a tangent line from the point of contact along the liquid-vapor interface in the profile of the droplet. The contact angle (θ) can also be defined as the quantitative measure of a liquid-solid interaction when a liquid is placed against a solid. The angle θ represents the angle made between the normal to the solid and liquid surface (measured in the gas) along the three-phase interline at the point of interest (Figure 7.9). It is the identical to the angle formed between the solid surface and the tangent to the liquid-gas surface (drawn in the liquid) in the plane perpendicular to the interline at the point of interest. The contact angle may be related to the surface energy (γ's) of the three interfaces by Young's equation:

$$Cos \ \theta = (\gamma_{sv} - \gamma_{sl})/\gamma_{lv}$$

where:

γ_{sv} = effective boundary tension of the solid vapor interface (or solid/vapor interfacial energy),

γ_{sl} = effective boundary tension of the solid liquid interface (or solid/liquid interfacial energy),

γ_{lv} = liquid surface tension.

7.6.1.1 Case I) When $\theta = 0°$

The liquid wets the entire solid and spreads out more to form a monomolecular film. As per Young's equation, this case is favored by high γ_{sv} and low γ_{sl} and γ_{lv}.

7.6.1.2 Case II) When $0° < \theta < 90°$

The solid is not completely wet by the liquid. To some extent, the solid prefers to be covered by the liquid as to oppose the gas.

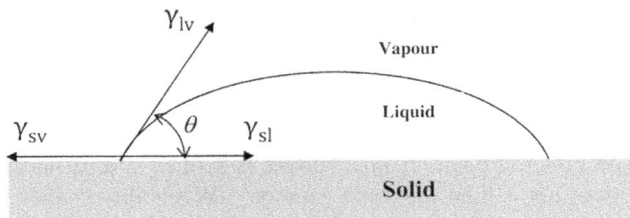

FIGURE 7.9 Contact angle as a measure of wettability.

7.6.1.3 Case III) When 90° < θ < 180°

In such conditions, the liquid does not wet the solid. This condition is favored by liquids with high surface tension on solids with low surface energy.

A limiting case occurs at $\theta \rightarrow 180°$ (Velde & Kiekens, 2000).

7.6.2 CONTACT ANGLE HYSTERESIS

A unique contact angle, θ_Y, can be determined using three thermodynamic parameters, γ_{lv}, γ_{sv}, and γ_{sl}, by applying Young's equation to any specific liquid-solid system. But in the real world, there is an existence of many metastable states of a droplet on a solid. These metastable states are responsible for a deviation in contact angle measured (θ_Y) to that of contact angle observed. The process of wetting cannot be limited to just a static state because the liquid moves constantly to expose its fresh surface and in turn it wets the fresh surface of the solid. Hence, a single static contact angle measurement for characterizing the wetting behavior became obsolete. If the contact line involving three phases, as discussed previously, is in real motion, the contact angle formed is known as a dynamic contact angle. The contact angle can also be classified as advancing (θ_a) when the angle is formed by expanding the liquid and receding (θ_r) if the angle is formed by contracting the liquid (Figure 7.10). There is a certain range under which these angles lie, whose maximum is defined by the advancing angle and the minimum value is governed by the receding angle. The measurement of dynamic contact angles depends on rates of speed. The measurement at lower speed gives an accurate value almost similar to the one achieved by static measurement. Hysteresis (H) can be defined as the difference between the advancing angle and the receding angle:

$$H = \theta_a - \theta_r$$

Extensive studies have been done on the contact angle hysteresis, which proves that the basic cause of it is surface roughness and/or heterogeneity. For non-homogeneous surfaces, there are domains that act as a barrier to the motion of the contact line. Consider this as an example; hydrophobic domains limit the motion of the water front as it advances, which causes an increase in the contact angle observed. The same domains hold back water front's contracting motion when the water recedes, causing a decrease in the observed contact angle. In several cases, the surface roughness is responsible for the generation of hysteresis. Barriers are

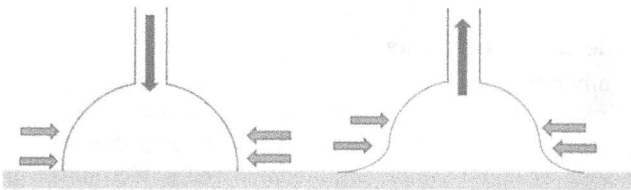

FIGURE 7.10 Advancing and receding angle illustration.

created due to the actual microscopic variations of slope on the surface, which limits the contact line motion and hence changes the macroscopic contact angles. Interpretation of such contact angle data according to Young's equation can be futile since the equation does not consider the surface topography.

This ambiguity lying in the contact angle phenomena is attributed to a significant difference between the experimentally observed contact angle and the Young's contact angle, θ_Y. However, there is no hysteresis involved in the contact angle on ideal solid surfaces. In such case, the contact angle measured by Young's equation (θ_Y) is equal to the observed contact angle, which denies any possibility of contact angle hysteresis. Smooth solid surfaces that are chemically heterogeneous show a difference in contact angle measured experimentally to that of contact angle observed. The experimental advancing contact angle, θ_a, can be approximated as θ_Y, instead of experimental receding angle θ_r, which exhibits poor reproducibility because of liquid sorption or solid swelling. But again on solid surfaces that are rough, no relation can be established between θ_a and θ_Y. It can be concluded that contact angles on rough surfaces are of no use as per Young's equation. The thermodynamic equilibrium contact angles on rough surfaces are called Wenzel angles, whereas the contact angle on heterogeneous surfaces is termed a Cassie-Baxter angle. These angles bear no equivalence to Young's contact angle (Yuan & Lee, 2013).

How smooth the surface (minimum roughness) should be to not have any effect on the contact angle still remains unanswerable. But literature tells that the solid surface prepared should be as smooth as possible and should be inert to the liquids.

7.6.3 MEASUREMENT OF CONTACT ANGLE

In the context of FRP composites, it is highly recommended to gather sufficient information on the wettability of the natural fibers, such as degree of wetting by the liquid or the penetration of liquid inside the fibers, before actually using them as reinforcements in composites. This purpose can be achieved through measurement of the contact angle. Hence, the contact angle measurement is a very common but key method of wettability measurements (Chen et al., 2012). There are several techniques present in the literature to measure the contact angle, which include sessile drop technique, captive bubble method, tilted plate method, wilhelmy balance method, capillary rise method, individual fiber technique, capillary tube method, capillary penetration method, and capillary bridge method. Among all these methods, the most common direct measurement technique is known as the sessile drop technique, as well as an indirect measurement technique in the form of the wilhelmy balance method will be discussed in detail.

7.6.3.1 Sessile Drop Technique

This is the commonly used technique to measure the contact angle. This method involves the direct evaluation of the tangent angle, which is formed at the three-phase junction point on a sessile drop profile. The setup comprises a horizontal stage where a solid or liquid sample are mounted and a micrometer pipette is also provided to form a liquid droplet. Apart from that illumination source, along with a telescope equipped, is also there. The evaluation involves the alignment of the tangent line of the sessile drop profile at the junction point with the surface and the protractor is then read with the help of an eyepiece. A high-speed camera can also be included to take images of the drop profile in case the contact angle needs to be measured later. High magnifications help in accurate examination of the intersection profile. A motor-driven syringe provide an easy way to control the rate of addition or removal of liquid, which makes the study of advancing, receding, or dynamic contact angles easier (Yuan & Lee, 2013). To measure the advancing contact angle, the sessile drop is very slowly added to make it grow up to 5 mm approximately with a micrometer syringe provided. To avoid any unwanted vibration, the needle should remain inside the liquid drop during the whole analysis. The diameter of the needle must be as small as possible so that the shape of the drop does not get distorted. The measurement must be done on both sides of the liquid drop profile and results should be averaged to cancel the effect of asymmetry in the drop, if any present. Measurement should be done on multiple points in case of a larger substrate and the results should be averaged so that the entire surface of the substrate can be incorporated in the measurement. There are certain advantages and limitations associated with this method. It is known as the most convenient method unless high accuracy is not required because it requires only a very small liquid quantity with the sample size as small as a few square millimeters. But, as said, that is advantageous only when a high accuracy is not required because its application gets limited when small contact angles (below 20°) have to be measured. The accuracy of this technique becomes questionable in such cases due to the uncertainty in various steps like assigning a tangent line when the drop profile is nearly flat. Also, only the largest meridian section of the sessile drop is focused on by its imaging device; hence, the profile image will only reflect the contact angle at the point in which the meridian plane is intersecting the three-phase line. Though the accuracy of this method is reported as ±2°, the amount of precision required is huge in this technique. As the sessile drop technique involves contact angle measurement by a sessile drop on a flat surface, there are several other system methods that also include measurements of contact angles on solid samples with several other geometric forms including plates, fibers, granules, or powders.

7.6.3.2   Wilhelmy Balance Method

In the analysis of natural fibers, a direct estimation of the contact angle creates uncertainties, and so their wetting characteristics are difficult to analyse. The Wilhelmy method is a reliable and widely used method to study the wetting behavior. This technique involves indirect measurement of a contact angle on a solid sample. It involves the detection of a change in weight by a balance when a thin and smooth vertical plate is made to contact a liquid. The change in force detected by the balance is basically a combination of buoyancy and wetting force, assuming that the gravity force remains the same.

The wetting force f can be defined as (Figure 7.11)

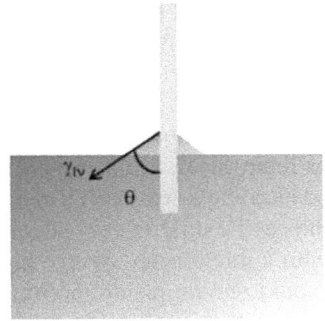

FIGURE 7.11 Wilhelmy balance method (diagrammatic representation).

$$f = \gamma_{lv} p \cos\theta$$

where:

γ_{lv} = the surface tension of the liquid,

p = the perimeter of contact line (same as the perimeter of solid sample),

θ = the contact angle.

The total change in force F detected on the balance can be given as:

$$F = \gamma_{lv} p \cos \theta - V\Delta\rho g$$

where:

V = the volume of the liquid displaced by the solid,

$\Delta\rho$ = density difference between the liquid and air (or a second liquid),

g = the acceleration of gravity.

Various trends of the curve force vs depth (1–4) can be analyzed as follows:

1. The sample is still approaching towards the liquid; hence, the force per unit length is zero.
2. The sample comes in contact with the surface of a liquid, forming a contact angle $\theta < 90°$. A positive wetting force is created with a rise in liquid level.
3. The sample is further immersed into the liquid, causing an increase of buoyancy that further causes a decrease in the force detected on the balance. The force measured on the balance is for the advancing angle.
4. The sample is pulled out of the liquid after reaching the desired depth. The force measured on this occasion is for the receding angle.

So, it can be concluded that if the liquid surface tension along with the wetting perimeter are known, the value of the contact angle can be calculated. In a very rare situation where the contact angle can be made zero with a known wetting perimeter, the force measured by balance can directly be called the surface tension of the liquid. In 1969, Princen (Princen, 1969) developed such a technique where the contact angle was zero to calculate the surface tension of liquid using the Wilhelmy balance method. Depending on whether the solid specimen is being pushed in or pulled out of the liquid, advancing or receding, the contact angle is established. The whole process

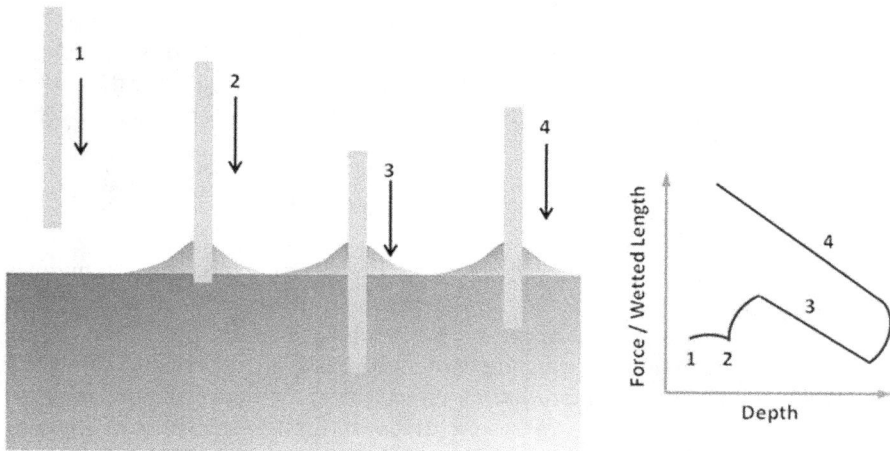

FIGURE 7.12 A submersion cycle for the Wilhelmy balance measurement.

can be summarized as per Figure 7.12. Unlike the sessile drop technique, the Wilhelmy balance method is an indirect force method. This method has distinct advantages over the conventional optical approach. First, this technique measures the contact angle by measuring weight and length. Hence, it is considered to be highly accurate and less subjective. Secondly, the force measured will always be an average value at any given depth of immersion. Although this feature does not go well when it comes to evaluating the heterogeneity, still it automatically gives a more accurate value of contact angle reflecting the entire sample's property. In addition, the graph produced by this method is quite helpful in the study of dynamic contact angles and contact angle hysteresis at different wetting speeds. Heterogeneity in a solid sample indicated by the smoothness of the curve. A repeated submersion circle can make the study of absorption or surface reorientation possible. Like every other method, this method also suffers from different drawbacks. The cross section of the solid sample must be uniform along the submersion direction. Another limitation lies for all those samples where the determination of wetted perimeter is a complex task. Samples like rods, plates, and fibers with known perimeters are the only ideal samples for this technique. Apart from regular geometries, the sample must have the identical composition and topography at all sides, which can be a herculean task, sometimes especially when anisotropic systems or films are investigated. The requirement of sufficient quantity of liquid in the method might cause a swelling in the solid sample or can cause an undesirable absorption of vapor by the sample.

7.6.4 APPLICATIONS OF CONTACT ANGLE MEASUREMENT AND WETTABILITY ANALYSIS IN COMPOSITES

Contact angle measurement and wettability analysis play a very important role in the field of natural fiber reinforced polymer composites as the matrix is hydrophobic and natural fibers are hydrophilic, so wettability analysis gives an idea of water absorption characteristics of the composite material. With the help of this analysis, we can calculate the surface energy of the composites before and after the surface treatment of natural fibers. Work of adhesion can also be calculated by this phenomena.

7.7   RECENT DEVELOPMENTS

After discussing all the traditional or conventional characterization techniques in detail, let's try to focus on some of the emerging trends in the field of natural fiber–reinforced polymer composites. New dimensions are ever demanding and their need for understanding the structural basis of performance of NFCs cannot be denied. Hence, there are significant developments marked in some of the discussed techniques in recent decades. Finite element modeling of natural fiber–reinforced polymer composites are one of them. Apart from that, advanced integral analysis is also used to understand the degradation behavior and calculation of the activation energy analysis of these composites (Sbirrazzuoli et al., 2009).

CONCLUSION

With the increasing use of natural fiber–reinforced polymer composites, the accurate characterization is very important in order to have a better application of these materials in diverse fields. The traditional and advanced characterization techniques and their fundamental theories have been discussed in this compilation. This will facilitate the commercialization of these novel products.

REFERENCES

Abilash, N., & Sivapragash, M. (2013). Environmental benefits of eco-friendly natural fiber reinforced polymeric composites materials. *International Journal of Application or Innovation in Engineering & Management, 2*(1), 2319–4847.
Akil, H. M., Cheng, L. W., Mohd Ishak, Z. A., Abu Bakar, A., & Abd Rahman, M. A. (2009). Water absorption study on pultruded jute fibre reinforced unsaturated polyester composites. *Composites Science and Technology, 69*(11–12), 1942–1948. 10.1016/j.compscitech.2009.04.014
Alix, S., Lebrun, L., Morvan, C., & Marais, S. (2011). Study of water behavior of chemically treated flax fibres-based composites: A way to approach the hydric interface. *Composites Science and Technology, 71*(6), 893–899. 10.1016/j.compscitech.2011.02.004
ASTM. (2017). ASTM D7264/D7264M-21. *Annual Book of ASTM Standards.*
ASTM. (2014). Astm D3039/D3039M. *Annual Book of ASTM Standards,* 1–13. 10.1520/D3039
ASTM International. (2019). ASTM D7337/D7337M-12.
Barton, J. M. (1986). Differential scanning calorimetry cure studies of tetra-N-glycidyldiaminodiphenylmethane epoxy resins. *Part 1 – Reaction with 4,4'–Diaminodiphenylsulphone. 18*(1), 37–43.

Beckermann, G. W., & Pickering, K. L. (2008). Engineering and evaluation of hemp fibre reinforced polypropylene composites: Fibre treatment and matrix modification. *Composites Part A: Applied Science and Manufacturing, 39*(6), 979–988. 10.1016/j.compositesa.2008.03.010

Célino, A., Fréour, S., Jacquemin, F., & Casari, P. (2013). Characterization and modeling of the moisture diffusion behavior of natural fibers. *Journal of Applied Polymer Science, 130*(1), 297–306. 10.1002/app.39148

Chen, H., Fei, B., Wang, G., & Cheng, H. (2012). Contact angles of single fibers measured in different temperature and related humidity 1 6. *Proceedings of the 55th International Convention of Society of Wood Science and Technology August 27–31.* Beijing, China.

Devnani, G. L., Mittal, V., & Sinha, S. (2018). Mathematical modelling of water absorption behavior of bagasse fiber reinforced epoxy composite material. *Materials Today: Proceedings, 5*(9), 16912–16918. 10.1016/j.matpr.2018.04.094

Dhakal, H. N., Zhang, Z. Y., & Richardson, M. O. W. (2007). Effect of water absorption on the mechanical properties of hemp fibre reinforced unsaturated polyester composites. *Composites Science and Technology, 67*(7–8), 1674–1683. 10.1016/j.compscitech.2006.06.019

Dos Santos, P. A., Giriolli, J. C., Amarasekera, J., & Moraes, G. (2008). Natural fibers plastic composites for automotive applications. *SPE Automotive and Composites Division - 8th Annual Automotive Composites Conference and Exhibition, ACCE 2008 - The Road to Lightweight Performance, 1,* 492–500.

El Hachem, Z., Célino, A., Challita, G., Moya, M. J., & Fréour, S. (2019). Hygroscopic multi-scale behavior of polypropylene matrix reinforced with flax fibers. *Industrial Crops and Products, 140*(July), 111634. 10.1016/j.indcrop.2019.111634

Faruk, O., Bledzki, A. K., Fink, H. P., & Sain, M. (2014). Progress report on natural fiber reinforced composites. *Macromolecular Materials and Engineering, 299*(1), 9–26. 10.1002/mame.201300008

Fraga, A. N., Frullloni, E., De La Osa, O., Kenny, J. M., & Vázquez, A. (2006). Relationship between water absorption and dielectric behavior of natural fibre composite materials. *Polymer Testing, 25*(2), 181–187. 10.1016/j.polymertesting.2005.11.002

Fuentes, C. A., Beckers, K., Pfeiffer, H., Tran, L. Q. N., & Dupont-Gillain, C. (2014). Colloids and surfaces A: Physicochemical and engineering aspects equilibrium contact angle measurements of natural fibers by an acoustic vibration technique. *Colloids and Surfaces A: Physicochemical and Engineering Aspects, 455,* 164–173. 10.1016/j.colsurfa.2014.04.054

Ghosh, R., Ramakrishna, A., Reena, G., Ravindra, A., & Verma, A. (2014). Water absorption kinetics and mechanical properties of ultrasonic treated banana fiber reinforced-vinyl ester composites. *Procedia Materials Science, 5,* 311–315. 10.1016/j.mspro.2014.07.272

Gupta, M. K. (2018). Water absorption and its effect on mechanical properties of sisal composite. *Journal of the Chinese Advanced Materials Society, 6*(4), 561–572. 10.1080/22243682.2018.1522600

Heidi, P., Bo, M., Roberts, J., & Kalle, N. (2011). *The Influence of Biocomposite Processing and Composition on Natural Fiber Length, Dispersion and Orientation. 1,* 190–198.

Ho, M., Wang, H., Lee, J., Ho, C., Lau, K., Leng, J., & Hui, D. (2012). Critical factors on manufacturing processes of natural fibre composites. *Composites Part B, 43*(8), 3549–3562. 10.1016/j.compositesb.2011.10.001

Holbery, J., & Houston, D. (2006). Natural-fiber-reinforced polymer composites in automotive applications. *Jom, 58*(11), 80–86. 10.1007/s11837-006-0234-2

Hosseinihashemi, S. K., Arwinfar, F., Najafi, A., Nemli, G., & Ayrilmis, N. (2016). Long-term water absorption behavior of thermoplastic composites produced with thermally treated wood. *Measurement: Journal of the International Measurement Confederation*, *86*, 202–208. 10.1016/j.measurement.2016.02.058

Hu, R. H., Sun, M. Young, & Lim, J. K. (2010). Moisture absorption, tensile strength and microstructure evolution of short jute fiber/polylactide composite in hygrothermal environment. *Materials and Design*, *31*(7), 3167–3173. 10.1016/j.matdes.2010.02.030

Jacob, M., Francis, B., & Thomas, S. (2006). Dynamical mechanical analysis of sisal/oil palm hybrid fiber-reinforced natural rubber composites.*Polymer Composites*, *27*(6), 671–680.10.1002/pc

Jawaid, M., & Khalil, H. P. S. A. (2011). Effect of layering pattern on the dynamic mechanical properties and thermal degradation of oil palm-jute fibers reinforced epoxy hybrid composite. *BioResources*, *6*, 2309–2322.

Jawaid, M., Khalil, H. P. S. A., Hassan, A., Dungani, R., & Hadiyane, A. (2015). Effect of jute fibre loading on tensile and dynamic mechanical properties of oil palm epoxy composites. *Composites Part B*, *45*(1), 619–624. 10.1016/j.compositesb.2012.04.068

Jiang, L., & Hinrichsen, G. (1999). Flax and cotton fiber reinforced biodegradable polyester amide composites. Die Angewandte Makromolekulare Chemie, *268*(4649), 13–17.

Joseph, P. V., Joseph, K., & Thomas, S. (1999). Effect of processing variables on the mechanical properties of sisal-fiber-reinforced polypropylene composites. *Composites Science and Technology*, *59*(11), 1625–1640.

Kim, J. K., & Mai, Y. W. (1998). *Engineered Interfaces in Fibre Reinforced Composites.* Elsevier.

Kim, H. J., & Seo, D. W. (2006). Effect of water absorption fatigue on mechanical properties of sisal textile-reinforced composites. *International Journal of Fatigue*, *28*(10), 1307–1314. 10.1016/j.ijfatigue.2006.02.018

Laminates, T. (2000). *Standard Test Method for Short-Beam Strength of Polymer Matrix Composite Materials and Their Laminate*, D 2344/D 2344M.

Lau, K. Tak, Hung, P. Yan, Zhu, M. H., & Hui, D. (2018). Properties of natural fibre composites for structural engineering applications. *Composites Part B: Engineering*, *136*(September 2017), 222–233. 10.1016/j.compositesb.2017.10.038

Lila, M. K., Komal, U. K., & Singh, I. (2019). Characterization techniques of reinforced polymer composites. In P. K. Bajpai (Ed.), *Reinforced Polymer Composites: Processing, Characterization and Post Life Cycle Assessment*, 119–145. Wiley. 10.1002/9783527820979.ch7

Madsen, B., & Lilholt, H. (2003). Physical and mechanical properties of unidirectional plant fibre composites-an evaluation of the influence of porosity. *Composites Science and Technology*, *63*(9), 1265–1272. 10.1016/S0266-3538(03)00097-6

Madsen, B., Thygesen, A., & Lilholt, H. (2009). Plant fibre composites - porosity and stiffness. *Composites Science and Technology*, *69*(7–8), 1057–1069. 10.1016/j.compscitech.2009.01.016

Malkapuram, R., Kumar, V., & Negi, Y. S. (2009). Recent development in natural fiber reinforced polypropylene composites. *Journal of Reinforced Plastics and Recent Development in Natural Fiber*, *28*(10), 1169–1189. 10.1177/0731684407087759

Masoodi, R., & Pillai, K. M. (2012). A study on moisture absorption and swelling in bio-based jute-epoxy composites, *Journal of Reinforced Plastics and Composites*, *31*(5), 285–294. 10.1177/0731684411434654

Mrad, H., Alix, S., Migneault, S., Koubaa, A., & Perré, P. (2018). Numerical and experimental assessment of water absorption of wood-polymer composites. *Measurement: Journal of the International Measurement Confederation*, *115*(August 2016), 197–203. 10.1016/j.measurement.2017.10.011

Muñoz, E., & García-Manrique, J. A. (2015). Water absorption behavior and its effect on the mechanical properties of flax fibre reinforced bioepoxy composites. *International Journal of Polymer Science*, *2015*, 16–18. 10.1155/2015/390275

Mustafa, A., Abdollah, M. F. Bin, Shuhimi, F. F., Ismail, N., Amiruddin, H., & Umehara, N. (2015). Selection and verification of kenaf fibres as an alternative friction material using Weighted Decision Matrix method. *Materials and Design*, *67*, 577–582. 10.101 6/j.matdes.2014.10.091

Nascimento, L. F. C., Monteiro, S. N., Louro, L. H. L., Luz, F. S. Da, Santos, J. L. Dos, Braga, F. D. O., & Marçal, R. L. S. B. (2018). Charpy impact test of epoxy composites reinforced with untreated and mercerized mallow fibers. *Journal of Materials Research and Technology*, *7*(4), 520–527. 10.1016/j.jmrt.2018.03.008

Oke, S. R., Omotoyinbo, J. A., & Alaneme, K. K. (2013). Water absorption characteristics of polyester matrix composites reinforced with oil palm ash and oil palm fibre. *Usak University Journal of Material Sciences*, *2*(2), 109–120. 10.12748/uujms.201324253

Princen, H. M. (1969). Capillary phenomena in assemblies of parallel cylinders: I. Capillary rise between two cylinders. *Journal of Colloid and Interface Science*, *30*(1), 69–75. https://doi.org/10.1016/0021-9797(69)90379-8

Raja, T., Anand, P., Karthik, M., & Sundaraj, M. (2017). Evaluation of mechanical properties of natural fibre reinforced composites – A review. *International Journal of Mechanical Engineering and Technology*, *8*(7), 915–924.

Romanzini, D., Lavoratti, A., Ornaghi, H. L., Amico, S. C., & Zattera, A. J. (2013). Influence of fiber content on the mechanical and dynamic mechanical properties of glass/ramie polymer composites. *Materials and Design*, *47*, 9–15. 10.1016/j.matdes.2012.12.029

Saba, N., Jawaid, M., Alothman, O. Y., & Paridah, M. T. (2016). A review on dynamic mechanical properties of natural fibre reinforced polymer composites. *Construction and Building Materials*, *106*, 149–159. 10.1016/j.conbuildmat.2015.12.075

Sanadi, A. R., Caulfield, D. F., & Jacobson, R. E. (1996). Agro-fiber thermoplastic composites. In R. M. Rowell & J. Rowell (Eds.), *Paper and Composites From Agro-Based Resources* (pp. 378–399). CRC Press.

Sarkar, J. K., & Wang, Q. (2020). Characterization of pyrolysis products and kinetic analysis of waste jute stick biomass. *Processes*, *8*(7), 837. https://doi.org/10.3390/pr8070837

Sbirrazzuoli, N., Vincent, L., Mija, A., & Guigo, N. (2009). Integral, differential and advanced isoconversional methods. Complex mechanisms and isothermal predicted conversion-time curves. *Chemometrics and Intelligent Laboratory Systems*, *96*(2), 219–226. 10.1016/j.chemolab.2009.02.002

Shah, D. U., Porter, D., & Vollrath, F. (2014). Can silk become an effective reinforcing fibre? A property comparison with flax and glass reinforced composites. *Composites Science and Technology*, *101*, 173–183. 10.1016/j.compscitech.2014.07.015

Sreekala, M. S., & Thomas, S. (2005). Dynamic mechanical properties of oil palm fiber/phenol formaldehyde and oil palm fiber/glass hybrid phenol formaldehyde composites*Polymer Composites*, *26*(3), 388–400. 10.1002/pc.20095

Summerscales, J., Dissanayake, N. P. J., Virk, A. S., & Hall, W. (2010). A review of bast fibres and their composites. Part 1 - Fibres as reinforcements. *Composites Part A: Applied Science and Manufacturing*, *41*(10), 1329–1335. 10.1016/j.compositesa.2010. 06.001

Thiruchitrambalam, M., Athijayamani, A., Sathiyamurthy, S., & Syed Abu Thaheer, A. (2010). A review on the natural fiber-reinforced polymer composites for the development of roselle fiber-reinforced polyester composite. *Journal of Natural Fibers*, *7*(4), 307–323. 10.1080/15440478.2010.529299

Velde, K. Van De, & Kiekens, P. (2000). Wettability and surface analysis of glass fibres. *Indian Journal of Fibre and Textile Research*, *25*(March), 8–13.

Venkatesan, K., & Bhaskar, G. B. (2020). Evaluation and Comparison of Mechanical Properties of Natural Fiber Abaca-sisal Composite. *Fibers and Polymers*, *21*(7), 1523–1534. 10.1007/s12221-020-9532-5

Venkateshappa, S. C., Jayadevappa, S. Y., Kumar, P., & Puttiah, W. (2011). Mechanical behavior of areca fiber reinforced epoxy composites. *Advances in Polymer Technology*, 1–12. 10.1002/adv

Wu, X., & Dzenis, Y. A. (2006). Droplet on a fiber: Geometrical shape and contact angle. *Acta Mechanica*, *185*(2006), 215–225. 10.1007/s00707-006-0349-0

Yuan, Y., & Lee, T. R. (2013). *Contact Angle and Wetting Properties*. Springer. 10.1007/978-3-642-34243-1

Zhang, Z., Wang, P., & Wu, J. (2012). Dynamic mechanical properties of EVA polymer-modified cement paste at early age. *Physics Procedia*, *25*, 305–310. 10.1016/j.phpro.2012.03.088

8 Thermoset Polymer Matrix–Based Natural Fiber Composites

Ayushi Kushwaha
Department of Chemical Engineering, Harcourt Butler
Technical University, Kanpur, U.P., India

INTRODUCTION

With the rapid development in the manufacturing industries, the necessity is to have materials with better properties like strength, lower cost, toughness, and low moisture absorption properties with great sustainability. Over time, natural fibers have become better replacement materials for reinforcement. Presently, the researchers have been inclined towards the aspect of green technology, so the eco-friendly fibers exhibit good mechanical and thermal properties, are cheaper, and easily accessible are extensively used as reinforcement materials. There are three categories of matrices for composite materials: metal matrix composite, ceramic matrix composite, and polymer matrix composite. The polymer matrix, which is used to prepare composites, is classified into elastomer, thermoplastic, and thermosetting materials. Polymers matrix composites are composed of fibers held together by a polymer matrix. They are predominantly used due to low-price raw materials with good binding properties. The purpose of the matrix is to hold the fibers together in an order and protect it from the environment as well. A polymer matrix composite helps to transfer the loads among fibers. The thermosetting matrix is more impregnable than thermoplastics due to their cross-linked density and network structure. They exhibit higher stiffness, resistance to thermal degradation, durability, surface texture, etc. The natural fiber–reinforced thermosetting matrix is now used in various applications such as aerospace, automobiles, marine, construction materials, and electronic industries due to the improvement in their surface as well as mechanical properties. The advantage of the polymer over conventional reinforcement material is mainly ease of processing, lower cost, lightweight, recyclability, sustainability, and mechanical properties.

8.1 NATURAL FIBERS, THERMOSETS, AND TYPES

Natural fibers are thread-like formations which are inherited from the plants and animals. They contain cellulose, lignin, protein, wax, and pectin. Pectin and wax are in low contents. These are quite longer than wider. The main resources to inherit

DOI: 10.1201/9781003201724-8

natural fibers are plant parts like bast, leaf, stem, and fruit such as hemp, kenaf, abaca, sisal, jute, banana, palm fiber, bagasse, flax, areca, cotton, etc. and animal fur like wool, silk, etc. The resources for natural fiber are much more than synthetic fiber. The resources for natural fiber are unlimited and easily available. Now, the scope of using natural fiber is increasing day by day. The easier the availability, the cheaper the cost (Thyavihalli Girijappa et al., 2019). The main advantages of selecting natural fiber over synthetic fiber is its cheaper price, availability, high specific properties like moisture sensitivity, thermal sensitivity, and recyclability are better than synthetic fiber. Natural fibers are eco-friendly in nature, which is an important step for green technology. The various industrial applications where natural fibers that can be used with additives are ceramics, automobiles, aerospace, construction sites, packaging materials, furniture, roofing sheets, etc. There are diverse applications where it has further scope. The nature of natural fiber is hydrophilic due to the presence of lignocellulosic content. Cellulose is semi-crystalline in nature, which is responsible for the hydrophilic nature of fiber. The moisture absorption competency is very high in them. The hydrophilic nature of natural fibers can be understood through diffusion theory. The increasing moisture content results in low mechanical properties like hardness, resilience, toughness, and stiffness. The hydrophilic nature of natural fiber may result in imperfect adhesion between a natural fiber and thermosetting matrix (Wang et al., 2006). Thermosets are the class of materials that possess low viscosity and low molecular weight. Thermosets are durable and substantial due to a three-dimensional cross-linking network. These materials have good mechanical properties like high mechanical strength, hardness, and thermal resistivity. Thermosets experience permanent deformation and thus these cannot be reshaped or remelted back once these are cured because curing results in less mobility of molecules of thermosets and increases viscosity, so resin forms a solid (Sanjay et al., 2018). Curing is a chemical process that produces toughening of composites. It is a thermally driven process, thermosets generally have a long processing time. A thermoset polymer composite matrix mainly includes epoxy resins, polyesters, vinyl esters, polyimides, polyurethanes, cyanate esters, and phenolics, etc. Mainly epoxies are used as the dominant resin for lower temperature application (up to 135°C). Vinyl esters and polyester can also be used for low temperatures, but these are quite lower in properties; thus, they are not high-performance materials. Polyimides are used for higher temperature applications (up to 315°C). Cyanate esters were designed to replace epoxies due to the advantage of lower moisture absorption, but the cost is quite higher. Phenolics can also be used for higher temperatures but they yield char due to the presence of carbon-carbon bonding. Thermosets ensure great rigidity, high stability, good resistance to creep, and the highest ability. The thermoset matrix is made by mixing a resin and a hardener, which undergo curing at room temperature (Figure 8.1).

8.1.1 PHENOLIC RESIN

Phenolic resin can be considered as the first polymer product produced from simple compounds with lower molecular weight. The phenolic resin was invented by Dr. Leo Baekeland in 1907. These are produced by the reaction of phenols with aldehydes. They are characterized by great properties like higher temperature

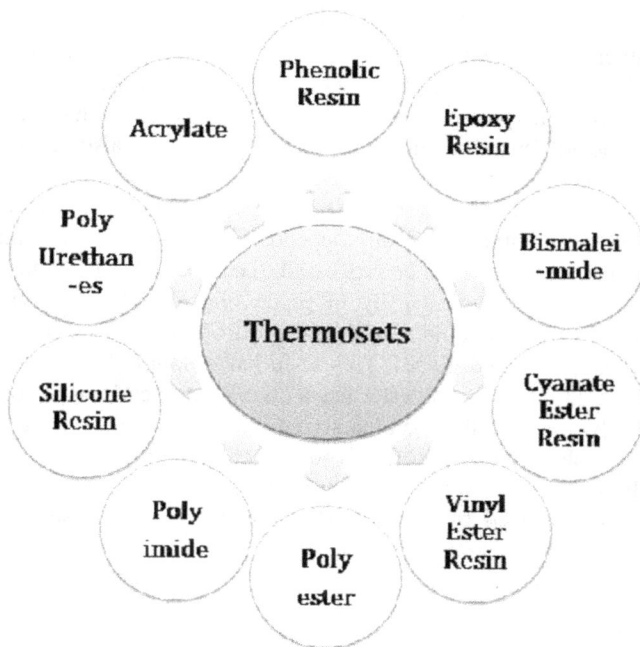

FIGURE 8.1 Types of thermosetting polymers.

stability up to 350°C, less toxic, greater friction, higher chemical stability, excellent thermal resistance, and high-performance polymers. The phenolic resins are widely used in a variety of applications like binders, adhesive, matrix, and surface coating. Phenolic resin is utilized for inducting and mass transit (Allen & Ishida, 2001). Phenolic resin is prepared by the condensation reaction of formaldehyde with phenol which gives water as a by-product. Phenolic resins are classified as resoles and novolac. Resoles are formed when the excess of formaldehyde is reacted with phenol in the presence of a base catalyst, the product is a low molecular weight liquid, whereas novolac is formed when the excess of phenol is reacted with formaldehyde in the presence of an acidic catalyst; the product is solid Novolac. Resoles are used for phenolic pre-impregnated material like epoxies. Phenolic resins are one of the better options for producing high thermal resistance composite. The phenolic is pyrolyzed to develop a carbon matrix, but this process develops a porous structure due to the moisture ability of phenolic resin. Some examples of phenolic resins are bakelite, novolac, and catalin.

8.1.2 POLYESTERS

Polyester was discovered in the DuPont lab in the 1930s but W. H. Caruthers set it to work on newly invented nylon in 1941. The first commercial polyester terylene was made in 1941. They are mainly used in commercial applications, but have certain limitations for use in the high-performance composite. The curing of polyester is done by an additional reaction in which unsaturated double bonds appear wherever cross-linking occurs. Polyester mainly contains three parts: a polyester chain; a cross-linking agent such as styrene that helps in lowering the viscosity to improve the process; and an initiator, mainly peroxides such as benzoyl peroxide. Styrene not only works as a cross-linking agent but also helps to lower the viscosity to attain better processability. Certain materials like chloro styrene, methyl methacrylate, and diallyl phthalate are used as pre-impregnation. The characteristic of formed polyester composites mainly depends on curing or a cross-linking agent used for impregnation. These can be processed to cure at either room temperature or high temperature, showing versatility in processing (Kandelbauer et al., 2014). Polyester resin is a viscous liquid that is prepared from the condensation reaction of unsaturated dibasic acid and glycol. This resin has a double bond between carbon atoms. The basic structure of polyester has a double bond carbon-carbon group and ester group. Polyester usually contains styrene as a monomer, which reduces the viscosity of the solution and also reacts with a chain to form a cross-linked rigid structure. Polyester has a lower temperature ability and less mechanical properties but they are characterized by lower cost, great processing versatility, and high performance.

8.1.3 EPOXY RESIN

Epoxy resin was made by Dr. Sylvan Greenie in the United States and Dr. Pierre Castan in Switzerland in the 1930s. Epoxy resin is a polymer that contains more than one epoxide group. They are the most utilized matrix material to produce adhesives and high-performance composites. Days passes, it retains with great strength, good adhesion property, low shrinking, and versatility (Massingill & Bauer, 2000). Commercial matrix of epoxy is one epoxy and a single curing agent. Matrices either contain major or minor epoxies. The minor ones are used to improve thermal properties, reduce moisture absorption, and improve toughness. The major ones used in the aerospace sector:

- DGEBA (diglycidyl ether of bisphenol A)
- TGMDA (tetraglycidyl methylene dianiline)

DGEBA is difunctional epoxy that can either be solid or liquid which can be used for adhesion or pultrusion, while TGMDA is a higher functionality resin based on aromatic amine and is used for preparing composite matrices. Due to their higher functionality, they provide a more cross-linked structure that demonstrates higher temperature resistance, rigidity, and high strength (Gibson, 2017). The site of cross-linking of epoxy resin is the oxirane or epoxy ring. This epoxy group is present as glycidyl amine or aliphatic ring. The curing of these resins depends on cross-linking of the curing agent and opening of the oxirane ring. Glycidyl amines are used for composite matrices whereas aliphatics are for electrical appliances. Aliphatic amines are highly reactive, which can be cured at room temperature and shows rigidity and toughness, whereas aromatic amines require higher temperature for curing and show higher strength but less toughness. Epoxy resin is viscous and it is characterized by its good aggressive property with reinforcement, good moisture resistance, outstanding mechanical properties, good fatigue resistance, and strong durability at different temperatures. Epoxy resin is transparent in color and is mixed with suitable hardener being darker in color to polymerize fibers to get an irreversible composite.

8.1.4 Polyimide

Polyimide was discovered by Bogert and Rensitshaw in 1908. However, the polyimide used for commercial purposes was made in 1950. Polyimide is a high-temperature engineered polymer that is used for temperatures up to 315°C. The curing system can either be condensation or addition. In addition to the curing system, polyimides can be made with solvents like dimethylformamide (DMF) and dimethylacetamide (DMAC). which are removed during the curing process. The condensation curing system gives water as a by-product, which results in serious volatile management issues during curing. To remove the lowering in mechanical properties of the matrix, the volatiles should be removed before resin gelation (McKeen, 2021). They are difficult to process compared to epoxy because they need high processing temperature and higher pressure. The volatile impedes the potential while processing polyimide. They have great resistance power and excellent flexural strength. PMR-15 is one of the best addition-cured polyimides, which means polymeric monomer that contains three monomers and has a molecular weight of 1,500. Another promising resin PETI-5 was developed after screening the materials. PETI stands for phenylethynyl terminated imide. NMP, which stands for N-methyl pyrrolidone, is a major retainer during curing to manage volatiles and voids. They are characterized by an exceptional combination of mechanical toughness, thermal stability, outstanding dielectric property, and chemical resistance.

8.1.5 Vinyl ester Resin

Vinyl ester resin was first introduced by shell chemicals in the 1960s with the name of Epocrgl and another brand named Derakane was introduced by Dow Chemical in 1966.

Curing of vinyl ester resin can be done at elevated or room temperature. The curing condition affects mechanical properties. Vinyl ester resins are assortments of methacrylate epoxy compound with styrene. By altering the content of styrene and molecular weight of ether of the resin, viscosity can be controlled. The curing of this is done by free radical copolymerization where styrene acts as diluent and vinyl ester acts as cross-linking agent. The primary curing initiates with induction. The availability of inhibitors result in micro-gel formation. These micro-gels delimited the area of cross-linking density disseminates with unreactive monomers. These micro-gels when coagulated into clusters result in the cured resin (Ziaee & Palmese, 1999). Vinyl ester resin is characterized by a specific type of denatured epoxy resin that is obtained by reacting bisphenol with methacrylic acid. Vinyl ester resin possesses great properties like great curability, good corrosion resistance, fracture toughness, good strength, and thermal resistance. Vinyl ester resin has mainly been used for fiberglass-reinforced composites, commercial, and military applications.

8.1.6 POLYURETHANE RESIN

Polyurethane was discovered by professor Dr. Otto Bayer in 1937. He is known as the father of the polyurethanes industry. The polyurethane was synthesized from 1,6-hexamethylene diisocyanate with di- or tri-isocyanate and polyester polyol by the polymer addition method.

Urethanes can be produced by the reaction of alcohols with isocyanates. Then isocyanates react with water to form carbamide acid, which readily decomposes into carbon dioxide and amine as it is unstable. The carbon dioxide formed after decomposition will help in the foaming of polyurethane and amine reacts with another isocyanate to form a substituted urea (Atiqah et al., 2017). They can be cured at higher temperatures. Aliphatic amines are more reactive than alcohols and aromatic amines. Hence, we can produce a large variety of polyurethanes by reaction of aliphatic amines with diisocyanate. They show two mechanisms for adhesion. Chemical bonding takes place due to the presence of active hydrogen and isocyanate that react with the surface to form a covalent bond between fiber and polyurethane, whereas physical bonding takes place due to close contact of the surface of the fiber and adhesive film (Engels, 2012). Polyurethanes are characterized by some specific properties like high resilience, flexible segments, rigidity, great resistance to moisture, and durability. Polyurethanes are now widely used in the biomedical sector, automotive sector, and shoe manufacturing.

8.1.7 BISMALEIMIDE RESIN

Bismaleimide resin was made as a bridge to cover the temperature gap between polyimides and epoxy resin. The curing is done by an additional reaction at a higher temperature. They set free post-curing to accomplish polymerization reaction to obtain high thermal properties. They can undergo filament winding, autoclave curing, as well as resin transfer molding process. Their flip and mantle are better due to the presence of a liquid component of reactant. Bismaleimides are formed in two steps. The first step is the reaction of di-amines with maleic anhydride to form bismaleic acid. This is exothermic and fast and the second step is done by imidization. The most prevailing resin with base monomer for adhesives is 4,4'- bismaleimide diphenylmethane (Bao et al., 2001). Curing of bismaleimide can be done by either homopolymerization, which results in brittle fabrication or by copolymerization, which results in flexible and extended fabrication when reacted with diamine and toughening when reacted with olefins. Bismaleimide resin can be prepared by resin transfer molding and autoclave curing process. They are characterized by their properties like high-temperature polymerization, flexibility, toughening, and good thermal-oxidative behavior. A copolymerization reaction with discyanobidiscyan a resulted in BT resin, where B is bismaleimide and T is triazine. BCB (benzocyclobutene) exhibits toughness and good thermal behavior. Earlier, they were considered as less processing and low toughness, and with the modification they show better flip and good toughness.

8.1.8 Silicone Resin

Silicone resins are silicon material that has good thermal resistance up to 350°C. Silicones are prepared by the Muller-Rochow method using methyl chlorosilanes at a temperature range of 250°C–300°C and 2 to 5 bar of pressure. To improve the activity, the copper catalyst can be used along with metals such as zinc, cadmium, antimony, tin, and aluminium. The homogeneous mixture of copper and silicon is entered into a fluidized bed reactor, which is fluidized using gaseous methyl chloride. In the separation unit, reactants are set apart from solid, and primitive liquid silane is formed on cooling. The heat is released during the reaction; thus, it is an exothermic reaction that needs accurate control. The main product formed is dimethyl dichlorosilane, along with some more products at different yields are methyl dichlorosilane, methyl trichlorosilane, and trimethylchlorosilane (Fink, 2013). They contain oxygen, silicon, and organic groups. They possess a branch of alternating silicon and oxygen atoms in a three-dimensional network that can be dissolved in an organic solvent. Silicone resins are characterized by the properties like great resistance to tearing, hydrophobicity, and hardness (Z. Liu et al., 2020). They are used in pressure-sensitive adhesives and as water repellent due to their hydrophobicity nature. Silicones are non-toxic in nature; hence, they are used in the medical sector and pharmaceutical industry. They have good resistance to corrosion and heat resistance and thus are used for surface coatings. They can also be used as adhesives due to their excellent flip and high thermal resistance. They exhibit low surface tension, which makes them anti-foaming agents and also an essential constituent in the polymerization of PVC.

8.1.9 Cyanate Ester Resin

Cyanate ester is a great substitute for bismaleimides and epoxies. They are quite expensive due to less usage. They are mainly used in such applications which require low dielectric dissipation. Their adhesive nature is inferior to bismaleimides or epoxies and they demonstrate lower moisture absorption than epoxies and bismaleimides and are less thermal resistant. In the curing process, the impregnation is less reliable due to moisture pickup, which produces carbon dioxide. They are the derivatives of bisphenol having a ring that forms a cyanate functional group. A three-dimensional cross-link network of oxygen-linked triazine ring and bisphenol is formed through an addition reaction during the curing process. Benzene and triazine rings exhibit higher content of aromaticity to provide high glass temperature. The more the cross-linking density, the more will be the toughness. Bonding of single oxygen atoms helps in the reduction of local stress on the load. They can be cured at higher temperatures or at lower temperatures with the help of a suitable catalyst like transition metal complexes of cobalt, copper, manganese, and zinc (Nair et al., n.d.). Due to the absence of hydrogen bonding and dipoles balanced, the results are excellent electrical properties like dissipation factor and low dielectric constant. Water/moisture resistance property of cyanate ester resin is due to lack of polarity on the symmetry of triazine ring makes it more water resistant than bismaleimides and epoxies.

8.1.10 ACRYLATES

Acrylates were first innovated by DuPont during the 1970s through chloro-sulfonated polyethylene substrate; later in the mid-1990s, improved acrylates were made by changing the substrate as well as curing technique (Engels, 2012). Structural acrylates are generally two parts available in different mixture ratios. These two parts are adhesive and activator. The curing is done by a free radical addition polymerization reaction using redox-active agents. This redox-active agent is incorporated with a reducing agent and an oxidizing agent. The use of metal catalysts depends on the process. They are versatile and showcase good adhesion with different substrates. Acrylate composition is methacrylate monomer, a toughener, a resin or cross-linking agent, a reducing agent, and an oxidizing agent that acts as a source of filler and free radical. Methacrylate monomer is selected from alkyl, cycloalkyl, or alkoxy alkyl groups present in methacrylic acid. The monomer derived from the alkyl chain provides significant volatility, lower cost, and good adhesion surface but it exhibits a peculiar odor and flammability. Some acidic monomers include methacrylic and acrylic acids, monoester of phosphoric acid, semi-ester of maleic and succinic acid required for adhesion. For the fabrication of a hybrid curing system, an acidic monomer is used along with glycidyl epoxy resin to provide better thermal property. Some second-generation acrylates, such as partially reduced pyridine, and PDHP (1,2-dihydropyridine) are also used as reducing agents. Third-generation acrylates are based on organoborane chemistry, which consists of boro hydrides and trialkyl borane amine salts. Acrylates are used in the automobile sector, marine industry, and metal bonding applications due to their characteristics of hardness, tensile strength, and tough bonding of different substrates and toughness.

8.2 NATURAL FIBER–BASED THERMOSET COMPOSITES

The primary object to use natural fiber as reinforced material in a thermoset composite is to improve the elasticity and strength. The natural fibers that are reinforcement material need to be modified in various steps for their adaptability with composites. These fibers are either in short or in long form (Shekar & Ramachandra, 2018). During extraction, the fibers are quite longer and they can be changed into a desired size. The natural fibers are mainly inherited from plants through various steps such as decortication, scraping, and crushing. The extraction method allows the changes of mechanical properties in natural fibers. The primary contents of natural fiber are hemicellulose, cellulose, lignin, wax, and pectin. The percentage of these components differs from one fiber to another due to their species. Cellulose is the most rich component found in natural fiber. It is a linearly structured polymer made by many glucose molecules that are recognized as the monomeric units. The molecular formula of cellulose is $(C_6H_{10}O_5)_n$. Hemicellulose is attached to various kinds of sugars such as mannose, xylose, glucose, etc., unlike cellulose, hemicellulose is a non-crystalline in nature and results in hydrophilic nature of fibers that is highly soluble in an alkaline solution (Komuraiah et al., 2014). Lignin is the second richest component that is a phenolic three-dimensional polymer. It provides rigidity and is amorphous in nature. It provides rough and stiff nature to natural fibers. It can easily dissolve in an alkaline solution. Thus, different components exhibit different functions, although they have similar structures. Researchers have classified the natural fibers in various types according to their inherited parts.

8.3 CHEMICAL TREATMENTS OF NATURAL FIBER

Due to the hydrophilic nature of natural fibers, whereas thermoset is hydrophobic in nature, the adhesion between natural fibers and thermoset matrix is incompatible and that results in imperfect composite materials (Li et al., 2007). To reduce this nature, it is essential to follow various chemical modifications of the surface of natural fiber. The various procedures to carry chemical treatments are in the figure (Figure 8.2).

The chemical treatment results in removal of impurities, grafting of coupling agents to establish good adhesion between the matrix and fiber and modify mechanical properties of composites, their moisture, resistance, and strength.

8.4 ADVANTAGES AND DISADVANTAGES OF NATURAL FIBERS

8.4.1 ADVANTAGES

Natural fibers are in abundance so they are easily accessible at a cheaper cost. The ultimate advantage of natural fiber is its recyclability. Compared to synthetic fibers, natural fibers are non-toxic and non-corrosive in nature and have no adverse effect

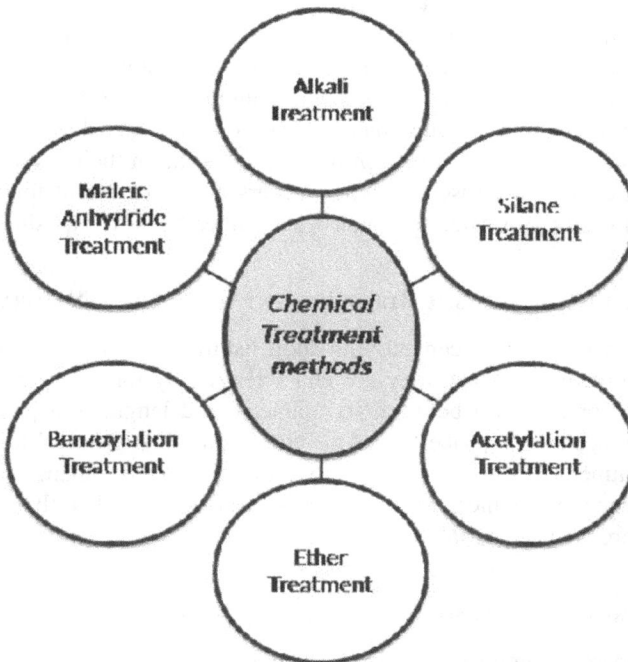

FIGURE 8.2 Chemical treatment methods of natural fibers.

on health. Natural fibers incorporated with a thermoset matrix produce composites with better mechanical properties, such as tensile strength, resistance to fracture, stiffness, etc.

8.4.2 DISADVANTAGES

The main drawback of natural fiber is its low thermal resistance, hydrophilic nature, and non-uniformity in properties. The incompatibility between the matrix and fiber can be modified by physical and chemical treatment of natural fiber to increase adhesion between the matrix and fiber.

8.5 THERMOSET COMPOSITES

Thermoset composites can be fabricated using unlike kinds of reinforcement, mainly natural fibers. The short and long natural fibers that are used as reinforcement depend upon the properties required to meet the target applications. There are many methods to process the thermosetting composite reinforced with natural fiber (Campbell, 2004).

Different thermoset matrices are used for particular natural fibers for various applications.

8.5.1 THERMOSET COMPOSITE REINFORCEMENT WITH LONG NATURAL FIBERS

Thermoset composite reinforced with long natural fibers are of two types: woven and non-woven chopped string. The woven fiber prerequisite is a specific preference, whereas the non-woven chopped fiber has a random preference. They are made using different hands and automatic techniques for weaving include backstrap, tablet weaving, and looming machine. Thermoset composites reinforced with woven fibers are stiffer in the direction of fiber but weak in their reverse direction, yet thermoset composites based on woven fibers are more useful for various applications and easily fabricated than non-woven fiber-based composite.

8.5.2 THERMOSET COMPOSITE REINFORCEMENT WITH SHORT NATURAL FIBERS

Thermoset composite reinforcement with short natural fibers are utilized when the load is omnidirectional. Thus, they are characterized by random preference. The length of fiber or the ratio between its diameter and length is a parameter that influences the mechanical properties of the composite. The chemical treatment like the alkali treatment of fiber reduces its hydrophilicity and enhances mechanical properties. Hence, mechanical properties are more effective when the short fiber is used as reinforcement material.

8.5.3 THERMOSET COMPOSITE PREPARATION TECHNIQUES

When we prepare a composite matrix, first select a suitable thermoset matrix then natural fiber reinforcement and the key step is the selection of processing because it has an impact on the final product. There are diverse techniques to prepare thermoset composite such as hand layup, resin transfer molding, and autoclave molding techniques (Campbell, 2004) (Figure 8.3).

8.5.3.1 Hand Layup Technique

The hand layup technique is the simplest and oldest open molding method of fabrication of the thermoset composite. The process contains these steps:

- Coat the mold with an anti-adhesive agent to avert the sticking of molded part to the surface.
- Form the primer layer of the piece of work by gel coating.
- Apply a layer of fiber-reinforced matrix.

It is suitable for a layover of pre-impregnated layers of a composite of thickness in the range of 0.125 to 0.30 mm.

Although it is a time-consuming process and quality completely depends upon the way of processing, this technique is widely used to produce the large and low-cost tool. The curing process includes co-operation between vacuum and temperature to meet two main purposes: first is to acquire the correct viscosity level of the thermoset resin for its consistent distribution via laminate and second is to remove entrapped air gaps. This technique provides the fabrication of reinforced material with optimal characteristics.

8.5.3.2 Autoclave Molding Technique

Autoclave molding is an advanced composite preparation technique that is an open molding process.

The process is done in the following steps:

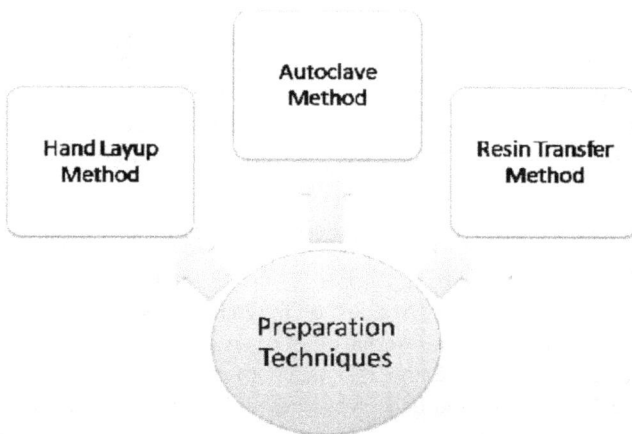

FIGURE 8.3   Thermoset composite preparation techniques.

- In this technique, reinforced material is stacked in a two-sided mold set. The lower side is rigid mold, whereas the upper side is made of a flexible membrane.
- The whole assembly is vacuum bagged to remove the entrapped air.
- After removal of air, the assembly is transferred to an autoclave where curing of the molded part is done by applying heat, vacuum, and pressure of inert gas.
- After uniform distribution of matrix, it is cooled to a definite rate, and the composite part is taken out from the mold.
- The mold surface must be coated with anti-adhesive gel to avoid sticking to the matrix.

This technique is mainly used in the aerospace industry to fabricate high-strength fiber composite products. This process is quite expensive but it produces high-quality products.

8.5.3.3   Resin Transfer Molding Technique

The resin transfer molding technique is in intermediate volume and the closed molding process is used for fabrication of composites for aircraft and automotive components mainly. This process is quite popular as it is cheaper and has a high capacity to produce miniature components to a large component in good numbers.

The process is done in the following steps:

- The mold is coated with anti adhesive gel.
- The core material is placed in the mold and then the mold is closed.
- For curing the reason is introduced in the mold under pressure using meter injection in a mold cavity.
- The cured reinforced material is done separately and then it is placed in the mold to attain shape.

Resin transfer molding techniques can be done at room temperature but a heated mold can give better results. This molding technique can make three-dimensional complex parts with great tolerance and the composites made from this technique possess excellent mechanical properties and good surface finish.

8.5.4   Toughening of Thermosets and Approaches for Toughening

The toughening approaches are developed to understand the relationship between fracture behavior and the microstructure of toughened epoxy resins. A toughened thermoset mainly contains a thermoplastic domain which is dispersed from end to end of the matrix to enhance the resistance to crack. Brittle and single-phase thermosets are usually unmodified, whereas multiphase systems are tough and thermosets. When thermoplastic domains are dispersed distinctly from end to end in the thermoset matrix, toughness can be enhanced. In the toughening process, it is suitable to minimize the mechanical and thermal properties of the matrix material (Brostow & Singh, 2004). The toughness delimitation of the thermoset matrix is the result of its highly cross-linked and rigid structure that is formed during curing. These rigid structures have several advantages and drawbacks. The main advantage of the structure is its high-temperature efficiency and capability to stabilize reinforced fiber during compression. The main disadvantage is less reliability to delamination during impact (Mullins et al., 2018). Over the past years, major efforts have been put into developing such resin systems that are tougher and less prone to damage due to impact. During the mid-1980s, two major matrix components emerged:

- Damage-resistant thermoplastic composite
- Toughened thermosets.

Despite the fundamental differences in the chemical structure, the thermosets show almost similar properties. Over the years, their resistance to impact damage has been improved, which results in the great capability to carry load after set to impact. As compared to a stiff thermoset, the above matrix component shows less resin modulus, which results in less compression strength. The stiffness of thermoset is more complex than homogeneous metallic material due to their non-homogeneous nature where the variation in properties is multi-directional. In composites, loading in-plane is restricted basically by reinforced fiber, whereas loading out-of-plane is restricted by attributes of the resin matrix. Hence, structures of the composite are designed purposely so that they are basically in-plane and the path of the load is stiff. Out-of-plane loadings are developed during the process of in-plane compression as they are convinced by various design features. Normal stresses and interlaminar shear stress can be developed at those locations even when exposed to normal in-plane loading. Some indirect loads like air pressure can act with out-of-plane loads. When an out-of-plane load turns out to be large, delaminations are formed, but the design criteria are specified to resist small delaminations. Some design details that result in out-of-plane loading are:

- Free edge
- Notch
- Ply drop
- Bonded joint
- Bolted joint

Cross-linking is the bonding between polymer chains that provides thermal resistance, rigidity, and strength to the cured polymer. The lesser the length of the polymer chain and the more the density of cross-linking, the more will be the stiffness and thermal resistance. The stiffness can be brought using a stiffener in the main polymer chain. However, rigidity causes brittleness and a poor impact on properties. The structures having higher cross-linking show good thermal stability. Since the bonds formed in cross-linking are covalent in nature, they hold a major portion of the strength when the temperature increases. Thus, higher cross-linking polymers exhibit moderate to higher strength, excellent temperature resistance, and stiffness. However, rigidity causes brittleness and is more prone to impact damage.

The molecular structure of thermosets decides its processing and its consequent properties. Some properties that depend upon molecular structure are moisture/water absorption, toughness, strength, elongation, and modulus. The performance of thermosets can be altered by changing their molecular structure. The molecular structure can be controlled by the main polymer chain (backbone structure) and the number of cross-linking (network structure). The type of curing agent used in the curing process of particular thermoset influences the network structure. To improve the toughness of thermosets, resin formulators are widely used in a significant amount with polymers to formulate a new molecular structure (Campbell, 2004). To contribute to higher toughness to the thermoset polymers, different approaches have been developed that are used to enhance toughness. Some of the toughness approaches are (Figure 8.4):

• Network alteration
• Thermoplastic elastomer toughening
• Rubber elastomer toughening
• Inter layering

8.5.5 Network Alteration

The higher cross-linking density causes brittleness in thermoset polymers. One process of toughening can be lowering of cross-linking density. Decreasing the cross-linking density can also result in the reduction of desirable properties like resin modulus. There are two popular methods to reduce the cross-linking density. The first method is to change the main backbone chain by reducing the

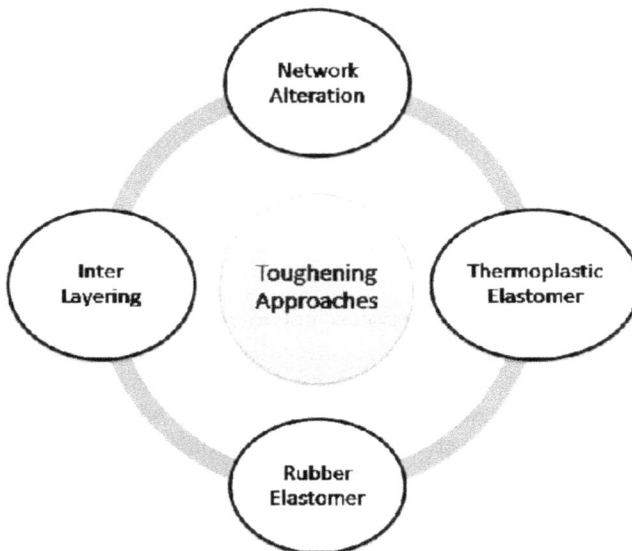

FIGURE 8.4 Toughening approaches.

cross-linking density with a long-chain monomer. The decreasing glass transition temperature can be set off with the help of fabricating a long-chain monomer that has rigidity. The reduced mobility of the polymeric chain will remunerate the loss in transition temperature. The second method is to reduce the functionality of the monomer. Thermosets that exhibit high cross-linking for reactive end groups show functionality of four end groups that react and form the cross-link. If the part of a polymer has a monomer of functionality two, then during the curing there will be fewer active sites for cross-linking, which will result in improvement in toughness as the cross-linking density will reduce but the thermal resistance will be affected due to lower transition temperature of the monomer of functionality.

8.5.6 RUBBER ELASTOMER SECOND-PHASE TOUGHENING

When the cracking occurs in a brittle solid, it needs minimal energy to distribute. In a fiber-reinforced polymer matrix composite, reinforced fiber will forbid crack growth. If the crack is between the plies (interlaminar), the fibers will not contribute to the prevention of crack growth. To reduce the crack growth, a second-phase elastomer can be used. Distinct rubber particles assist to weaken the crack growth by upholding higher plastic flow at the crack site. The crucial factor that determines the micro-deformation process that curbs toughening is the size of the elastomer domain. Rubber particles that initiate shear yielding have a small domain diameter between $100–1,000°A$, whereas for matrix crazing they have large domain diameters $10,000–20,000°A$. However, matrix crazing does not occur in a more cross-linked system because of low tensile elongation. Small domains assist to increase shear deformation process. Toughening can be doubled if cross-linked density allows utilizing a large domain of elastomers resulting in both shear and crazing. In an epoxy-based matrix composite, reactive liquid polymers like carboxyl terminated butadiene acrylonitrile rubber are frequently used to impart the desirable compatibility characteristics. The bonding between continuous resin phase and elastomer domain is essential. The poor bonding can result in debonding of elastomer from resin while cooling, which can form voids. Elastomers should have fine rubbery qualities to prevent crack growth in the matrix, if the elastomer is rich in domain and the transition temperature is lower than $-100°F$, Then the domain will still perform as an elastomer, although the crack growth occurs through the matrix. The additional specification is that the elastomer should be thermally oxidatively stable. If thermally oxidatively unstable rubber is used, it can degrade on oxidation, which would cause brittleness of the elastomer domain. Rubber elastomer second-phase toughening is efficient enough if the resin has high cross-linking. When the cross-linking density is reduced, the advantages of second-phase toughening increase rapidly.

8.5.7 THERMOPLASTIC ELASTOMERIC TOUGHENING

Some necessarily available thermoset composites depend on thermoplastic toughening. In cured composites, thermoplastic toughening indicates four apparent morphologies:

- Single-phase (homogeneous)
- Co continuous (thermoset and thermoplastic are continuous)
- Particulate (thermoplastic particle suspended in thermoset matrix)
- Phase inverted (thermoset is discontinuous and thermoplastic is continuous).

The outcome of continuous morphology is the toughness refinement. In this morphology, both thermosets and thermoplastics are in the continuous phase. The thermoplastic helps to enhance the toughness, whereas thermoset cross-linking assists to preserve high transition temperature and performance. Thermoplastic is selected on the basis of its thermal stability, heat resistance, and chemical compatibility. The viscosity of resin should be controlled to keep it low.

8.5.8 INTER-LAYERING

An engineered procedure to acquire toughness can be attained by assimilating a tough ductile resin between the distinct impregnated sheets. This layer is 0.01 inches or less and it should be impartially separated during the curing operation. The tough interlayer exhibits greater strain-to-failure, which assists interlaminar shear force that can influence delaminations. Recently, distinct toughening particles have been adjoined to the surface in the course of the pre-impregnation process. These particles are quite larger than reinforced fiber and persist on interlayers throughout pre-impregnation and curing. This method can be employed with either a toughened or brittle resin system. A toughened resin system is utilized in entering layering furnishes good impact resistance as compared to a brittle resin system.

8.5.9 TOUGHENING MECHANISMS

There are distinct toughening techniques that depend on the characteristics of the toughening agent, stress condition, material, and the main polymer, which are operated in the toughening system (J (Daniel) Liu et al., 2010) (Table 8.1).

The toughening phase maintains the mechanical and physical properties of the matrix. The toughening agent is selected on the basis of several applications:

- The type of elastic phase
- Interface
- The dispersion level
- The phase morphology
- The size of particles

The matrix has an essential role in toughening. The matrix is selected on the basis of the following properties:

- Yield stress
- Toughening characteristic
- The cross-linking density

TABLE 8.1

Toughening Techniques in Various Polymers with Their Effects

S.No.	Toughening Mechanisms	Toughening Effects
1.	Crazing	Modification in fracture toughness up to several layers.
2.	Croiding	Modification in fracture toughness up to the order of magnitude.
3.	Shear banding/yielding	Improvement in toughness up to the order of magnitude.
4.	Crack bridging	Increment in modification in fracture toughness.
5.	Crack deflection	Modifications in fracture toughness up to twice.
6.	Crack bifurcation	Modifications in fracture toughness occurs fractionally.
7.	Crack pinning	Improvement in fracture toughness occurs fractionally.
8.	Segmental crack growth	Modifications in fracture toughness occurs fractionally.

- Mobility of particles
- Morphology and crystallinity
- Thermal and mechanical properties

Cavitation of the elastomer phase is one of the most efficient techniques for a toughening mechanism as it advances the shear banding in thermoplastic and thermosetting matrices under stress conditions. It has been illustrated that rubber-modified polymers undergo cavitation of particles for a toughening mechanism. The rubber-modified plastic can have vicinity due to hydrostatic tension. Thus, the cavitation facilitates the shear banding of the matrix. The toughening of rubber-modified epoxy is affected by particle size but particle size greater than 20 mm is not effective for epoxy matrix. Various techniques are present for the toughening of the thermoset matrix. The addition of some toughening agents has a great impact on the toughening mechanism.

8.5.10 Mechanical Properties of Thermoset Polymers

The mechanical property of thermosets is an essential and fundamental characteristic to explain their mechanical behavior. The mechanical properties of thermoset composites can be determined by various characterization techniques. The vicinity in the thermoset composite can lead to reduction in mechanical performance. Some of the key mechanical properties that determine the character of a thermoset matrix are as follows.

8.5.10.1 Elastic Deformation

Thermosets are glassy at lower temperature with Young's modulus of approximately 3 GPa. With increasing temperature, Young's modulus goes down promptly throughout the zone of transition temperature where the thermoset depends on temperature and Young's modulus rate and is viscoelastic. They become rubbery at high temperatures. Thus, for thermosets, Young's modulus must be almost a constant 10 Pa (Ouarhim et al., 2019). The molecular structure, test temperature, and cross-linking density affect the tensile behavior of thermosets. It is evident that any alteration in the property can alter the tensile behavior.

8.5.10.2 Plastic Deformation

There are some thermosets that experience plastic deformation at higher strains even if they did not face Britain fracture. There are two primary techniques of plastic deformation, which are crazing and shear yielding; basically shear yielding is observed in thermosets.

Shear yielding exhibits a change at constant volume in sample shape. This is closely related to polymer deformation. Von Mises and Tresca are two criteria to predict the outbreak of yielding.

Shear yielding plays an essential role in the initiation and impregnation of cracks in the thermoset matrix. Shear yielding is expanded uniformly over the domain of a crack and ductile failure will take place and material will show high toughness. Thus, shear yielding is a productive toughening technique for polymers (Pickering et al., 2016).

8.5.10.3 Static Tensile Property

The tensile characteristic of thermoset polymers can be increased when induced with natural fibers to optimal value. The Young's modulus for composites of natural fibers based on a thermoset polymer matrix can be increased with increasing fiber content. The different fibers embedded with thermosets exhibit different tensile strengths. The chemically treated fibers have better tensile strength and Young's modulus than the untreated natural fibers due to fine interfacial bonding of fiber and matrix.

8.5.10.4 Static Torsional and Shear Properties

The torsional property is deliberate by conducting a torsional test carried out in ARES-LS in a machine. The characterization of viscoelastic property must be determined as it depends on the deformity of a natural fiber–based thermoset. The torsional domain can be upgraded by enhancing fiber content but the elastic property of composites reduces due to chemically treated fiber content resulting in reduction of shear property.

8.5.10.5 Dynamic Mechanical Behavior

The dynamic mechanical behavior such as damping and stiffness can be emphasized by dynamic mechanical analysis. Strong modulus value is upgraded when chemically treated natural fiber–based thermoset composite is used. The fiber-based thermoset composites are more thermally stable than a neat matrix. Hence, the dynamic mechanical behavior can be improved if chemically treated natural fiber–based composites are used.

8.5.10.6 Fracture Behavior

Griffith's theory is an essential technique to provide detail of the fracture action is the energy balance theory given by Griffith to guide the brittle fracture of polymers with foregoing flaws. These flaws can be a notch, a sharp crack, or a scratch. The extent of fracture relies on the geometrical restraints. Any of the three principal strain values are zero in the plane strain. It is frequently acquired in the deformation throughout the crack inside an adhesive joint. Deformation in sheets results in any stress condition where two principal stresses are finite, whereas the third one is zero.

8.5.10.7 Linear Elastic Fracture Mechanics

When glass polymers undergo brittle fracture, then a linear elastic fracture mechanism has pertained. In this mechanism, any yielding is limited to a small domain throughout the crack and the material is elastic linearly. Crack displacement occurs in three distinct approaches. The first mode is an opening mode, and the fracture side dissociates consistently concerning the crack plane. The second mode is sliding, and the fracture side slip consistently concerning normal, whereas unsymmetrically with respect to the crack plane. The third mode is the tearing mode, and the fracture side slips unsymmetrically concerning crack plain and its normal. The first mode is commonly applicable, although the fracture surface may require any sort of crack displacement mode. Therefore, fracture energy and toughness for the first mode is specified for most of the conditions. The stress state of the crack in the unit is normally in-plane strain condition.

8.5.10.8 Elastic-Plastic Fracture Mechanics

When large plastic deformation or elastic deformation is developed in the vicinity of the tip, then the linear elastic fracture mechanics approach is not sufficient to apply. Thus, the concept of elastic-plastic fracture mechanism was developed, which calculates the energy criterion for plastic materials.

8.5.10.9 Essential Work of Fracture

When the material is exceptionally ductile, then the elastic-plastic fracture mechanism cannot be certainly carried out. Hence, a new approach to the essential work of fracture is considered. In this method, the deformation region is divided into two different regions around the crack up. The regions are the inner process region and the outer plastic region. Thus, the total work of fracture is classified into two parts as the plastic work of fracture and the essential work of fracture.

8.5.11 Thermal Properties of Thermoset Polymers

The thermoset polymers cannot be remolded once they are cured. Thermosets exhibit excellent resistance to mechanical deformations and heat. Thermoset polymers like epoxies, phenolic resin, cyanate ester, polyimide, and vinyl ester are used for high-temperature usages due to the greater thermal stability. The thermal stability property of a thermoset strongly depends upon molecular structure, polymerization degree, curing conditions, and thermal history (Vengatesan et al., 2018).

8.5.11.1 Thermal Stability of Epoxy Resin

The consequent thermal properties and curing mechanism of epoxies can be determined by curing conditions like temperature and time and type of resin and hardener. The selection of the type of hardener used for curing is important due to the difference in thermal properties. It is evident that the curing system based on the aromatic amine has a higher curing temperature than the aliphatic amine. The molecular structure of epoxy has an impact on thermal stability and transition temperature. It is necessary to keep the appropriate temperature and curing conditions to get an absolute cured epoxy polymer. The thermal properties can be enhanced by substantial abatements like mixing with other polymers. The flame-retardant property of epoxy determines its high-temperature applications.

8.5.11.2 Thermal Stability of Polyester Resin

Polyester resin is found to be more available for structural composite materials. The polyester goes through cross-linking copolymerization of monomer styrene, which results in the initiation of cross-linking network structure connected by monomeric units. The curing reactions in the polyester pre-polymer are controlled by the desired curing temperature and a cross-linking agent. A system having more cross-linking agent concentration shows a higher conversion rate. The length of cross-linking of styrene is reduced by high-curing temperature due to easier molecule orientation of active sites. The enhancement in the curing rate of polyester resin with an increasing amount of radical initiator can be noticed from the rate of temperature rise and exotherm peak. This resin has been mixed with a thermally stable group to improve its overall performance by altering the amount of the activation energy that is necessary to initiate thermal degradation. Fully cured polyester resin exhibits poor thermal stability for temperature above 300°C. To improve the flame-retardant property of polyester resin, the improvement of a polyester backbone with fire-retardant group that partly replaces styrene with a fire-retardant group and allows to cross-link with this reason and mix with thermal stable thermoset resin.

8.5.11.3 Thermal Stability of Cyanate Ester Resin

Cyanate ester resin is capitulated from mono- and difunctional cyanate. The derived cyanate ester resin dominates an excellent combination of degradation temperature above 400°C and transition temperature above 200°C. The flexibility can be changed either by incorporating a functional group or altering the backbone structure. The structural changes directly simulate the thermal properties, such as thermal stability and transition temperature and their curing behavior. This resin can also be mixed with other thermoset polymers like bismaleimide and epoxy to improve their thermal behavior and curing property. This reason exhibits outstanding thermal stability in the high-temperature domain, but it has less flame-retardant property. The flame-retardant property can be increased with the increase in cyclic matrix network and cross-linking density.

8.5.11.4 Thermal Stability of Polyimide

Polyimides are outstanding thermoset compounds exhibiting excellent high-temperature effects. The molecular formation of the polyimides unit is an essential criterion to determine the thermal stability and transition temperature. Polyimide monomeric units with rigid structures show higher transition temperature, whereas units with flexible structure exhibit a lower transition temperature. The transition temperature of polyimides changes with the solubility criterion. Thus, the higher transition temperature polyimides are hardly soluble in the organic solvent.

Some side or pendant groups like sulfonyl, alkyl, and phthalonitriles are introduced to restructure polyimides and the property of the side or pendant group is a subsequent part that decides the thermal properties. Many kinds of research have concluded the effect of different pendant groups on a restructuring of polyimides and their thermal stabilities. They show great thermal conductivity when introduced with nanomaterials.

8.5.11.5   Thermal Stability of Vinyl Ester Resin

Vinyl ester resin is an excellent type of thermoset that shows the associated characteristics of unsaturated polyesters and epoxy resin. The thermal properties and the curing process of vinyl ester resin can be measured employing MEKP (methyl ethyl ketone peroxide) catalyst and a cocatalyst, cobalt octoate. The resin has been mixed with more polymers to review their impact on the thermal properties of vinyl ester resin. The nano- and micro-composites have been prepared to increase the flame-retardant characteristic and thermal property of vinyl ester resin.

The thermal stability polymer based on vinyl ester resin can be examined by the inclusion of graphite. The presence of graphite enhanced the flame-retardant characteristic. The composites of natural fiber and vinyl ester resin show higher thermal stability initially and it can be increased by incorporation of nano clay. The association of different materials with vinyl ester resin shows the enhancement of thermal properties and stability. The processing of vinyl ester resin nanocomposites results in increased thermal instability reprocessing of the polymer.

8.5.11.6   Thermal Stability of Phenolic Resin

Phenolic resins are also known as phenol-formaldehyde (PF) resin. These are obtained in novolac and resole forms depending on the catalyst used and molar proportion of formaldehyde and phenol. Thermally activated cross-linking of novolac or resole resin can convert into a higher cross-linking structure density with excellent properties. The phenolic resins exhibit excellent adhesive properties. Thus, they are used for adhesive applications, electronic applications, composites, laminates, and abrasives. The thermal property and flame-retardant behavior can be enhanced through the modification of their structure and composition or association of nano- and micro-materials. The thermal stability is higher for addition-cured phenolic resin than conventional phenolic resole resin. The phenolic resin incorporated with lignin exhibits greater thermal stability than pure phenolic resin. The thermal stability of phenolic resin can be enhanced through nanocomposite induced with POSS, which is an inorganic nanofiller and it is non-toxic in nature. Thermoset polymers have become better replacement materials than conventional and structural materials. As a result of their excellent mechanical and thermal properties, the improvement in thermal properties of thermosets and their incorporating materials have been analyzed and developed the growing recognition to enhance their performance at higher temperatures. The thermal properties of important thermosets that are glass transition temperature, flame retardancy, thermal stability, and thermal cure properties can be increased by modifications in compositions and structure. Thermosets exhibiting excellent thermal performances have versatile application potential in high-temperature processes with no adverse effect on the physical and mechanical properties.

8.5.12 REINFORCEMENT OF THERMOSETS

8.5.12.1 Reinforcing Methods

The mechanical properties of strengthening polymers such as stiffness and strength can be altered by rigid phase reinforcement. On the basis of a kind of reinforced agent, polymer composites can be categorized as one of three distinct groups: fiber reinforced, filler reinforced, and structural composites. Fillers are used to enhance stiffness and lower cost, while fibers are used to enhance fatigue performance and strength (Chaudhary & Ahmad, 2020) (Figure 8.5).

The fiber-reinforced composites mainly contain continuous fiber with an orientation in the polymer matrix. Cellulosic, glass, and carbon are some prominent fiber materials. Oriented fiber-based composites show non-linearity in properties. The stiffness and strength in the direction parallel to the fiber orientation can be modified with the inclusion of fiber in the polymer matrix, whereas the strength in the opposite direction to fiber orientation can be extremely low. The filler-reinforced composites are basically random dispersions of hard, small pillars in any polymer matrix. These fillers are platelets (zirconium, graphene, phosphate etc.), spherical molecules (zinc oxide, aluminium oxide, silicon dioxide, etc.), and random-oriented tubes (carbon nanotubes). An identical refinement in modulus and strength can be done and other properties like flame retardancy and thermal stability can also be modified. The polymer nanocomposites are eminently used in various applications due to their lightweight and excellent performance. The structural composites are those in which incorporating material have multi-dimensional structure. Interlayered composites are structural composites that require overlay sheets of fabric medium like paper, wood, carbon, or woven glass fiber impacted in a polymer matrix. The final property of the composite

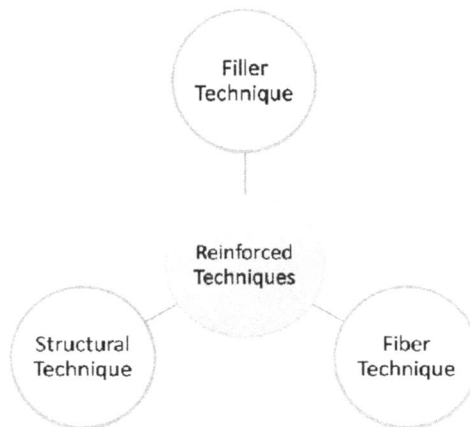

FIGURE 8.5   Different reinforcement techniques.

depends not only on the geometric shape of the structure but also on the physical property of fabric material. The structural composites fabricate composites with high stiffness and strength due to load transmission and efficiency. The transformation technique of structural composites is an essential step because inappropriate processing techniques can give critical defects in composites like voids and delamination.

Structural as well as fiber-reinforced composites require similar methods for crack growth. Initiation of a crack generally takes place at the fiber-matrix interface. Crack impregnation takes place across the interface that leads to debonding, fiber breakage, and fiber extract. This propagation of a crack can expand through the matrix to pull it to a stop; thus, cavitation and deformation can lead to reduction in mechanical properties.

8.5.13 APPLICATIONS OF THERMOSETTING MATRIX NATURAL FIBER

The various applications proposed for natural fiber–reinforced thermoset matrix-based composites have replaced many sythentic fibers in various industries. The cost optimization is a fundamental basis for industrial applications. Now, cross-woven and unidirectional fibers both are potential replacements of others due to their properties (Mohammed et al., 2015) (Figure 8.6).

8.5.13.1 Automotive Applications

Natural fiber–based thermoset composites are extensively used in automotive applications due to their sustainability and mechanical properties. Natural fibers like sisal, hemp, and flax are used to make door cladding, floor panels, and seat back linings. Coconut fiber is processed to make head restraints, back cushions, and seat bottoms, whereas sound proofing is provided by cotton fiber. Acaba fiber is processed for under floor body panels (Holbery & Houston, 2006).

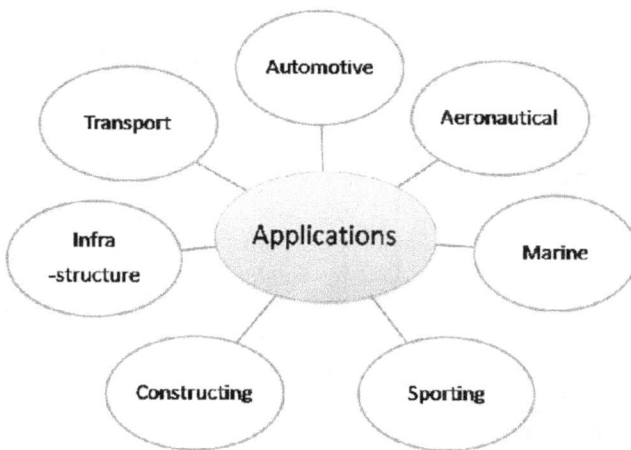

FIGURE 8.6 Applications of natural fiber–based thermoset composites.

Many eminent automobile industrial group like Chevrolet, BMW, Ford, and Toyota are extensively using natural fiber–reinforced thermosetting polymer composites to manufacturer various parts of their automotive vehicles. The BMW group embedded significant amount of natural fibers to produce different parts. Ford sets up tires that were made from incorporating corn with a matrix that has improved the fuel economy. With the help of wood fiber, food made sliding doors for their automobiles. Toyota desired to make shelves using kenaf fiber with a polymer matrix. In the last few years, the automotive sector has experienced rapid growth in the use of natural fiber–based polymer composites and they attracted the automotive sector globally due to their cheaper cost, performance, recyclability, and ease of availability. Some efforts are required to modify some technical challenges like interface compatibility and moisture stability.

8.5.13.2   Infrastructure Applications

Natural fiber composites are now used in infrastructural and structural applications for features such as eco-friendly, lower cost, and moderate strength are important. In the year 1986, a study reported about the use of polyester/coir-based composites were used to manufacture projector covers, mirror casing, mailbox, and roofing materials. In recent years, with the improvement of mechanical properties of natural fiber–based thermoset matrix composites, they are extensively used in load bearing materials like roofs, beams, and water tanks. A beam is the basic component of infrastructure and it is a structural element that exhibits flexural mode. Conventionally, beams were made from reinforced concrete, timber, and steel, but recent advances show the possible advantages like cost, installation, time, and weight for using natural fiber composite beams (Al-Azad et al., 2021). The roofing sector is the second most important part of the infrastructure. This sector requires moderate strength that can be attained by the use of natural fiber composites. Roofing materials such as a woven mat is prepared by cashew nut shell liquid/sisal fiber. The sector of a structural rehabilitation uses jute composite for reconstructing underground water pipes and drain pipes. There are some drawbacks of natural fibers that are related to the variation in their properties due to different compositions. Thus, major concerns are required to develop modified structural elements using chemical treatments and optimization.

8.5.13.3   Aeronautical Applications

Natural fibers like kenaf, hemp, sisal, jute, banana, flax, etc. are used in aerospace industries to develop various components. With the recent advances, some components in civil aircraft are prepared by natural fiber composites that are better alternatives of synthetic fiber due to their specific properties. Many aerospace industries such as Boeing 757, 767 and Airbus A310, A320 and many more airlines have significantly used natural fiber composites. Many aircraft components such as a blade to reduce weight and secondary structure and some machinery components are developed by incorporating natural fiber polymer composites (Asim et al., 2018). Some defense aerospace industries are recognizing the potential of natural fiber composites to obscure it from radar signals. In aeronautical applications, natural fiber–reinforced composites are significantly used due to cost effectiveness and low density. Natural fiber composites should have impact on damage behavior, as some components of aircraft such as wing leading edge, tail plane, etc. have to resist low velocity impact damage. Bast fibers such as hemp, kenaf, ramie etc. exhibit low velocity impact characteristics that make them a potential for their applications in the aeronautical section.

8.5.13.4 Marine Applications

The natural fiber–based composites are used to prepare marine plywood with the help of woven natural fiber reinforced in a matrix. This matrix contains 95% of vegetable oil and it results in sustainable composites that help in manufacturing boats and surfboards. Flax/epoxy composite are used to prepare various marine components like yachts and canoes. These natural fibers furnish the composite with good moistening properties. The external framework is manufactured from natural fiber and structural foam incorporated in a sandwich overlay (Fragassa, 2017). The boat manufacturing requires two main features: (i) reducing the weight of the boat and (ii) constructing parts easier to operate. Superstructures are important parts that are obtained by incorporating natural fiber that provides light weight. The flanges for an eco-friendly pulley are manufactured with the help of natural fiber composites. The potential of natural fibers can be increased with the modifications in their constituents and chemical treatments to treat their surface for proper adhesion and improved mechanical performance.

8.5.13.5 Construction Applications

Natural fiber composites are popularly used for various construction bases. In ancient times, houses were constructed with bricks that were prepared by mixing clay with pieces of sawdust or husk fiber. Generally massive structures constructed by concrete are hard to repair. The construction industry looked for an alternative material that was easy to repair. Natural fiber composites are used to reduce the waste. Hemp/basalt fiber composite is significantly used in different construction applications. Another natural material, bamboo, is employed for construction form and also to provide support to the framework. Bamboo-based hybrid composites are utilized to create high-strength panels. These panels are utilized as artificial walls, manufacturing furniture, etc.; composites of fibers like jute and sisal incorporated with polyester are used as false ceilings. Roofing panels and post boxes are manufactured from coir/polyester composites. Jute and phenolic resin are incorporated together to manufacture door frames. The researchers are doing a lot of work on these natural fibers to increase their potential and areas of application (Jariwala & Jain, 2019).

8.5.13.6 Sporting Applications

The use of natural fiber composites has grown rapidly in the manufacturing of sport equipments such as fishing poles, bicycle frames, tennis racket, badminton, and golf clubs. Hybrid composites based on epoxy and flax are used to manufacture racing bikes, bicycle frameworks, and spares of bicycles. Many sports industries are assembling sport helmets based on coir fiber sports gear such as golf clubs and fishing poles based on flax fiber and hemp is used to manufacture snowboards and skate boards.

TABLE 8.2

The Thermosetting Matrix Material Used for Different Fibers and Their Applications

Fiber	Matrix	Technique	Applications
Hemp	Poly urethane	Compression molding	Automotive sector
Kenaf	Epoxy resin	Compression molding	Tooling, bearings
Sisal	Epoxy resin	Hand layup	Roofing sheets
Jute	Polyester	Compression molding	Door panels
Banana	Epoxy resin	Hand layup	Automotive

8.5.13.7 Transport Applications

Many hybrid composites prepared from the incorporation of bamboo or wood with glass are employed to construct boats that are used in the bodies of trucks, buses, and cars. The seats of trucks, buses, cars, and rails are prepared from coir-based composites. Various internal parts of buses and trucks are manufactured by incorporating different natural fibers such as kenaf, sisal, hemp, etc. with a matrix. Low-pressure tanks for storage and transport are also manufactured by mixing jute with MF resin. Many researchers are going on to modify the performance of these natural fiber composites Table 8.2 summarizes the applications.

CONCLUSION

The natural fibers are significantly that are used as reinforced material matrix hold the reinforced materials due to adhesion force in a particular orientation. The transfer of loading to fiber helps to protect the microcracking of composites. The advantage of the polymer over conventional reinforcement material is mainly ease of processing, cheaper cost, light weight, sustainability, degradability, recyclability, and mechanical properties. The main drawback of natural fiber is its hydrophilic nature and lignocellulosic contents. The nature of fiber and variation in its properties depends upon the content of the constituents. With the help of various chemical treatements, the surface and mechanical properties of fiber can be improved. Thermosets are the class of materials that possess low viscosity and low molecular weight. They are durable and substantial due to a three-dimensional cross-linking network and possess better mechanical and thermal properties. They cannot be remolded back once they are molded. The thermoset matrix reinforced with natural fiber exhibits high strength and toughness. There are many toughening approaches that impact the matrix. Huge efforts are being made to modify natural fiber composites and hybrid composites to create high-performance parts for automotive, transport, marine, and aerospace industries. The modifications and enduring efforts to prepare hybrid composites using different natural fibers will be an advantage in the sustainable and eco-friendly application in various industries. For the past decades, the application and consumption of natural fibers has grown rapidly. These efforts will lead to sustainable, eco-friendly, recyclable, and biodegradable components manufactured for different industries without any adverse effect on the environment.

REFERENCES

Al-Azad, N., Asril, M. F. M., & Shah, M. K. M. (2021). A review on development of natural fibre composites for construction applications. *Journal of Materials Science and Chemical Engineering*, *9*(07). 1–9, 10.4236/msce.2021.97001

Allen, D. J., & Ishida, H. (2001). Thermosets: Phenolics, novolacs, and benzoxazine. In *Encyclopedia of Materials: Science and Technology*. Elsevier. 9226-9229, 10.1016/B0-08-043152-6/01662-4

Asim, M., Saba, N., Jawaid, M., & Nasir, M. (2018). Potential of natural fiber/biomass filler-reinforced polymer composites in aerospace applications. In *Sustainable Composites for Aerospace Applications*. Edited by: Mohammad Jawaid and Mohamed Thariq Elsevier. 253-268, 10.1016/B978-0-08-102131-6.00012-8

Atiqah, A., Mastura, M., Ali, B., Jawaid, M., & Sapuan, S (2017). A review on polyurethane and its polymer composites. *Current Organic Synthesis*, *14*(2). 233-248 10.2174/1570179413666160831124749

Bao, L.-R., Yee, A. F., & Lee, C. Y.-C. (2001). Moisture absorption and hygrothermal aging in a bismaleimide resin. *Polymer*, *42*(17). 7327-7333 10.1016/S0032-3861(01)00238-5

Brostow, W., & Singh, R. P. (2004). Mechanical performance of plastics. In *Encyclopedia of Polymer Science and Technology*. John Wiley & Sons. 10.1002/0471440264.pst451

Campbell, F. C. (2004). Thermoset resins: The glue that holds the strings together. In *Manufacturing Processes for Advanced Composites*. Elsevier 63-101. 10.1016/B978-1-85617415-2/50004-6

Chaudhary, V., & Ahmad, F (2020). A review on plant fiber reinforced thermoset polymers for structural and frictional composites. *Polymer Testing*, *91*. 106792 10.1016/j.polymertesting.2020.106792

Engels, T (2012). 228-253 Thermoset adhesives: Epoxy resins, acrylates and polyurethanes. In *Thermosets*. edited by Qipeng Guo Elsevier. 10.1533/9780857097637.2.228

Fink, J. K. (2013). Silicones. In *Reactive Polymers Fundamentals and Applications*. 303-323 Elsevier. 10.1016/B978-1-4557-3149-7.00008-5

Fragassa, C (2017). Marine applications of natural fibre-reinforced composites: A manufacturing case study. In *Advances in Applications of Industrial Biomaterials*. edited by Eva PellicerDanilo NikolicJordi SortMaria BaróFatima ZivicNenad GrujovicRadoslav Grujic Svetlana Pelemis Springer International Publishing. 21-47 10.1007/978-3-319-62767-0_2

Gibson, G (2017). Epoxy resins. In *Brydson's Plastics Materials*. Edited by: Marianne Gilbert Elsevier. 773-797 10.1016/B978-0-323-35824-8.00027-X

Holbery, J., & Houston, D (2006). Natural-fiber-reinforced polymer composites in automotive applications. *JOM*, *58*(11). 80-86 10.1007/s11837-006-0234-2

Jariwala, H., & Jain, P (2019). A review on mechanical behavior of natural fiber reinforced polymer composites and its applications. *Journal of Reinforced Plastics and Composites*, *38*(10). 441-453 10.1177/0731684419828524

Kandelbauer, A., Tondi, G., Zaske, O. C., & Goodman, S. H. (2014). Unsaturated Polyesters and Vinyl esters. In *Handbook of Thermoset Plastics*. Elsevier. 10.1016/B978-1-4557-3107-7.00006-3

Komuraiah, A., Kumar, N. S., & Prasad, B. D. (2014). Chemical Composition of Natural Fibers and its Influence on their Mechanical Properties. *Mechanics of Composite Materials*, *50*(3). 359-376 10.1007/s11029-014-9422-2

Li, X., Tabil, L. G., & Panigrahi, S (2007). Chemical Treatments of Natural Fiber for Use in Natural Fiber-Reinforced Composites: A Review. *Journal of Polymers and the Environment, 15*(1). 25-33 10.1007/s10924-006-0042-3

Liu, Z., Huang, Y., & Deng, S (2020). Synthesis and characterization of thermosetting polyacetylene-terminated silicone resins. *Journal of Applied Polymer Science, 137*(23). 10.1002/app.48783

Liu, J (Daniel), Thompson, Z. J., Sue, H.-J., Bates, F. S., Hillmyer, M. A., Dettloff, M., Jacob, G., Verghese, N., & Pham, H (2010). Toughening of epoxies with block co-polymer micelles of wormlike morphology. *Macromolecules, 43*(17). 7238-7243 10.1 021/ma902471g

Massingill, J. L., & Bauer, R. S. (2000). Epoxy resins. In *Applied Polymer Science: 21st Century*. Elsevier. 10.1016/B978-008043417-9/50023-4

McKeen, L (2021). The effect of heat aging on the properties of polyimides. In *The Effect of Long Term Thermal Exposure on Plastics and Elastomers*. Elsevier. 10.1016/B978-0-323-85436-8.00008-4

Mohammed, L., Ansari, M. N. M., Pua, G., Jawaid, M., & Islam, M. S. (2015). A review on natural fiber reinforced polymer composite and its applications. *International Journal of Polymer Science, 2015*. 10.1155/2015/243947

Mullins, M. J., Liu, D., & Sue, H.-J. (2018). Mechanical properties of thermosets. In *Thermosets*. Edited by: Qipeng Guo Elsevier. 35-68 10.1016/B978-0-08-101021-1.00002-2

Nair, C. P. R., Mathew, D., & Ninan, K. N. (n.d.). Cyanate ester resins, recent developments. In *New Polymerization Techniques and Synthetic Methodologies*. Springer Berlin Heidelberg. 1-99 10.1007/3-540-44473-4_1

Ouarhim, W., Zari, N., Bouhfid, R., & Qaiss, A. El kacem (2019). Mechanical performance of natural fibers–based thermosetting composites. In *Mechanical and Physical Testing of Biocomposites, Fibre-Reinforced Composites and Hybrid Composites*. Edited by Mohammad Jawaid, Mohamed Thariq and Naheed Saba Elsevier. 43-60 10.1016/B978-0-08-102292-4.00003-5

Pickering, K. L., Efendy, M. G. A., & Le, T. M. (2016). A review of recent developments in natural fibre composites and their mechanical performance. *Composites Part A: Applied Science and Manufacturing, 83*. 98-112 10.1016/j.compositesa.2015.08.038

Sanjay, M. R., Madhu, P., Jawaid, M., Senthamaraikannan, P., Senthil, S., & Pradeep, S (2018). Characterization and properties of natural fiber polymer composites: A comprehensive review. *Journal of Cleaner Production, 172*. 566-581 10.1016/j.jclepro.2017.10.101

Shekar, H. S. S., & Ramachandra, M (2018). Green composites: A review. *Materials Today: Proceedings, 5*(1). 2518-2526 10.1016/j.matpr.2017.11.034

Thyavihalli Girijappa, Y. G., Mavinkere Rangappa, S., Parameswaranpillai, J., & Siengchin, S (2019). Natural fibers as sustainable and renewable resource for development of eco-friendly composites: A comprehensive review. *Frontiers in Materials, 6*. 10.3389/fmats.2019.00226

Vengatesan, M. R., Varghese, A. M., & Mittal, V (2018). Thermal properties of thermoset polymers. In *Thermosets*. Edited by: Qipeng Guo Elsevier. 69-114 10.1016/B978-0-08-101021-1.00003-4

Wang, W., Sain, M., & Cooper, P (2006). Study of moisture absorption in natural fiber plastic composites. *Composites Science and Technology, 66*(3–4). 379-386 10.1016/j.compscitech.2005.07.027

Ziaee, S., & Palmese, G. R. (1999). Effects of temperature on cure kinetics and mechanical properties of vinyl-ester resins. *Journal of Polymer Science Part B: Polymer Physics, 37*(7). 725-744 10.1002/(SICI)1099-0488(19990401)37:7<725 ::AID-POLB23>3.0.CO;2-E

9 Thermo Polymer Matrix–Based Natural Fiber Composite

Manan Tyagi, G. L. Devnani, and Raj Verma
Department of Chemical Engineering, Harcourt Butler
Technical University, Kanpur, India

INTRODUCTION

With the broader industrial applications of natural fiber–reinforced thermoplastic polymer matrix, it increased the attention of researchers and academicians. If there is no need of much higher structural performance, they can even beat the majority of synthetic fiber–reinforced polymer blend. The wider industrial application is due to their phenomenal properties, such as biodegradability, in spite of good strength to weight ratio, good tensile strength, and economical as they are much cheaper than synthetic fibers and many more. As aforementioned, natural fibers are used in fields where acceptable performance is required with reduced weight, like in the automobile and aerospace industries. Y. K. Kim & Chalivendra, (2020) reviewed the same and discussed that using a natural fiber–reinforced epoxy matrix in an Audi car door panel reduced the energy utilization by 45%. Along with less energy utilization, it was also ecologically beneficial, as it also reduced greenhouse gas emissions.

As we all know, a natural fiber thermoplastic composite is not comparable to synthetic fibers, such as glass or carbon. Selvaraj et al., (2019) validate that using natural additives will eventually increase their various mechanical properties like tensile strength and flexure strength. The composite with additives showed better mechanical properties than that the one without additives. However, this paper also states that using natural additives will increase their water adsorption tendency because of the presence of natural additives on the surface of a composite polymer matrix. This paper validates that adding additives is one of best methods in order to increase the mechanical properties of a composite. Natural fiber is also used extensively in cellular foam manufacturing industries (Kuranchie et al., 2021). This paper checks out how the reinforcement of natural fibers in polyurethane is beneficial in resulting properties as result of which natural fibers are getting more known in cellular foam industries. This paper also discusses that natural fibers obtained from agricultural waste can be reinforced with TPU_s which proves to be a biodegradable and sustainable material. So, due to easily available and less manufacturing cost, natural fibers are being given more impetus. Also, they also confirm that the final product obtained from reinforcement is better than virgin polyurethane foams in terms of properties.

DOI: 10.1201/9781003201724-9

Reinforcement of natural fibers always has ecological benefits whether it is biodegradability or recyclability (Pegoretti, 2021); they throw light on the recycle concepts of thermoplastic reinforced with natural fibers. This paper discusses how the recycling of composites aid the environment by using less resources, utilizing less energy, and limiting the emission of CO_2 during production of such kinds of composites. They also validate the challenges to overcome for profitable recyclability, such as mechanical and chemical recycling, use of recycled fibers, or utilizing waste matrices (Wirawan, 2020); this paper also discusses some techniques that can be used to recycle natural fiber thermoplastics. It also states that there are still limitations that should be overcome in this aspect. During fabrication, impregnation of resin for the reinforcement is one of major issues (Awais et al., 2019). This symposium gives impetus to the so-called comingled fabrics. Comingled fabrics are the intermediate materials that are used in the fabrication process of continuous fiber-reinforced thermoplastics. Comingled fabrics make sure that matrix fibers and natural fibers such as jute and flax are in the same alignment. They also studied the effect of commingling on mechanical properties of composites. This paper validates that knitted comingled composites show better mechanical properties than that of woven comingled composites.

Hence, as aforementioned, natural fibers are those versatile materials that are better alternatives (if there is not any structural performance concern) of synthetic fibers; they are even good for the environment and nature in terms of recyclability, biodegradation, easily available at lesser cost, and better than virgin metals in terms of specific strength. So this paper focuses exclusively on natural fiber reinforcement in thermoplastic polymers as it there that there is still a lot of limitations and drawbacks of using natural fibers in thermoplastic polymers. For awareness, this symposium tries to cover every aspect of the mentioned topic.

9.1 NATURAL FIBERS

Natural fibers are among those precious yet readily available resources that are very useful in today's era for sustainable development. Due to increments in demand on petroleum-based products, petroleum resources have started depleting. Also, global warming is a major issue that is caused by using petroleum-based products (Dunne et al., 2016). Hence, here comes this renewable resource that is much cheaper than that. They are generally classified into two categories (i) plant fiber and (ii) animal fiber. If we talk about plant fibers, jute, flax, and hemp come into the picture. When these biodegradable fibers are reinforced with polymers they emerge as an excellent material for decent structural performance at a reasonable cost (per kg) (Parbin et al., 2019). So, conserving the environment along with sustainable development comes the utilization of natural fibers (Dixit et al., 2017). Animal fibers are protein based; they are generally hairs of animals like wool or secretion of animals like silk. The most abundant polymer in nature is cellulose, of which all plant fibers consist. Along with cellulose, waxes, gums, pectin, and hemicellulose are all bound together to provide strength to plants. The polymeric chain of cellulose contains carbon, hydrogen, and oxygen atoms. Plants can be extracted from various parts of plants like a leaf, fruit like in coconut trees, and even seeds. Here we will discuss various types of plants and animal fibers that are popular nowadays.

9.2 TYPES OF FIBERS

9.2.1 WOOL FIBER

As we all know, wool is among the oldest animal fibers that are obtained from the hairs of goats, alpacas, and sheep. China, New Zealand, and Australia are the major countries where production of wool is quite numerous. The properties of wool depend upon the length and diameter of the wool. The wool extracted from the hairs of sheep from their shoulders, stomach, and back is said to be higher quality than obtained from the legs (Mcgregor & Butler, 2016). The sheep's breed can also be a factor for wool quality.

9.2.2 SILK FIBER

As we all know, silk fiber is known as the "queen of textile" as it provides warmth in winter and comfort in summer. Firstly, it was originated from China; later, India and Japan also started to produce silk. Mulberry silk is the only commercial available silk. Silk is produced by the secretions of certain insects to build their cocoons or webs (Hao et al., 2018). Silk consists of two binders that have different roles for silk. They are called fibroin and sericin. The former is responsible for providing stiffness and strength, and fiber structure is the responsibility of the latter (Hao et al., 2018). Silk fibers are triangular and there is non-uniformity in their shape. Silk fibers are known to be anti-bacterial; hence, they are as hygienic as they are dust-proof. Pharmaceutical, cosmetic, automotive, dietary, etc. are the industrial implementation of silk fiber. It is also used in tissue engineering (Chen et al., 2019).

9.2.3 COTTON FIBER

Cotton fiber contains around 90% (Hsieh, 2007) of cellulose. called the purest form of cellulose. It is a seed fiber. Its scientific name is *Gossypium hirsutum L.* India, Pakistan, Brazil, the United States, and China collectively produce 80% of the cotton produce globally, although cotton is being cultivated in more than 50 countries (Nix et al., 2017; Raina et al., 2015).

Due to their hollow nature, they are breathable and good absorbers (Hosseini Ravandi & Valizadeh, 2011). Cotton has a characteristic property of absorbing water of weight 24–27 times the weight of itself (Hosseini Ravandi & Valizadeh, 2011). If we see a cross-sectional view, we see a shape akin to beans, which turn circular when absorbing water. They are used in home furnishings, apparel, etc. Cotton fiber length ranges from 20 mm to 32 mm.

9.2.4 JUTE FIBER

Jute is obtained from *Corchorus* named genus plants. It has over 100 species (Samanta et al., 2016). After cotton, it is the second most vital fiber in terms of usability and cost and also the highest-produced bast type of fiber than any other bast-type fiber. The favorable condition for the growth of jute plants exists in India, China, Myanmar, and Bangladesh. And therefore they are the leading producer of jute fiber (Faruk et al., 2012). The process used for extracting jute from a jute plant is water retting. The length of jute fiber ranges from 1 m to 3.5 m. The shape of jute fiber is a polygon with round corners (Bhowmick, 2015).

Despite the strength and stiffness of jute fiber, there is drawback, too i.e., extensibility. Around 90–120 days are required for cultivating the jute plants and the retting process. If has various uses like packaging bags for clothes, large storage bags for agriculture products like manure and fertilizers or seeds, cements, etc. (Morin-Crini et al., 2019).

9.2.5   HEMP FIBER

After jute, one of the important bast fibers is hemp fiber. It is cheap as well as strong, stiff, and lightweight. These properties make hemp fiber useful for chelating material as well (Morin-Crini et al., 2019). It is obtained from fast-growing plants of the genus type *Cannabis sativa*. As compared to cotton bleaching, it is more difficult for hemp fiber due to its coarser and dark-colored nature. Hemp fiber can be 2 m in length. This cylindrical-shaped fiber has cracks and joints on its surface (Morin-Crini et al., 2019). It is used in textiles, composites, paper, animal bedding, etc. China, the United States, France, and Canada are the leading countries that grow hemp fiber.

9.2.6   FLAX FIBER

Flax fiber is one of the oldest fibers, found around 4,500 years ago, in the Stone Age. It is also known as linen, and it is the first fiber of bast type. Flax fiber is derived from the stem of the *Linum usitatissimum* plant species. It is cultivated from March to July. One hundred days are required for its complete cultivation. The length of flax fiber is about 90 cm. France, Russia, Canada, Belgium, and China are the main countries cultivating flax fiber; the shade of soft and lustrous flax fiber depends upon how it is being retted (Debnath, 2017).

9.2.7   BAMBOO FIBER

Bamboo is also known as a natural glass fiber because of its good tensile strength. It is often used in place of steel directly when there is a need for tensile strength (Muhammad et al., 2019). It has over 1,000 species with different properties that depend upon the various geographical conditions like temperature, moisture of soil, and habitat. This bast fiber can grow as fast as 3 cm per hour. Bamboo length can reach up to 40 m. Bamboo fiber is extracted from plants by various mechanical or chemical or both methods. In general, steam explosion is considered to be the best method for extraction of fiber from plants, as it has high specific strength and reinforced composites used in automotive, protection, aerospace, and leisure applications (A. U. M. Shah et al., 2016).

9.2.8 Asbestos Fiber

Asbestos fiber is a heat resistant and inflammable fiber that is obtained from various silicate mineral rocks. The most common silicates that have a fibrous crystalline nature are serpentine, anthophyllite, and amphibole. The largest producer of asbestos is Canada. Asbestos fiber has a shape similar to glass fiber's smooth shape. The fiber length of asbestos is between 12–300 mm. In a cross-sectional view, it circular or polygon in shape. It has various extraordinary properties like alkali or acid resistance and corrosion resistance and that's why it being used in industrial packing material, electrical windings, conveyor belts, etc. Pure asbestos also has serious drawbacks too; long and continuous exposure to it can trigger cancer in human beings.

9.3 PROPERTIES OF PLANT FIBERS

Cellulose, lignin, and hemicellulose are the main constituents of the plant fibers. Pectin and wax are the minor components. The amount of each these components in a plant fiber influences the properties of plant fibers. Also, chemical composition, degree of polymerization, fiber surface morphology, microfibrillar angle, and defects also impact mechanical properties. Cellulose generally controls the strength and stiffness of the plant fiber. Cellulose's crystallinity is modulated by hydrogen bonding in glucose units present in cellulose as result of which it impacts physical properties. The biodegradability, which is a characteristic property of plant fibers, is controlled by a hemicellulose component. Cellulose is embedded in hemicellulose like in a matrix. Hemicellulose also impacts thermal properties of plant fibers. Rigidity and moisture absorption in plant fibers are controlled by aromatic hydrocarbon lignin (Dhakal et al., 2007; Komuraiah et al., 2014; Saidane et al., 2016).

From production to processing stage, there are significant factors that also impact properties. These factors can be processing techniques, testing methods, and fiber maturity level as among bamboo fiber only some species can mature over 3 years or some mature over a period of 6 to 7 years, and last but not least, environmental conditions like temperature and the nature of soil in which the plant is cultivated (Faruk et al., 2012).

9.4 METHODS FOR REINFORCEMENT IN COMPOSITES

There are distinct forms available for reinforcement of natural fibers with composites so that we can modulate the manufacturing cost, resulting strength, and properties as per need. There are 1D, 2D, and 3D fiber-reinforced composite. They are categorized in terms of alignment of fiber. To design certain fabric architecture, some conventional textile technologies are available in published literature. Continuous fibers are being used in woven, knitted, and braided fabrics, whereas short fibers are being used in non-woven fabrics.

9.4.1 WOVEN FABRICS

Woven fabrics are manufactured by the oldest technique of weaving in which two sets of yarn are interlaced perpendicularly to each other like a grid. The two sets are known as warp, which is laid lengthwise and the other is weft, which is laid breadth wise. The term *weave design* is defined here as an interlacement pattern on which properties of woven fabric depend. Generally, there are three basic interlace patterns, plain, twill, and satin, for the manufacturing of woven fabric.

Yarn is a basic building block for weaving; it has to go through some preparatory processes so that it is suitable for weaving machines. Weave design, type of fiber from which yarn is made of, and number of weft and warp per unit distance are all factors that influence properties of woven fabric. During fabric formation, warp yarns are required to handle higher stresses than weft yarns as a result of which warp requires some extra preparation. For further processes, one type of package is transferred into another package, done through winding. Winding is also responsible for removing defects (if any) in yarn. As the warp yarn requires it to be more strong, extensible, and smooth, it is done by a protective coating of polymer. This process is called sizing. After sizing, warp yarn is stronger, it becomes abrasion resistant, its hairiness is reduced, and the weaver beam gets the sufficient length of yarn.

After that, the drawing-in process occurs through weaving elements in these manually or automatically warp yarn enters. Now this whole system is put on a weaving machine. Now the whole plain in which the warp is there is divided into two layers; this process is called shedding. Shedding is done as per the weave design. After shedding, weft yarn is inserted and pulled back by the reed. And this whole woven fabric gets rolled at the cloth beam.

9.4.2 BRAIDED FABRICS

Braided fabrics are manufactured when three or more yarns are interlaced diagonally, making 30 to 80 angles with the product axis. The angle that the yarns make with the product axis is known as the braiding angle. The cross section of braided fabrics can be of constant area or variable, and it is controlled by a braiding machine. The development of braided fabrics is either in open structure or in closed structure. The number of yarn carriers can be changed for the formation of different types of braids.

9.4.3 NON-WOVEN FABRICS

As the name suggests, non-woven fabrics are developed directly from fibers rather than converting into threads (Rawal et al., 2016). In this technique, firstly, fiber is made ready so that it can be directly used in the form of web or batt and then it is consolidated with various mechanical or chemical, thermal, or by means of solvent. There are different ways to do these processes; these different ways define the final properties of the product. Because of the anisotropic nature of non-woven fibers, the orientation of fibers is random. For the first process, i.e., conversion of fiber into suitable batt or web, various methods are being used, such as spunmelt, drylaid, and wetlaid. In wetlaid and drylaid, there is only a difference of state in which fibers are opened and carded. In drylaid, the process occurs through air and in the wetlaid process it occurs through water. And in the spunmelt process, there is melting of polymers' continuous filaments; after that, they are collected over a conveyor belt. After preparation of web or batt, it is required to be bonded, because the properties of the final product are determined by bonding (Pourmohammadi, 2013). As aforementioned, bonding of webs is done by mechanical bonding such as stitch bonding, needle punching, chemical bonding, and thermal bonding (Mao & Russell, 2015). As compared to woven fabrics, the strength and stiffness of non-woven fabrics is lowered. Bond failure, fiber sliding, and orientation are currently the main drawbacks of non-woven fabrics. But on the bright side of non-woven fabrics, high deformability and energy absorption capability is exhibited by non-woven fabrics that are at a lower cost.

9.4.4 KNITTED FABRICS

In this technique, loops made by bending or curving of one set of yarn or minimum of one yarn is interconnected (Ashraf et al., 2015). In this process, a continuous length of yarn is required so that by the interloping method this continuous length of yarn gets converted into thick yet flexible knitted fabrics. The flexibility and shape retention along with comfort are due to these loops (Karthikeyan et al., 2016) (Hasani et al., 2017). In terms of cost, design flexibility, and processiblity knitted fabrics are better than woven but woven fabrics are better than knitted fabrics in durability (Ashraf et al., 2015). Knitting is classified into two types: weft knitting and warp knitting. They differ to each other by their direction of feeding of yarn with respect to the fabric formation. In weft, yarn is fed perpendicularly to the direction of fabric formation, while in warp, it is parallel. Weft is a more useful and common technique than weft for the formation of fabrics because in warp there is a requisite for the multiple yarns in form beams, while in the weft technique only a single yarn can work. Also, low capital cost, low stock requirements of yarns, less space requirement, amenity to change pattern quickly, versatility, and higher production all are due to the weft knitting technique (Karaduman et al., 2017) (Figure 9.1) shows the process for preparing yarn for weaving.

9.5 THERMOPLASTIC POLYMER MATRIX COMPOSITES

High molecular weight resins are used to fabricate thermal polymer matrix composites. After melting of resins, they can undergo reinforcement without cross-

FIGURE 9.1 Processes for preparing yarns for weaving.

linking because molecular chains are bound together due to the weak secondary bonds. Because of their ability not to cross-link on reheating at their processing temperature, they can be recycled into new, different shapes. As compared to thermoset composites, thermoplastics are better in terms of their automated, fast and clean production, fracture toughness, and reduced processing time (Awais et al., 2019, 2020; Yassin & Hojjati, 2018). Due to the higher molecular weight of thermoplastic composite resins, they are more viscous than thermoset composites as a result of which reinforcement of long fibers is found to be difficult (Mallick, 2010a). With various advancements in fabrication techniques, development of new fibers, eco-sustainability, and incredible final properties led to the success and popularity of thermoplastic composites (L. Mohammed et al., 2015).

There are various types of thermoplastic resins available like acrylonitrile butadiene (ABS), polypropylene (PP), polyethylene (PE), polyamide (PA), polyetheretherketone (PEEK), etc. (Mallick, 2010a; Yao et al., 2018). Among all these, PE is mostly used in thermoplastics globally. Polymerization and condensation of ethylene forms polyethylene (Patel, 2016). If we talk about this was-like plastic mechanical properties, moderate tensile strength, density less than water, melted between 80°C–130°C. In the case of thin films, PE is transparent; otherwise, it is translucent or opaque (Ronca, 2017).

Because of its electrical insulation, resistance to chemical, non-poisonous nature, and odorless, it has extensive use in wire coverings, pipes, sheets, and packaging material like water bottles, but it has drawbacks like its softens at low temperatures so it cannot be used in the places where extensive heat is present, it is not resistance to scratches, and possibility of being oxidized (Sastri, 2014).

Polypropylene (PP) is one of a kind, having incredible properties at moderate cost. It has a high strength to weight ratio that makes it suitable for automotive industries. The melting range of PP is about 160°C–170°C and PP offers better heat resistance than any other thermoplastic resin in same range of cost. PP is also chemical resistant but up to a certain temperature. In spite of the opaque nature of PP, we can see different colors of products made in PP (Park & Seo, 2011). Various methods like injection molding, compression molding, blow molding extrusion, and thermoforming can be used to used to process PP and PE, so that they can be ready to fabricate thermoplastic composites (Calhoun, 2010; Gopanna et al., 2019).

But some thermoplastic are biodegradable too, so that they can be benign to environment. Poly (lactic acid) PLA is one of most popular biopolymer that is produced by heating lactic acid under vacuum. It was discovered in 1932 (Mehta et al., 2005). Lack of cross-sectional adoption is due to its lower molecular weight. Higher molecular weight PLA can be synthesized by two techniques: direct polymerization and ring-opening (Farah et al., 2016) (Table 9.1) is showing the comparison of properties of biodegradable polymers.

9.6   FABRICATION TECHNIQUES FOR THERMOPLASTIC COMPOSITES

The fabrication process is defined as the process in which raw polymers are converted into certain and usable shapes. The selection of fabrication techniques is

TABLE 9.1

A Comparative Study of Properties of PP, PLA, and PL (Calhoun, 2010; Farah et al., 2016; Gopanna et al., 2019; Mehta et al., 2005; Mukherjee & Kao, 2011; Patel, 2016; Pretula et al., 2016; Su et al., 2019)

Properties of Polymer	Polyethylene (HDPE)	Polypropylene (PP)	Polylactic Acid (PLA)
Hardness (MPa)	55–67	60–100	82–88
Tensile strength (MPa)	20–30	21–40	21–70
Density (g/cm^3)	0.945–0.97	0.9–0.94	1.26–1.4
Melting temperature ($^\circ$C)			150–175
Tensile modulus (GPa)	0.9–1.1	0.1–0.6	3.5–4.1
Failure strain (%)	500–700	100–300	4–7

based on the size, shape, and geometry and softening temperature of the final product (Bourmaud et al., 2020). During the fabrication process, the three primary processes are flowing, shaping, and hardening of resin. There are various types of techniques like injection molding, filament winding compression molding, vacuum infusion, extrusion, pultrusion, melt electrospinning, and resin transfer molding that are used to fabricate thermoplastic composites. But to fabricate continuous fiber-reinforced thermoplastic composites, mainly pultrusion, resin transfer molding, filament winding, and compression molding techniques are used.

9.6.1 COMPRESSION MOLDING

The oldest technique that is suitable for fabricating complex, large, or small parts of a product is compression molding. In this process, thermoplastic resin undergoes pressurized heating in a closed mold cavity until it is properly cured. The performance of final product is determined by the temperature, pressure, and the holding time during compression molding. These are the condemning parameters for compression molding. Molding temperature is selected on the basis on fiber degradation temperature as well as polymer matrix melting temperatures. If the lower temperature is selected than the appropriate temperature, then fibers are not wetted sufficiently, and in opposite, higher temperature leads to the degradation of fiber.

Interfacial bonding between the matrix and fibers are refined by the pressure taken during compression molding (D. J. Kim et al., 2019). Although higher pressure can demolish the hollow fiber structure, as a result, the mechanical properties will be downgraded (Medina et al., 2009). The benefits of using compression molding are less post-production scrap, high production, product consistency and better dimensional control (Gopanna et al., 2019). But there are also limitations too like restrictions in design and higher cost of equipment. This technique is more beneficial than any other technique due to the lesser cost, raw material utilization, and rate of production (Jaafar et al., 2019; Tatara, 2011).

9.6.2 Filament Winding

In this technique, there is high accuracy in the positioning of the fiber, this process is automated, and has high fiber volume content and low void space. When the resin is saturated with fiber, then this infused resin is further stiffened by a mandrel shape that is rotating with a geometric path. This technique is used in high-profile manufacturing like aircraft fuselage, marine and automotive industries, in aeronautics (rocket engines), and pressure vessels. The vital limiting factors that determine the final properties are wall thickness, winding angle, and filament winding pattern (Dun et al., 2019). As aforementioned, high fiber volume fractions can be achieved readily along with less cost in process and material cost as compared to pre-peg composites and high repeatability. And because of all of these attributes, it is the future potential techniques. Mandrel is an acute part in filament winding so extra cost is required for the mandrel in order to fabricate complex shapes.

9.6.3 Resin Transfer Molding (RTM)

When we talk about a feasible alternative to the autoclave process and an advanced liquid injection procedure, then RTM comes into the picture. It was deployed by industries in 1970 (Miller, 1990). In this process, resin is first heated. Resin's viscosity should be low, and then is injected into a pre-decided shape or form in a closed mold for dry reinforcement (Jamir et al., 2018). The resin is injected at pressure between 0.4 to 1 MPa for the superior impregnation of matrix into reinforcement. To obtain certain performance, the resin is cured under pressure (Mallick, 2010b). The advantages of RTM include less tooling and setup cost, less amounts of defects, and excellent finish on both surfaces. Using the RTM method, rudder tips, ribs, and panels of airplanes are fabricated (Meola et al., 2017).

9.6.4 Pultrusion

In this technique, fibers are heated first, then pulled through a puller to make to pass through impregnated equipment for the purpose of wetting and to pass through a cooling die to accomplish the requisite finish, size, and shape (Verma & Fortunati, 2019). After that, cutting of composites into final products is governed by a pelletising system. High fiber volume fractions, excellent reinforcement alignment, precise control of resin and fiber, good dimensional control, low scrap rates, and highly automated continuous and cost-effective process are the main advantages of pultrusion. Although skilled labor and initially high investment for setup, the industry is essential for pultrusion (Gopanna et al., 2019).

9.6.5 Vacuum Infusion

In this technique, vacuum pressure is used for helping the resin flow into reinforcement after the resin is being injected into the closed mold cavity. Here, vacuum pressure is a driving force for the movement of resin into laminate (Spasojevic, 2019). But there is a limitation of high viscosity of melted resin in this process, but certain thermoplastic polymers such as polyamide-6, polyamide-12, and polybutylene terephthalate can be fabricated at uplifted temperatures of a mold cavity. However, fabrication can be possible at room temperature also after the development of reactive thermoplastic resins like Arkema's acrylic-based Elium (Obande et al., 2019). Vacuum infusion requires lower capital investment and less resin for utilization. Superior quality products can be accomplished using vacuum infusion because of fewer voids and uniformity in consumption of resin (Zin et al., 2016). Although vacuum infusion also has some key challenges for its further improvement like surface finish and high consumable costs, for large structures, gelation before impregnation of resin, low cycle time, struggling to accomplish vacuum (Hammami & Gebart, 2000).

9.7 MECHANICAL PROPERTIES OF NFR THERMOPLASTIC COMPOSITE

Mechanical properties of NFR thermoplastic composite are determined by the properties of a fiber and polymer matrix (Shubhra et al., 2013), their ratio in composite, orientation of fiber in matrix, geometry of fiber in matrix, and fiber matrix adhesion (Rahman & Putra, 2018). The adhesion between fiber and matrix is considered to be significant in determining mechanical properties. Adhesion between fiber and matrix can be enhanced by improving impregnation by consuming low viscous matrix or lowering the distance of matrix melt flow (George et al., 2001). The mechanical properties of NFR thermoplastic composite that have major consideration are flexural strength, tensile strength, impact, and short beam strength.

9.7.1 TENSILE STRENGTH

For the structural performance of composites, tensile strength is one of the most vital mechanical properties. Composites depict their tensile properties when a composite is under a tension due to a load. For checking the ability of a composite and whether it resists or deforms under a load, tensile tests are conducted. A tensile test gives information of tested material about modulus, tensile strength, elongation at yield, and break (Rahman & Putra, 2018). All the aforementioned limiting factors together determine whether the composite is suitable for a particular application or not. Academicians and researchers are getting concerned on the influence of fiber reinforcement into a polymer matrix because of the enhancement of properties of a polymer matrix (Malkapuram et al., 2009; Pailoor et al., 2019).

It is well documented that tensile properties can be improved by various techniques such as post-curing, sandwiching and additive or filler addition, and chemical treatments (Selvaraj et al., 2019). Published research articles also claim that inclusion of nanofillers and micro-fillers can upgrade tensile properties (Awais et al., 2019; Selver, 2020; Shoja et al., 2020). Also, accompaniment of numerous organic and inorganic fillers can also refine properties of NF thermoplastic composites (Balan et al., 2017). For enhancement of performance of NF composites, the most suitable additive is hollow glass microspheres (HMG) (Charière et al., 2020; Kang et al., 2017; Kumar et al., 2017).

As we all know, heating temperature, curing time, and pressure are the crucial limiting factors in the compression molding technique. Optimization of these parameters ameliorates tensile properties (Yallew et al., 2020). The direction of fibers with respect to the direction of stress applied is very crucial in tensile properties of composites, like if the composite bear uniaxial load then unidirectional structures are considered to be the most suitable structure (Akonda et al., 2018). Higher yarn twist reduces the tensile strength because of obliquity effect and limiting the perforation of resin into yarns (Bar et al., 2019).

Another enthralling factor is fiber architecture. It is stated in published literature that knitted laminates offer less tensile strength but better elongation compared to woven laminates. In fact, in woven laminates, the weaving pattern also influences tensile properties like twill and satin offers better tensile strength (Arju et al., 2015; Baghaei et al., 2015).

The high melt viscosity of thermoplastic resins is a major drawback that hampers their performance. However, to overcome this problem, a commingling technique is used (Patou et al., 2019) (Tables 9.2 and 9.3). is showing the comparison in properties as an effect of different fabrication procedure, architecture and reinforcement.

9.7.2 FLEXURAL STRENGTH

The test that is used to analyze flexural properties of composites is a three-point bending test. In this test, tested material is bent at a preset span until it is fractured and the speed of bending is also preset. Due to this, compressive force acts on the upper

TABLE 9.2

Different Tensile Strengths of Composites at Different Fabrication Techniques, Reinforcement, and Architecture

Reinforcement Architecture	Fabrication Technique	Reinforcement	Tensile Strength (MPa)	Reference
Short fibers	Compression molding	Sugar palm	9.9–18.4	(A. A. Mohammed et al., 2018)
Woven	Compression molding	Jute	69.3	(Kandola et al., 2018)
Non-woven commingled	Compression molding	Hemp	40.7–51.7	(Merotte et al., 2018)
Friction spun hybrid yarn	Compression molding	flax	80–128	(Hao et al., 2018)
Short fibers/particles	Injection molding	Bamboo/flax	42.43–48.6	(Kumar et al., 2017)
Short fibers	Injection molding	Sisal	38–40.8	(El-Sabbagh, 2014)
Short fibers	Injection molding	flax	40.1–44.1	(El-Sabbagh, 2014)

TABLE 9.3

Flexural Strengths of Some NFR Composites

Polymer Matrix	Reinforcement Architecture	Reinforcement	Fabrication Technique	Flexural Strength	Reference
PLA	Woven fabric	Hemp	Compression molding	77.9–118.7	(Baghaei et al., 2015; Durante et al., 2017)
PLA	Short fibers	Jute	Compression molding	180–225	(Goriparthi et al., 2012)
PP	Non-woven fabric	Flax	Compression molding	30–75	(John & Anandjiwala, 2009)
HDPE	Short fibers	Hemp	Compression molding	32.6–44.6	(Lu & Oza, 2013)
PP	Short fibers	Hemp	Compression molding	28–50	(Sullins et al., 2017)
PP	Short fibers	Flax	Injection molding	50.4–57	(Dickson et al., 2014)

surface of the specimen and the lower surface is subjected to tensile forces (Dong et al., 2012). The crucial parameters that determine the flexural strength are fibers waviness, mold temperature, stacking sequence, and fiber orientation (Azzam & Li, 2014). Void space and the magnitude of fiber loading also impact flexural properties (Hagstrand et al., 2005; Khan et al., 2018; Mahmud Zuhudi et al., 2016).

The adhesion between matrices and fillers greatly enhance flexural strength and of the modulus of composite due to the increment in interfacial area and rigidity (Donmez Cavdar et al., 2019). But the addition of fillers should be optimized e.g., up to 10% by weight of HMG, bending strength of composite will be improved but beyond that, bending strength will start decreasing (Kumar et al., 2017). Furthermore, reinforcement techniques also have a great impact on flexural strength as there is a 25% upgradation in flexural strength by using commingled twill fabric (Formisano et al., 2019).

9.7.3 IMPACT PROPERTIES

Impact strength is defined as the maximum force per unit area that a material can withstand without breaking or fracturing (Ramesh, 2019). In other words, it is a capability to absorb energy when suddenly an objects strikes or the material itself strikes a rigid surface. The service life of materials undergo various impacts so it is necessary to study their behavior under such various loads of impact. There are generally three types of impact classified on the basis of velocity of impact. First is low impact. Velocity range for low impact is less than 10 m/s, second one is medium (50–200 m/s), and the last one is high velocity impact. Its velocity range is

200–500 m/s. So in the automotive industry, thermoplastic composites are very useful on the account of their good impact strength. But on account of some impulse loads like crash strikes, bird and stone strikes, and tool drops, some internal damage can occur that makes them weak (Shetty & Sethuram, 2019). In fact, low velocity impact can be harmful because no external surface damage can be seen but internally it can be damaged due to the breakage of fibers or cracking of matrices. Therefore, its load-bearing capacity is hampered (S. Z. H. Shah et al., 2019); it is required to understand the behavior of a composite when they are subjected to these impacts, especially low velocity impact (Abir et al., 2017; Dogan & Arikan, 2017).

The prime factors responsible for ameliorate impact properties are resin system and fabric architecture (Dubary et al., 2017), whereas secondary factors are stacking sequence (Lebaupin et al., 2019), fiber and matrix hybridization, repeated impact, and hygrothermal conditions. Woven laminates have better impact properties than unidirectional structures (Baghaei et al., 2015; S. Z. H. Shah et al., 2019). Weave pattern also play a crucial role in determining impact properties as plain fabric laminates exhibited poor impact strength than twill fabric laminates because in twill fabric, pull out of fiber is less (Arju et al., 2015). The highest absorbed energy is shown by commingled non-woven structures when it is 30% fiber volume fraction (Ameer et al., 2019).

Furthermore, enhancement in impact properties is also influenced by the addition of micro-fillers and nano-fillers if and only if interfacial adhesion between fillers and matrices is eminent as 10% by weight of HGM leads to the reduction in impact strength on account of weak bonding (Kumar et al., 2017).

9.7.4 INTERLAMINAR SHEAR PROPERTIES

Interlaminar shear is an important property that is analyzed in order to study de-lamination characteristics of composite structures. A matrix polymer is primarily responsible for influencing ILSS. For scrutinizing the delamination, a short-beam shear (SBS) test is conducted. The advantage of SBS includes no-prerequisite obligation, employment is easy, and requirement of material is low (Espadas-Escalante & Isaksson, 2019). Mold temperature also affects ILSS of woven fabrics, and it is claimed that the highest interlaminar shear strength is shown by a composite when the mold temperature is 180°C (van Rijswijk et al., 2009). Also, the rate at which cooling of a composite is done also impacts ILSS; preferably a slow cooling rate due to the better ILSS in laminates.

Furthermore, fiber treatment can have a notable impact on ILSS of laminates. ILSS can be increased up to a certain limit by plasma treatment due to increased roughness of the surface. Published literature also discusses the reduction in ILSS after quenching treatment and little to enhancement after annealing (Yan et al., 2013).

9.8 IMPLEMENTATION OF NATURAL FIBER THERMOPLASTICS IN INDUSTRIES

Natural fiber–reinforced thermoplastics are a most viable alternative for metals or alloys or ceramic materials and that's why they are being used intensively in automobile industries, aerospace, in the manufacturing of sporting goods, electronic industries, and marine. As we know that NRF-reinforced thermoplastic is known for their incredible properties, also their properties can further be enhanced by adding additives and fillers. In automobile industries, a polypropylene matrix reinforced with natural fiber such as flax, jute, sisal, hemp, etc. is being used for manufacturing various parts. Due its high strength to weight ratio and marketing, it makes it useful in various manufacturing industries rather than technology norm. Due to low weight, less manufacturing cost, rigidness, high specific strength, and NRF thermoplastic composites are being used by German automobile companies like Mercedes Benz, Audi, and Volkswagen. Automobile companies use them for interior as well as exterior of car. The prime producer of these composites in the world is Germany. For a specific capacity engine, if the weight of the car is low, then it ameliorates fuel efficiency and eventually reductes emissions of carbon footprints. Also inside a car, these composites help in absorbing water content in the air and hence reduce humidity, which leads to more comfort.

For structural performance or as constituents for indoor housing, jute-reinforced thermoplastic composites are being used. They are also used in tooling appliances due to their nonabrasive nature. And on account of their nonabrasive nature, they are used in the interiors of automobiles as well as for interior paneling in aircraft for safety of commute (Rohit & Dixit, 2016). Casing mirrors, roofs, helmets, and electrical appliances like voltage stabilizers or projector covers are all fabricated of coir fiber–reinforced composites. As aforementioned, NFR composites are the potential and viable alternative to synthetic fiber composites, but still there is more space for their development and many limitations that have to be overcome in the near future. The adhesion between matrices and fibers is impacted by the hydrophilic nature of natural fibers. Absorbing moisture, large disparity in properties, less thermal stability, gap in composite fabrication temperature, and fiber degradation temperature are the prime challenges that are concerns for researchers and scientist. Also, environmental decay and vulnerability to microorganisms resist the utilization of them at full potential.

CONCLUSION

The ongoing research and recent developments make the utilization of natural fibers sustainable and environment benign. They are above the edge from synthetic fibers in terms of ecological and biodegradability. Also, incredible properties like low manufacturing cost, less density, light weight, and readily available in nature makes them the best possible for the utilization in parts for automotives and aircraft. But stated in various published literature, there are various shortcomings in the development as well as consumption in certain applications of NRF composites.

Due to the high melt viscosity of a thermoplastic polymer matrix, improper impregnation of a polymer matrix with fibers occurs due to the availability of natural fiber is mainly based on short fibers. This affects their structural performance. The solution of this issue is to lower the flow distance. However, the materials that are impregnated to a limited extent, such as fiber bundles, commingled yarns, and thermoplastic film stacking have notably scaled down the flow distance, but they are not of much use. Knitting and weaving may be seen as viable methods for fabrication of such composites without much compromising on structural performance. And among knitting and weaving, the former is a productive method for such development.

Also, due to their chemical nature of natural fibers, there is not sufficient adhesion between the matrix and fibers. Many additives and fillers are there to tackle this problem, but they have to be used in an optimized way as they also affect the mechanical properties of resulting material if they exceed their use limits.

REFERENCES

Abir, M. R., Tay, T. E., Ridha, M., & Lee, H. P. (2017). On the relationship between failure mechanism and compression after impact (CAI) strength in composites. *Composite Structures*, *182*(July), 242–250. 10.1016/j.compstruct.2017.09.038

Akonda, M., Alimuzzaman, S., Shah, D. U., & Rahman, A. N. M. M. (2018). Physico-mechanical, thermal and biodegradation performance of random flax/polylactic acid and unidirectional flax/polylactic acid biocomposites. *Fibers*, *6*(4). 10.3390/fib6040098

Ameer, M. H., Nawab, Y., Ali, Z., Imad, A., & Ahmad, S. (2019). Development and characterization of jute/polypropylene composite by using comingled nonwoven structures. *Journal of the Textile Institute*, *110*(11), 1652–1659. 10.1080/00405000.2 019.1612502

Arju, S. N., Afsar, A. M., Khan, M. A., & Das, D. K. (2015). Effects of jute fabric structures on the performance of jute-reinforced polypropylene composites. *Journal of Reinforced Plastics and Composites*, *34*(16), 1306–1314. 10.1177/0731684415589360

Ashraf, W., Nawab, Y., Maqsood, M., Khan, H., Awais, H., Ahmad, S., Ashraf, M., & Ahmad, S. (2015). Development of seersucker knitted fabric for better comfort properties and aesthetic appearance. *Fibers and Polymers*, *16*(3), 699–701. 10.1007/s12221-015-0699-0

Awais, H., Nawab, Y., Amjad, A., Anjang, A., Md Akil, H., & Zainol Abidin, M. S. (2019). Effect of comingling techniques on mechanical properties of natural fibre reinforced cross-ply thermoplastic composites. *Composites Part B: Engineering*, *177*, 107279. 10.1016/j.compositesb.2019.107279

Awais, H., Nawab, Y., Anjang, A., Md Akil, H., & Zainol Abidin, M. S. (2020). Effect of fabric architecture on the shear and impact properties of natural fibre reinforced composites. *Composites Part B: Engineering*, *195*, 108069. 10.1016/j.compositesb.2020.108069

Azzam, A., & Li, W. (2014). An experimental investigation on the three-point bending behavior of composite laminate. *IOP Conference Series: Materials Science and Engineering*, *62*(1). 10.1088/1757-899X/62/1/012016

Baghaei, B., Skrifvars, M., & Berglin, L. (2015). Characterization of thermoplastic natural fibre composites made from woven hybrid yarn prepregs with different weave pattern. *Composites Part A: Applied Science and Manufacturing*, *76*, 154–161. 10.1016/j.compositesa.2015.05.029

Balan, A. K., Mottakkunnu Parambil, S., Vakyath, S., Thulissery Velayudhan, J., Naduparambath, S., & Etathil, P. (2017). Coconut shell powder reinforced thermoplastic polyurethane/natural rubber blend-composites: Effect of silane coupling agents on the mechanical and thermal properties of the composites. *Journal of Materials Science*, *52*(11), 6712–6725. 10.1007/s10853-017-0907-y

Bar, M., Alagirusamy, R., & Das, A. (2019). Development of flax-PP based twist-less thermally bonded roving for thermoplastic composite reinforcement. *Journal of the Textile Institute*, *110*(10), 1369–1379. 10.1080/00405000.2019.1610997

Bhowmick, M. D. S. (2015). Mechanical properties of unidirectional jute-polyester composite. *Journal of Textile Science & Engineering, 5*(04). 10.4172/2165-8064.1000207

Bourmaud, A., Shah, D. U., Beaugrand, J., & Dhakal, H. N. (2020). Property changes in plant fibres during the processing of bio-based composites. *Industrial Crops and Products, 154*(June), 112705. 10.1016/j.indcrop.2020.112705

Calhoun, A. (2010). Polypropylene. In J. R. Wagner (Ed.), *Multilayer Flexible Packaging* (Sixth Edition). (pp. 31–36). William Andrew Publishing. 10.1016/B978-0-8155-2 021-4.10003-6

Charière, R., Marano, A., & Gélébart, L. (2020). Use of composite voxels in FFT based elastic simulations of hollow glass microspheres/polypropylene composites. *International Journal of Solids and Structures, 182–183*, 1–14. 10.1016/j.ijsolstr.2019.08.002

Chen, S., Liu, M., Huang, H., Cheng, L., & Zhao, H. P. (2019). Mechanical properties of Bombyx mori silkworm silk fibre and its corresponding silk fibroin filament: A comparative study. *Materials and Design, 181*, 108077. https://doi.org/10.1016/j.matdes.2019.108077

Debnath, S. (2017). Sustainable production of bast fibres. In S. Muthu (Ed.), *Sustainable Fibres and Textiles* (pp. 69–85). Elsevier. 10.1016/B978-0-08-102041-8.00003-2

Dhakal, H. N., Zhang, Z. Y., & Richardson, M. O. W. (2007). Effect of water absorption on the mechanical properties of hemp fibre reinforced unsaturated polyester composites. *Composites Science and Technology, 67*(7–8), 1674–1683. 10.1016/j.compscitech.2 006.06.019

Dickson, A. R., Even, D., Warnes, J. M., & Fernyhough, A. (2014). The effect of re-processing on the mechanical properties of polypropylene reinforced with wood pulp, flax or glass fibre. *Composites Part A: Applied Science and Manufacturing, 61*, 258–267. 10.1016/j.compositesa.2014.03.010

Dixit, S., Goel, R., Dubey, A., Shivhare, P. R., & Bhalavi, T. (2017). Natural fibre reinforced polymer composite materials – A review. *Polymers from Renewable Resources, 8*(2), 71–78. 10.1177/204124791700800203

Dogan, A., & Arikan, V. (2017). Low-velocity impact response of E-glass reinforced thermoset and thermoplastic based sandwich composites. *Composites Part B: Engineering, 127*, 63–69. 10.1016/j.compositesb.2017.06.027

Dong, C., Ranaweera-Jayawardena, H. A., & Davies, I. J. (2012). Flexural properties of hybrid composites reinforced by S-2 glass and T700S carbon fibres. *Composites Part B: Engineering, 43*(2), 573–581. 10.1016/j.compositesb.2011.09.001

Donmez Cavdar, A., Boran Torun, S., Ertas, M., & Mengeloglu, F. (2019). Ammonium zeolite and ammonium phosphate applied as fire retardants for microcrystalline cellulose filled thermoplastic composites. *Fire Safety Journal, 107*, 202–209. 10.1016/j.firesaf.2018.11.008

Dubary, N., Taconet, G., Bouvet, C., & Vieille, B. (2017). Influence of temperature on the impact behavior and damage tolerance of hybrid woven-ply thermoplastic laminates for aeronautical applications. *Composite Structures, 168*, 663–674. 10.1016/j.compstruct.2017.02.040

Dun, M., Hao, J., Wang, W., Wang, G., & Cheng, H. (2019). Sisal fiber reinforced high density polyethylene pre-preg for potential application in filament winding. *Composites Part B: Engineering, 159*, 369–377. 10.1016/j.compositesb.2018.09.090

Dunne, R., Desai, D., Sadiku, R., & Jayaramudu, J. (2016). A review of natural fibres, their sustainability and automotive applications. *Journal of Reinforced Plastics and Composites, 35*(13), 1041–1050. 10.1177/0731684416633898

Durante, M., Formisano, A., Boccarusso, L., Langella, A., & Carrino, L. (2017). Creep behaviour of polylactic acid reinforced by woven hemp fabric. *Composites Part B: Engineering, 124*, 16–22. 10.1016/j.compositesb.2017.05.038

El-Sabbagh, A. (2014). Effect of coupling agent on natural fibre in natural fibre/poly-propylene composites on mechanical and thermal behaviour. *Composites Part B: Engineering*, *57*, 126–135. 10.1016/j.compositesb.2013.09.047

Espadas-Escalante, J. J., & Isaksson, P. (2019). A study of induced delamination and failure in woven composite laminates subject to short-beam shear testing. *Engineering Fracture Mechanics*, *205*(October), 359–369. 10.1016/j.engfracmech.2018.10.015

Farah, S., Anderson, D. G., & Langer, R. (2016). Physical and mechanical properties of PLA, and their functions in widespread applications—A comprehensive review. *Advanced Drug Delivery Reviews*, *107*, 367–392. 10.1016/j.addr.2016.06.012

Faruk, O., Bledzki, A. K., Fink, H. P., & Sain, M. (2012). Biocomposites reinforced with natural fibers: 2000-2010. *Progress in Polymer Science*, *37*(11), 1552–1596. 10.1016/j.progpolymsci.2012.04.003

Formisano, A., Papa, I., Lopresto, V., & Langella, A. (2019). Influence of the manufacturing technology on impact and flexural properties of GF/PP commingled twill fabric la-minates. *Journal of Materials Processing Technology*, *274*(June), 116275. 10.1016/j.jmatprotec.2019.116275

George, J., Sreekala, M. S., & Thomas, S. (2001). A review on interface modification and characterization of natural fiber reinforced plastic composites. *Polymer Engineering and Science*, *41*(9), 1471–1485. 10.1002/pen.10846

Gopanna, A., Rajan, K. P., Thomas, S. P., & Chavali, M. (2019). Polyethylene and polypropylene matrix composites for biomedical applications. In V. Grumezescu, & A. M. Grumezescu (Eds.), *Materials for Biomedical Engineering: Thermoset and Thermoplastic Polymers*. Elsevier. 10.1016/B978-0-12-816874-5.00006-2

Goriparthi, B. K., Suman, K. N. S., & Mohan Rao, N. (2012). Effect of fiber surface treatments on mechanical and abrasive wear performance of polylactide/jute compo-sites. *Composites Part A: Applied Science and Manufacturing*, *43*(10), 1800–1808. 10.1016/j.compositesa.2012.05.007

Hagstrand, P. O., Bonjour, F., & Månson, J. A. E. (2005). The influence of void content on the structural flexural performance of unidirectional glass fibre reinforced poly-propylene composites. *Composites Part A: Applied Science and Manufacturing*, *36*(5), 705–714. 10.1016/j.compositesa.2004.03.007

Hammami, A., & Gebart, B. R. (2000). Analysis of the vacuum infusion molding process. *Polymer Composites*, *21*(1), 28–40. 10.1002/pc.10162

Hao, L. C., Sapuan, S. M., Hassan, M. R., & Sheltami, R. M. (2018). Natural fiber reinforced vinyl polymer composites. In S. M. Sapuan, & H. Ismail (Eds.), *Natural Fibre Reinforced Vinyl Ester and Vinyl Polymer Composites* (pp. 27–70). Elsevier. 10.1016/b978-0-08-102160-6.00002-0

Hasani, H., Hassanzadeh, S., Abghary, M. J., & Omrani, E. (2017). Biaxial weft-knitted fabrics as composite reinforcements: A review. *Journal of Industrial Textiles*, *46*(7), 1439–1473. 10.1177/1528083715624256

Hosseini Ravandi, S. A., & Valizadeh, M. (2011). Properties of fibers and fabrics that contribute to human comfort. In G. Song (Ed.), *Improving Comfort in Clothing* (pp. 61–78). Elsevier Masson SAS. 10.1533/9780857090645.1.61

Hsieh, Y. L. (2007). Chemical structure and properties of cotton. In S. Gordon & Y. L. Hsieh (Eds.), *Cotton: Science and Technology* (pp. 3–34). Woodhead Publishing. 10.1533/9781845692483.1.3

Jaafar, J., Siregar, J. P., Tezara, C., Hamdan, M. H. M., & Rihayat, T. (2019). A review of important considerations in the compression molding process of short natural fiber composites. *International Journal of Advanced Manufacturing Technology*, *105*(7–8), 3437–3450. 10.1007/s00170-019-04466-8

Jamir, M. R. M., Majid, M. S. A., & Khasri, A. (2018). Natural lightweight hybrid com-posites for aircraft structural applications. In M. Jawaid, & M. Thariq (Eds.),

Sustainable Composites for Aerospace Applications (pp. 155–170). Elsevier. 10.1016/B978-0-08-102131-6.00008-6

John, M. J., & Anandjiwala, R. D. (2009). Chemical modification of flax reinforced polypropylene composites. *Composites Part A: Applied Science and Manufacturing, 40*(4), 442–448. 10.1016/j.compositesa.2009.01.007

Kandola, B. K., Mistik, S. I., Pornwannachai, W., & Anand, S. C. (2018). Natural fibre-reinforced thermoplastic composites from woven-nonwoven textile preforms: Mechanical and fire performance study. *Composites Part B: Engineering, 153,* 456–464. 10.1016/j.compositesb.2018.09.013

Kang, D. H., Hwang, S. W., Jung, B. N., & Shim, J. K. (2017). Effect of hollow glass microsphere (HGM) on the dispersion state of single-walled carbon nanotube (SWNT). *Composites Part B: Engineering, 117,* 35–42. 10.1016/j.compositesb.2017.02.038

Karaduman, N. S., Karaduman, Y., Ozdemir, H., & Ozdemir, G. (2017). Textile reinforced structural composites for advanced applications. In B. Kumar & S. Thakur (Eds.), *Textiles for Advanced Applications.* IntechOpen. 10.5772/intechopen.68245

Karthikeyan, G., Nalankilli, G., Shanmugasundaram, O. L., & Prakash, C. (2016). Thermal comfort properties of bamboo tencel knitted fabrics. *International Journal of Clothing Science and Technology, 28*(4), 420–428. http://dx.doi.org/10.1108/IJCST-08-2015-0086%0Ahttp://dx.doi.org/10.1108/%0Ahttp://

Khan, M. Z. R., Srivastava, S. K., & Gupta, M. K. (2018). Tensile and flexural properties of natural fiber reinforced polymer composites: A review. *Journal of Reinforced Plastics and Composites, 37*(24), 1435–1455. 10.1177/0731684418799528

Kim, D. J., Yu, M. H., Lim, J., Nam, B., & Kim, H. S. (2019). Prediction of the mechanical behavior of fiber-reinforced composite structure considering its shear angle distribution generated during thermo-compression molding process. *Composite Structures, 220*(October 2018), 441–450. 10.1016/j.compstruct.2019.04.043

Kim, Y. K., & Chalivendra, V. (2020). Natural fibre composites (NFCs) for construction and automotive industries. In D. Verma, M. Sharma, K. L. Goh, S. Jain, & H. Sharma (Eds.), *Handbook of Natural Fibres: Volume 2: Processing and Applications* (pp. 113–128). Elsevier. 10.1016/B978-0-12-818782-1.00014-6

Komuraiah, A., Kumar, N. S., & Prasad, B. D. (2014). Chemical composition of natural fibers and its influence on their mechanical properties. *Mechanics of Composite Materials, 50*(3), 359–376. 10.1007/s11029-014-9422-2

Kumar, N., Mireja, S., Khandelwal, V., Arun, B., & Manik, G. (2017). Light-weight high-strength hollow glass microspheres and bamboo fiber based hybrid polypropylene composite: A strength analysis and morphological study. *Composites Part B: Engineering, 109,* 277–285. 10.1016/j.compositesb.2016.10.052

Kuranchie, C., Yaya, A., & Bensah, Y. D. (2021). The effect of natural fibre reinforcement on polyurethane composite foams – A review. *Scientific African, 11,* e00722. 10.1016/j.sciaf.2021.e00722

Lebaupin, Y., Hoang, T. Q. T., Chauvin, M., & Touchard, F. (2019). Influence of the stacking sequence on the low-energy impact resistance of flax/PA11 composite. *Journal of Composite Materials, 53*(22), 3187–3198. 10.1177/0021998319837339

Lu, N., & Oza, S. (2013). A comparative study of the mechanical properties of hemp fiber with virgin and recycled high density polyethylene matrix. *Composites Part B: Engineering, 45*(1), 1651–1656. 10.1016/j.compositesb.2012.09.076

Mahmud Zuhudi, N. Z., Jayaraman, K., & Lin, R. J. T. (2016). Mechanical, thermal and in-strumented impact properties of bamboo fabric-reinforced polypropylene composites. *Polymers and Polymer Composites, 24*(9), 755–766. 10.1177/096739111602400912

Malkapuram, R., Kumar, V., & Singh Negi, Y. (2009). Recent development in natural fiber reinforced polypropylene composites. *Journal of Reinforced Plastics and Composites, 28*(10), 1169–1189. 10.1177/0731684407087759

Mallick, P. K. (Ed.) (2010a). Thermoplastics and thermoplastic-matrix composites for lightweight automotive structures. In *Materials, Design and Manufacturing for Lightweight Vehicles* (pp. 174–207). Woodhead Publishing. 10.1533/9781845697822.1.174

Mallick, P. K. (Ed.) (2010b). Thermoset-matrix composites for lightweight automotive structures. In *Materials, Design and Manufacturing for Lightweight Vehicles* (pp. 208–231). Woodhead Publishing. 10.1533/9781845697822.1.208

Mao, N., & Russell, S. J. (2015). Fibre to fabric: Nonwoven fabrics. In R. Sinclair (Ed.), *Textiles and Fashion: Materials, Design and Technology* (pp. 307–335). Elsevier. 10.1016/B978-1-84569-931-4.00013-1

Mcgregor, B. A., & Butler, K. L. (2016). Coarser wool is not a necessary consequence of sheep aging: Allometric relationship between fibre diameter and fleece-free liveweight of Saxon Merino sheep. *Animal*, *10*(12), 2051–2060. 10.1017/S1751731116001038

Medina, L., Schledjewski, R., & Schlarb, A. K. (2009). Process related mechanical properties of press molded natural fiber reinforced polymers. *Special Issue on the 12th European Conference on Composite Materials, ECCM 2006*, *69*(9), 1404–1411. 10.1016/j.compscitech.2008.09.017

Mehta, R., Kumar, V., Bhunia, H., & Upadhyay, S. N. (2005). Synthesis of poly (lactic acid): A review. *Journal of Macromolecular Science –Polymer Reviews*, *45*(4), 325–349. 10.1080/15321790500304148

Meola, C., Boccardi, S., & Carlomagno, G. M. (Eds.). (2017). Composite materials in the aeronautical industry. In *Infrared Thermography in the Evaluation of Aerospace Composite Materials* (pp. 1–24). Woodhead Publishing. 10.1016/b978-1-78242-171-9.00001-2

Merotte, J., Le Duigou, A., Kervoelen, A., Bourmaud, A., Behlouli, K., Sire, O., & Baley, C. (2018). Flax and hemp nonwoven composites: The contribution of interfacial bonding to improving tensile properties. *Polymer Testing*, *66*, 303–311. 10.1016/j.polymertesting.2018.01.019

Miller, B. (1990). Resin transfer molding. *Plastics World*, *48*(6), 60–64. 10.1016/B978-0-08-099922-7.00010-X

Mohammed, L., Ansari, M. N. M., Pua, G., Jawaid, M., & Islam, M. S. (2015). A review on natural fiber reinforced polymer composite and its applications. *International Journal of Polymer Science*, *2015*. 10.1155/2015/243947

Mohammed, A. A., Bachtiar, D., Rejab, M. R. M., & Siregar, J. P. (2018). Effect of microwave treatment on tensile properties of sugar palm fibre reinforced thermoplastic polyurethane composites. *Defence Technology*, *14*(4), 287–290. 10.1016/j.dt.2018.05.008

Morin-Crini, N., Loiacono, S., Placet, V., Torri, G., Bradu, C., Kostić, M., Cosentino, C., Chanet, G., Martel, B., Lichtfouse, E., & Crini, G. (2019). Hemp-based adsorbents for sequestration of metals: A review. *Environmental Chemistry Letters*, *17*(1), 393–408. 10.1007/s10311-018-0812-x

Muhammad, A., Rahman, M. R., Hamdan, S., & Sanaullah, K. (2019). Recent developments in bamboo fiber-based composites: A review. *Polymer Bulletin*, *76*(5), 2655–2682. 10.1007/s00289-018-2493-9

Mukherjee, T., & Kao, N. (2011). PLA based biopolymer reinforced with natural fibre: A review. *Journal of Polymers and the Environment*, *19*(3), 714–725. 10.1007/s10924-011-0320-6

Nix, A., Paull, C., & Colgrave, M. (2017). Flavonoid profile of the cotton plant, Gossypium hirsutum: A review. *Plants*, *6*(4), 43. https://doi.org/10.3390/plants6040043

Obande, W., Mamalis, D., Ray, D., Yang, L., & Ó Brádaigh, C. M. (2019). Mechanical and thermomechanical characterisation of vacuum-infused thermoplastic- and thermoset-based composites. *Materials and Design*, *175*, 107828. 10.1016/j.matdes.2019.107828

Pailoor, S., Murthy, H. N. N., Hadimani, P., & Sreenivasa, T. N. (2019). Effect of chopped/ continuous fiber, coupling agent and fiber ratio on the mechanical properties of injection-molded jute/polypropylene composites. *Journal of Natural Fibers, 16*(1), 126–136. 10.1080/15440478.2017.1410510

Parbin, S., Waghmare, N. K., Singh, S. K., & Khan, S. (2019). Mechanical properties of natural fiber reinforced epoxy composites: A review. *Procedia Computer Science, 152*, 375–379. 10.1016/j.procs.2019.05.003

Park, S. J., & Seo, M. K. (Eds.). (2011). Element and processing. In *Interface Science and Technology* (Vol. 18) (pp. 431–499). Elsevier. 10.1016/B978-0-12-375049-5.00006-2

Patel, R. M. (2016). Polyethylene. In J. R. Wagner, Jr (Ed.), *Multilayer Flexible Packaging* (Second Edition) (pp. 17–34). William Andrew. 10.1016/B978-0-323-37100-1.00002-8

Patou, J., Bonnaire, R., De Luycker, E., & Bernhart, G. (2019). Influence of consolidation process on voids and mechanical properties of powdered and commingled carbon/PPS laminates. *Composites Part A: Applied Science and Manufacturing, 117*, 260–275. 10.1016/j.compositesa.2018.11.012

Pegoretti, A. (2021). Recycling concepts for short-fiber-reinforced and particle-filled thermoplastic composites: A review. *Advanced Industrial and Engineering Polymer Research, 4*(2), 93–104. 10.1016/j.aiepr.2021.03.004

Pourmohammadi, A. (2013). Nonwoven materials and joining techniques. In I. Jones & G. K. Stylios (Eds.), *Joining Textiles: Principles and Applications* (pp. 565–581). Woodhead Publishing. 10.1533/9780857093967.4.565

Pretula, J., Slomkowski, S., & Penczek, S. (2016). Polylactides—Methods of synthesis and characterization. *Advanced Drug Delivery Reviews, 107*, 3–16. 10.1016/j.addr.2016.05.002

Rahman, R., & Putra, S. Z. F. S. (2018). Tensile properties of natural and synthetic fiber-reinforced polymer composites. In *Mechanical and Physical Testing of Biocomposites, Fibre-Reinforced Composites and Hybrid Composites* (pp. 81–102). Elsevier. 10.1016/B978-0-08-102292-4.00005-9

Raina, M. A., Gloy, Y. S., & Gries, T. (2015). Weaving technologies for manufacturing denim. In *Denim: Manufacture, Finishing and Applications* (pp. 159–187). Elsevier. 10.1016/B978-0-85709-843-6.00006-8

Ramesh, M. (2019). Flax (Linum usitatissimum L.) fibre reinforced polymer composite materials: A review on preparation, properties and prospects. *Progress in Materials Science, 102*, 109–166. 10.1016/j.pmatsci.2018.12.004

Rawal, A., Shah, T. H., & Anand, S. C. (2016). Geotextiles in civil engineering. In A. Richard Horrocks, & S. C. Anand (Eds.), *Handbook of Technical Textiles* (Second Edition), (pp. 111–133). Woodhead Publishing.

Rohit, K., & Dixit, S. (2016). A review – Future aspect of natural fiber reinforced composite. *Polymers from Renewable Resources, 7*(2), 43–60. 10.1177/204124791600700202

Ronca, S. (2017). Polyethylene. In *Brydson's Plastics Materials: Eighth Edition* (pp. 247–278). Elsevier. 10.1016/B978-0-323-35824-8.00010-4

Saidane, E. H., Scida, D., Assarar, M., & Ayad, R. (2016). Assessment of 3D moisture diffusion parameters on flax/epoxy composites. *Composites Part A: Applied Science and Manufacturing, 80*, 53–60. 10.1016/j.compositesa.2015.10.008

Samanta, K. K., Basak, S., & Chattopadhyay, S. K. (2016). Potential of ligno-cellulosic and protein fibres in sustainable fashion. In S. Senthilkannan Muthu, & M. Gardetti (Eds.), *Environmental Footprints and Eco-Design of Products and Processes*. 10.1007/978-981-10-0566-4_5

Sastri, V. R. (2014). Commodity Thermoplastics. In *Plastics in Medical Devices* (pp. 73–120). William Andrew. 10.1016/b978-1-4557-3201-2.00006-9

Selvaraj, D. K., Silva, F. J. G., Campilho, R. D. S. G., Baptista, A., & Pinto, G. F. L. (2019). Influence of the natural additive on natural fiber reinforced thermoplastic composite. *Procedia Manufacturing, 38*(2019), 1121–1129. 10.1016/j.promfg.2020.01.200

Selver, E. (2020). Tensile and flexural properties of glass and carbon fibre composites re-inforced with silica nanoparticles and polyethylene glycol. *Journal of Industrial Textiles*, *49*(6), 809–832. 10.1177/1528083719827368

Shah, S. Z. H., Karuppanan, S., Megat-Yusoff, P. S. M., & Sajid, Z. (2019). Impact resistance and damage tolerance of fiber reinforced composites: A review. *Composite Structures*, *217*, 100–121. 10.1016/j.compstruct.2019.03.021

Shah, A. U. M., Sultan, M. T. H., Jawaid, M., Cardona, F., & Talib, A. R. A. (2016). A review on the tensile properties of bamboo fiber reinforced polymer composites. *BioResources*, *11*(4), 10654–10676. 10.15376/biores.11.4.Shah

Shetty, H., & Sethuram, D. (2019). Low velocity response of GFRP and hybrid laminates under impact testing. *Journal of Dynamic Behavior of Materials*, *5*(2), 150–160. 10.1 007/s40870-019-00194-y

Shoja, M., Mohammadi-Roshandeh, J., Hemmati, F., Zandi, A., & Farizeh, T. (2020). Plasticized starch-based biocomposites containing modified rice straw fillers with thermoplastic, thermoset-like and thermoset chemical structures. *International Journal of Biological Macromolecules*, *157*, 715–725. 10.1016/j.ijbiomac.2019.11.236

Shubhra, Q. T. H., Alam, A. K. M. M., & Quaiyyum, M. A. (2013). Mechanical properties of polypropylene composites: A review. *Journal of Thermoplastic Composite Materials*, *26*(3), 362–391. 10.1177/0892705711428659

Spasojevic, P. M. (2019). Thermal and rheological properties of unsaturated polyester resins-based composites. In S. Thomas, M. Hosur, & C. J. Chirayil (Eds.), *Unsaturated Polyester Resins: Fundamentals, Design, Fabrication, and Applications* (pp. 367–406). Elsevier. 10.1016/B978-0-12-816129-6.00015-6

Su, S., Kopitzky, R., Tolga, S., & Kabasci, S. (2019). Polylactide (PLA) and its blends with poly (butylene succinate) (PBS): A brief review. *Polymers*, *11*(7), 1–21. 10.3390/polym11071193

Sullins, T., Pillay, S., Komus, A., & Ning, H. (2017). Hemp fiber reinforced polypropylene composites: The effects of material treatments. *Composites Part B: Engineering*, *114*, 15–22. 10.1016/j.compositesb.2017.02.001

Tatara, R. A. (2011). Compression Molding. In M. Kutz (Ed.), *Applied Plastics Engineering Handbook* (pp. 289–309). Elsevier. 10.1016/B978-1-4377-3514-7.10017-0

van Rijswijk, K., van Geenen, A. A., & Bersee, H. E. N. (2009). Textile fiber-reinforced anionic polyamide-6 composites. Part II: Investigation on interfacial bond formation by short beam shear test. *Composites Part A: Applied Science and Manufacturing*, *40*(8), 1033–1043. 10.1016/j.compositesa.2009.02.018

Verma, D., & Fortunati, E. (2019). Biopolymer processing and its composites: An introduction. In D. Verma, E. Fortunati, S. Jain, & X. Zhang (Eds.), *Biomass, Biopolymer-Based Materials, and Bioenergy: Construction, Biomedical, and other Industrial Applications* (pp. 3–23). Elsevier. 10.1016/B978-0-08-102426-3.00001-1

Wirawan, R. (2020). Recyclability of natural fiber-filled thermoplastic composites. In S. Hashmi, & I. A. Choudhury (Eds.), *Encyclopedia of Renewable and Sustainable Materials*. Elsevier. 10.1016/b978-0-12-803581-8.11293-7

Yallew, T. B., Kassegn, E., Aregawi, S., & Gebresias, A. (2020). Study on effect of process parameters on tensile properties of compression molded natural fiber reinforced polymer composites. *SN Applied Sciences*, *2*(3), 338. 10.1007/s42452-020-2101-0

Yan, C., Li, H., Zhang, X., Zhu, Y., Fan, X., & Yu, L. (2013). Preparation and properties of continuous glass fiber reinforced anionic polyamide-6 thermoplastic composites. *Materials and Design*, *46*, 688–695. 10.1016/j.matdes.2012.11.034

Yao, S. S., Jin, F. L., Rhee, K. Y., Hui, D., & Park, S. J. (2018). Recent advances in carbon-fiber-reinforced thermoplastic composites: A review. *Composites Part B: Engineering*, *142*, 241–250. 10.1016/j.compositesb.2017.12.007

Yassin, K., & Hojjati, M. (2018). Processing of thermoplastic matrix composites through automated fiber placement and tape laying methods: A review. *Journal of Thermoplastic Composite Materials*, *31*(12), 1676–1725. 10.1177/0892705717738305

Zin, M. H., Razzi, M. F., Othman, S., Liew, K., Abdan, K., & Mazlan, N. (2016). A review on the fabrication method of bio-sourced hybrid composites for aerospace and automotive applications. *IOP Conference Series: Materials Science and Engineering*, *152*(1), 1–12. 10.1088/1757-899X/152/1/012041

10 Biodegradable Polymer-Based Natural Fiber Composites

Manash Protim Mudoi
Department of Chemical Engineering, Indian Institute of
Technology, Roorkee, Uttrakhand, India

Department of Chemical Engineering, University of
Petroleum and Energy Studies, Dehradun, Uttrakhand, India

Shishir Sinha
Department of Chemical Engineering, Indian Institute of
Technology, Roorkee, Uttrakhand, India

INTRODUCTION

Preserving natural resources and maintaining a clean environment for the eco-friendly existence of all living beings is the biggest priority of today's world. Intense market competition, demand for low-cost, high-performance products, and low energy utilization have led the industry to look for innovative and novel materials. The composite manufacturing industry is up against those challenges by developing the methods and techniques to prepare biodegradable composite materials. Synthetic fibers like glass, aramid, and ceramic, which are detrimental to the environment, are progressively replaced with natural fibers like flax, ramie, jute, sisal, coir, nettle, rich husk, and kenaf (Mahmud et al., 2021; Singh et al., 2020; Zaini et al., 2018). Natural fibers have been used since prehistoric ages, and various archeological evidence of natutal fiber usage in clothing and ropes are found (Hu et al., 2020; Muthu, 2016). In the last few decades, natural fibers are used extensively in the textiles and the highly engineered and technically complex sectors like automobile, aerospace, and spacecraft as reinforcement in biocomposite materials. Acceptance of natural fibers than synthetic fibers is growing due to the easy biodegradability and recyclability, low density, desired fiber aspect ratio, low abrasive behavior, sufficient thermal stability, tensile strength, good acoustic and electrical insulation, low cost, and marginal environmental hazard (Gholampour & Ozbakkaloglu, 2020; Khalid et al., 2021).

Synthetic polymers like polyethylene, polyesters, epoxy, and polypropylene have been widely investigated and used with natural fibers for composite synthesis for the last couple of decades. The objective of biodegradation and recyclability is hard to achieve with synthetic polymer-based biocomposites. In this regard, biodegradable

DOI: 10.1201/9781003201724-10

polymeric matrix and natural fiber reinforcement will be suitable combinations to
attain the objective of biodegradability. However, biodegradable polymers may not be
the ultimate solution to all problems of plastic pollution. Yet, their utilization as an
alternative to non-degradable conventional polymers will improve the environmental
deterioration to a large extent (Zumstein et al., 2019). The global production of
biodegradable bioplastics is expected to reach 1,800 metric tons in 2025 from 1,227
metric tons in 2020 (Figure 10.1).

The development of a suitable biodegradable polymer is crucial for the synthesis of
biocomposite for a specific application. Initially, biopolymers were used in farming,
packaging, and industries requiring low-strength material. High cost and limitation in
performance are significant issues hindering biopolymer's mass acceptance. For some
biopolymers, the low production volume is crucial for having a higher cost (Mohnty
et al., 2005). Avoiding the degradation of the biopolymers and natural fibers during
storage and usage, while achieving the desired degradation at the end of the life cycle
is a significant challenge when preparing the advanced biodegradable polymer
composites. Cellulose acetate (Asyraf et al., 2021), poly(hydroxyalkanoate) (Soon
et al., 2019), and poly(lactic acid) (Pan et al., 2021; Reddy et al., 2021) are some of the
biodegradable polymers. They are used with natural fibers like jute, hemp, kenaf,
bamboo, wood flour, abaca, and flax to prepare biocomposite materials (Visakh et al.,
2019). Even after having promising properties, biocomposites are subjected to
drawbacks like poor fiber wettability and poor interfacial adhesivity between the
hydrophobic polymer matrix and polar hydrophilic natural fiber. This incompatibility
results in low stiffness, strength, and moisture absorbance. Fiber surface modification
and introducing compatible coupling agents enhance the mechanical and physical

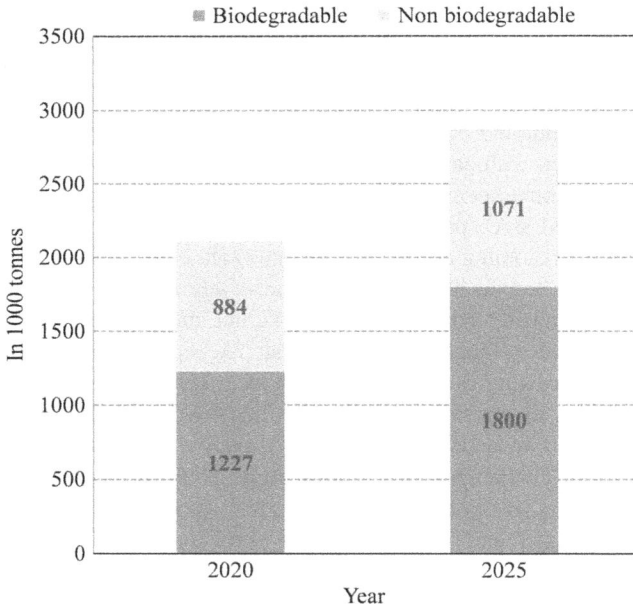

FIGURE 10.1 Global bioplastic production (European bioplastics, 2021).

properties of the biocomposite material. In this chapter, the current research trend and the applications of biodegradable composite material are discussed.

10.1 BIODEGRADABLE POLYMERS

Biocomposites are those materials where at least one of the materials is derived from the natural product. In most cases, the reinforcement material is naturally derived, and the matrix is derived from petroleum-based products termed *synthetic polymers*. Another class of material known as green composites (a subset of biocomposite) is prepared using a naturally derived polymeric matrix combining with natural fiber reinforcing. These green composites are easily degraded at the final stage of their life cycle without causing harmful effects to the environment (Gurunathan et al., 2015). The biodegradable biocomposites can be classified mainly into two broad categories – fully degradable and partially degradable composite material (Faruk et al., 2012). This classification is entirely based on the type of matrix material used. The degradable polymer can be derived mainly by three routes: (a) renewable source material, (b) petroleum derived, and (c) mixed source derived (Reddy et al., 2013). In the first route, the polymers are synthesized from renewable resources like animals or plants, including various products like proteins, lignin, cellulose, starch, chitosan, polyhydroxyalkanoates, polyhydroxy butyrate (PHB), and polylactic acid. In recent technological development, nylon, polypropylene, and polyethylene polymers can be derived from bioresources. In the second route, derived polymers are petroleum product-based, but they are subjected to biodegradation. Poly (butylene adipate-co-terephthalate) (PBAT) and polycaprolactone (PCL) are known polymers synthesized with this route. The third category includes the polymers produced from the mixture of petroleum and natural-based monomers like bio-thermosets (bio-epoxy) and poly (trimethylene terephthalate) (PTT). The PTT can be manufactured from terephthalic acid (petroleum-based) and 1, 3-propanediol (bio-derived). Figure 10.2 shows the various sources of degradable polymers.

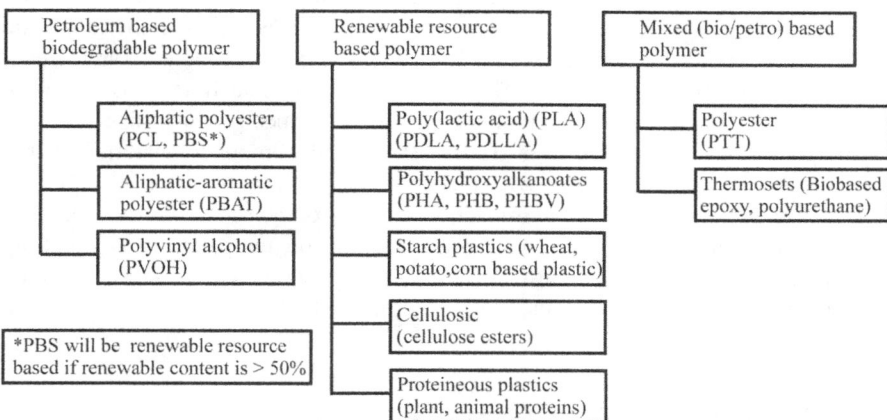

Petroleum based biodegradable polymer	Renewable resource based polymer	Mixed (bio/petro) based polymer
Aliphatic polyester (PCL, PBS*)	Poly(lactic acid) (PLA) (PDLA, PDLLA)	Polyester (PTT)
Aliphatic-aromatic polyester (PBAT)	Polyhydroxyalkanoates (PHA, PHB, PHBV)	Thermosets (Biobased epoxy, polyurethane)
Polyvinyl alcohol (PVOH)	Starch plastics (wheat, potato,corn based plastic)	
*PBS will be renewable resource based if renewable content is > 50%	Cellulosic (cellulose esters)	
	Proteineous plastics (plant, animal proteins)	

FIGURE 10.2 Classification of production routes for biodegradable polymers (Reddy et al., 2013).

Typically, the microbial and/or chemical polymerization route is used to synthesize these biopolymers (Chen & Patel, 2012; Gross & Kalra, 2002). The structure and synthesis routes of these materials are presented in Figure 10.3.

It is to be noted that not all bio-based polymers are biodegradable (Rai et al., 2021). Typically, the chemical structure, not the source, affects the degradability of these polymers. For instance, PBAT, which is petroleum-derived, is subjected to biodegradation, but naturally derived polyethylene (PE) is not degradable. The complex molecular structure of the polymer is attacked by the enzymes produced from the microorganisms and converted the polymers into simpler and smaller molecular structures like CO_2, CH_4, and biomass (Mohnty et al., 2005). Hydrolysis, oxidation, and photodegradation are the mechanisms for biodegradation (Satyanarayana et al., 2009). The biopolymers may not be degradable completely, but they show recyclable behavior (Arikan & Ozsoy, 2015). The degradation of these biopolymers occurs by the enzyme created by the microorganisms like algae, fungi, and bacteria (Wojnowska-Baryła et al., 2020). The polymers can be broken down by chemical hydrolysis, which is a non-enzymatic process. The ultimate degradation products are CH_4, CO_2, H_2O, humic matter, and biomass (Kumar et al., 2020). The recycling process of biodegradable polymers can be represented in Figure 10.4. The success of the biodegradation lies in the fact that the receiving environment should have the capacity to convert the polymer compounds into CH_4 and CO_2, and there should exist favorable incubation conditions like time, relative humidity, and temperature to catalyze the degradation process (Zumstein et al., 2019).

Bioplastics have proteins and polysaccharides as the main chemical constituents (Rai et al., 2021), which differ significantly from conventional plastics in composition and properties. Traditional plastics are superior in mechanical and physical character. Extensive research has been conducted to improve bioplastics' physical and chemical behavior to bring closure to non-degradable conventional plastics (Rai et al., 2021). It is now possible to reduce the carbon footprint of crops like banana, corn, stubble, husk, jute, vegetable waste, etc., by utilizing the garbage to synthesize more economically valuable bioplastic material. Lu et al., 2005 prepared a biodegradable plastic by blending polyurethane (castor oil derive) and thermoplastic starch, which exhibited improved tensile strength, Young's modulus, and breakpoint elongation (Lu et al., 2005). The material properties such as glass transition temperature (Tg), melting temperature (Tm), density, mechanical properties, and degradation time are presented in Table 10.1. Final composite material weight depends on the density of the matrix and reinforced fibrous material. The degradation time is essential for the impact on the environment. The thermal stability of the final composite depends on the glass transition temperature and melting temperature. Therefore, selecting the best-suited polymer matrix and reinforced fiber is crucial before applying the composite material for a specific purpose (Van de Velde & Kiekens, 2002).

10.2   NATURAL FIBER BIOCOMPOSITES

Composite materials are prepared by combining two or more compatible materials to get a final product with improved physical and mechanical properties. Generally,

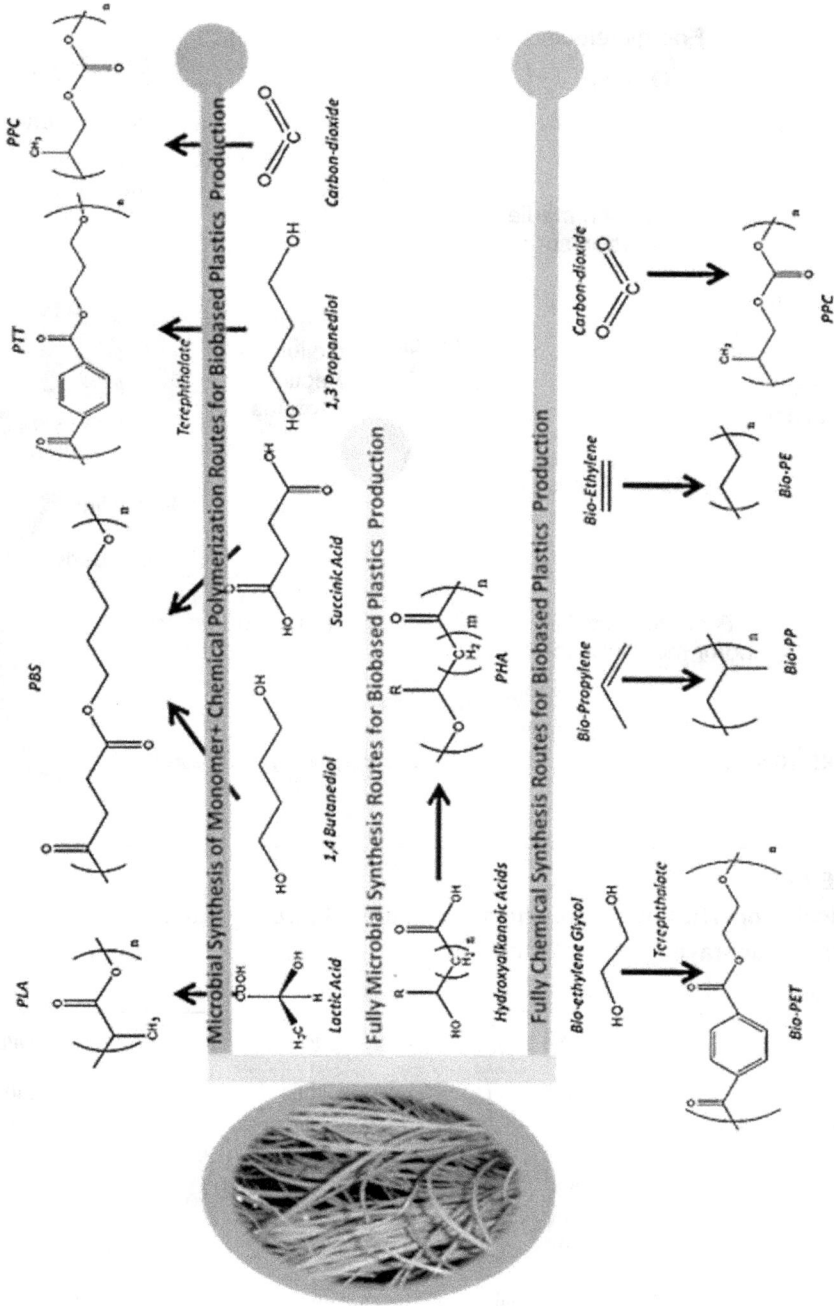

FIGURE 10.3 Various routes of production for biodegradable polymers. Reprinted with permission from (Reddy et al., 2013). Copyright © 2013 Elsevier.

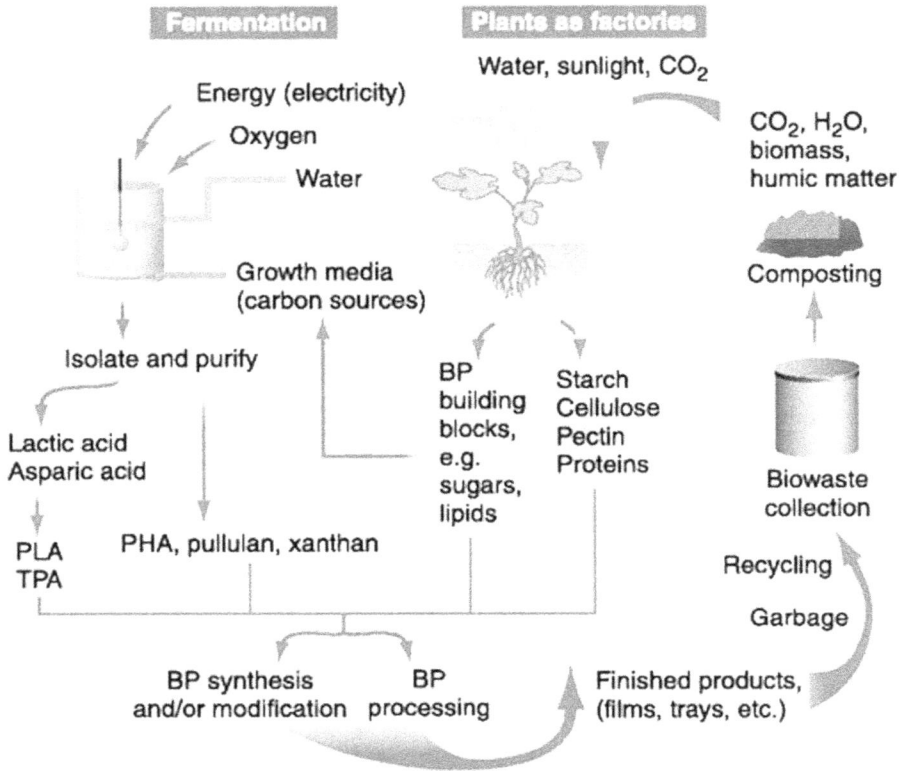

FIGURE 10.4 Recycling process of biodegradable polymers (Huyhua, 2010).

TABLE 10.1

Physical Properties of Biopolymers (Jawaid et al., 2016; Mochane et al., 2021; Satyanarayana et al., 2009)

Properties	Types of Biopolymer							
	Starch	PLA	PCL	PHB	PBS	PP	HDPE	LDPE
Density (kg/m^3)	–	1210	1110	1180	1260	920	952	940
T.S. (MPa)	5.0	21	20.7	40	34	33	28	10
Y.M. (GPa)	0.125	0.35	0.21	3.5	0.32	1.10	1.0	0.3
Elongation (%)	31	6	>50	5	560	415	700	300
Tg (°C)	–	45	–60	5	–32	–5	–120	–120
Tm (°C)	–	150	58	168	114	163	129	110

T.S.: tensile strength; Y.M.: Young's modulus; Tg: glass transition temperature; Tm: melting point, PLA: poly (lactide) acid; PCL: poly (ε-caprolactone); PHB: poly (hydroxybutyrate); PBS: poly (butylene succinate); PP: polypropylene; HDPE: high density polyethylene; LDPE: low-density polyethylene.

a polymer resin acts as a matrix and naturally derived fibers act as the reinforcement material in a biocomposite material. The mechanical properties of the neat polymer can be improved by creating composites with natural fiber reinforcement. Fiber properties such as morphology, modulus, aspect ratio, interface adhesiveness, and polymer properties are influencing factors for variation in biocomposite behavior. If the fiber property is not similar to a neat polymer, the final composite may not exhibit better physical characteristics (Meereboer et al., 2020). Biocomposites are widely used in consumer product, packaging, and automotive industries, where they help reducing weight and cost. Using biodegradable polymers for composite preparation has created new market opportunities for green products with the sustainable approach for life cycle assessment (John & Thomas, 2008; Netravali & Chabba, 2003; Satyanarayana et al., 2009). The problem of waste generation from the end use of synthetic plastics can be effectively controlled using biodegradable polymers in composite material (Plackett et al., 2003). The application suitability is dependent on the mechanical and degradation characteristics of the end product. The degradation time for most bio-based polymers and composites is effectively short, between 2 weeks to 6 months (Table 10.2).

Various factors such as good mixing of natural fiber, suitable biopolymer selection as matrix material, proper fabrication technique, and appropriate chemical and physical treatment must be recognized and considered to develop an efficient and biodegradable composite (Liu et al., 2007). Natural fibers provide benefits of

TABLE 10.2
Degradation Time for Various Materials (Satyanarayana et al., 2009)

Material	Time to Degrade in the Environment
Conventional copy paper	1 month
PHB–PHB/starch	1 month
WG	1 month
WG/SCB	1 month
WG/PVA	1 month
Cotton	1–5 months
PCL-g-MAH/starch	2 months
PCL-starch	2 months
Wool stocking	1 year
Bamboo stick	1–3 years
Chewing gum	5 years
Painted wood	13 years
Plastic	450 years
Glasses and tires	Uncertain time

MAH: maleic anhydride; WG: waste gelatin; PVA: poly (vinyl alcohol); SCB: sugar cane bagasse; g: grafting.

TABLE 10.3

Various Qualities Achieved by Biocomposite Materials (Satyanarayana et al., 2009)

Non-brittle fracture on impact	Energy management is efficient
Shows comparable performance	More shatter-resistant
Stronger (25%–30%) at a similar weight	Lower mold shrinkage
Lower cost than the matrix material	Easily colored
Easy recyclability	High flexural modulus – up to 5 times the resin
Reduced molding cycle time by 30% max	High tensile modulus – up to 5 times the resin
Non-abrasive to machinery	High notched impact – up to 2 times the resin
Lower thermal expansion coefficient	Lower processing energy requirements

low tool abrasiveness and a lower health hazard to the workers. Table 10.3 presents the qualities of natural fiber–reinforced composite materials.

10.2.1 Starch and Cellulose-Based Biocomposite

Starch (Hocking, 1992) and cellulose (Wilkinson, 2001) are the most famous natural sources for biodegradable plastic synthesis. Starch is one of the cheapest biomaterials in the global market. Starch is produced naturally in wheat, rice, potatoes, corn, etc., which is chemically a mixture of branched amylopectin (poly-α-1,4-D-glucopyranoside and α-1,6-D-glucopyranoside) and linear amylose (poly-α-1,4-D-glucopyranoside).

Biopolymer derived from starch is modified by functionalization, plasticization, and extrusion cooking processes to make suitable bioplastic. Starch-based thermoplastics find extensive applications in non-food industries. Plant-derived cellulose, on the other hand, can successfully substitute petroleum products to create cellulose plastics (Wilkinson, 2001) like cellulose acetate propionate (CAP), cellulose acetate butyrate (CAB), and cellulose acetate (CA). For example, the handles of the toothbrush and screwdriver are made with CAP and CAB material. Several natural fiber biocomposites are prepared in recent years with cellulosic plastic (Mathijsen, 2021; Mohanty et al., 2004) and starch-based plastic as matrix material (Ruhul Amin et al., 2019). The structure of starch and cellulose is shown in Figure 10.5.

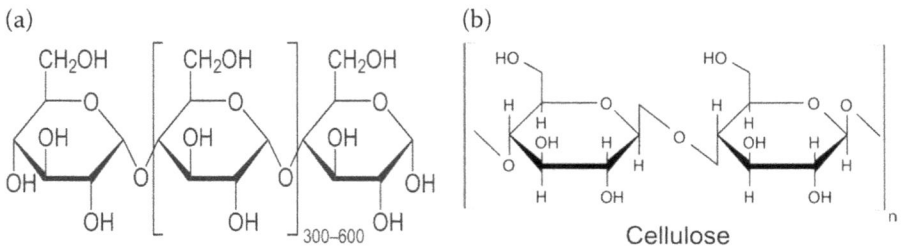

FIGURE 10.5 Chemical structure: (a) starch; (b) cellulose.

Cellulose esters can be used as matrix material with natural fiber reinforcement for biocomposite preparation (Toriz et al., 2006). Some grades of cellulose biodegrade readily, some grades can be "triggered" to "biodegrade," and some formulations (like butyrates, propionates, acetates) can be incinerated. However, all the cellulose grades are not equally degradable, but using suitable plasticizers, additives, and fillers improves the rate of degradability. Toriz et al. (2006) studied the properties of biocomposites prepared from CAP or CAB (45 wt%); bicomponent fibers (5 wt%); and wood, sisal, and flax fibers in a different composition. The tensile strength was more with fiber reinforcement composites than the neat cellulose ester. But impact strength was drastically reduced with fiber addition. Another polymer, cellulose propionate (CP), holds promising characters for use as a polymer matrix for composite synthesis (Mathijsen, 2021). From laboratory studies, it is found that CP-based composite is superior to the PP-based composite in many ways; however, CP-based natural fiber (flax)–reinforced composites are not commercialized yet due to the cost issues.

Researchers have studied different natural fibers and thermoplastic starch matrices for biocomposite synthesis (Espinach et al., 2014; Iman & Maji, 2015; Jiang et al., 2020). Such composites are susceptible to limitations like high water absorption and difficult processability, perhaps due to the starch's chemical composition and structure (Chi et al., 2008). Starch is highly hydrophilic due to the presence of hydroxyl groups in amylopectin and amylose (two main constituents of starch). The addition of plasticizers (mainly glycerine or glycerol) increases the starch's flexibility by helping starch to behave like a thermoplastic material. The tensile strength of starch is between 30–60 MPa, which is greater than polyethylene (around 30 MPa), but the higher elongation and fragile characters limit the widescale applications of starch. The mechanical properties can be enhanced with the reinforcement of natural fibers. Wollerdorfer & Bader (1998) reported a fourfold increase in the tensile property of wheat-based thermoplastic starch with cellulose fiber reinforcing (Jiang et al., 2020; Wollerdorfer & Bader, 1998). Corn starch is a widely used polymer. However, starch derived from potato, cassava, and wheat is also used in large quantities. Rojas-Bringas et al. (2021) studied the life cycle assessment (LCA) of starch-Brazil nut fiber composite by varying the plasticizer compositions (Rojas-Bringas et al., 2021). Table 10.4 represents the previous works on biocomposite preparation using a starch polymer as a matrix material.

10.2.2 Poly (Hydroxyalkanoates) (PHA)–Based Composite

PHAs are excellent biodegradable materials, having an aliphatic polyester chemical structure (Meereboer et al., 2020). PHAs are produced by bacterial and chemical synthesis routes. *Escherichia coli, Azotobacter, Ralstonia, Cupriavidus, Aeromonas, Clostridium, Syntrophomonas, Methylobacterium*, etc., are well-known microorganisms that can produce PHA (Ben Rebah et al., 2009; Choi & Lee, 1999). At the same time, the ring-opening polymerization reaction of β-butyrolactone is the chemical pathway for PHA synthesis (Vroman & Tighzert, 2009). Among the various PHAs, the most important PHAs are polyhydroxyvalerate (PHV), polyhydroxyhexanoate (PHH), and poly(3-hydroxybutyrate) (PHB). The chemical structure of PHA is shown in Figure 10.6.

TABLE 10.4

Works on Starch-Based Biopolymer

Polymer Matrix	Reinforced Natural Fiber and Plasticizer	Fiber Treatment and Preparation Method	Conclusion	Reference
Starch from sources like sweet potato, corn, and Andean potato	Brazil nut fiber (20 wt%), glycerol (15 wt%), and sorbitol (30 wt%) as a plasticizer	No chemical treatment/injection molding method	1. Cradle-to-gate LCA was studied for the prepared composite. 2. The starch-based biocomposite showed less impact than PP glass fiber and PLA-Brazil nut fiber composites. 3. Plasticizer and starch type influenced the load on the environment for the composite production. 4. The impact of sorbitol was more than glycerol due to the complex processing and wt% of sorbitol. 5. The study established BSF as sustainable reinforced material having less burden on the environment.	(Rojas-Bringas et al., 2021)
Tapioca starch	Bamboo fiber (10 wt%), PVA as s stabilizer, glycerol as plasticizer	Treatment of bamboo fiber with alkali (6%) and permanganate (1%) solution mixing method	1. The biodegradability and mechanical properties of the biocomposite were investigated. 2. Chemical treatment improved the mechanical and biodegradability of all the samples than the untreated composite samples. 3. Alkali-treated fiber composite showed better tensile, flexural strength, tensile, and	(Yusof et al., 2019)

Wheat starch (varying composition)	Leafwood and PPF as fiber. Glycerol as plasticizer, polycaprolactone and polyesteramide as biodegradable polyesters	No chemical treatment dry blend, single screw extrusion, and then injection molding

1. Studied the variation of mechanical, thermal degradation, thermo–mechanical behavior of the composites with respect to filler content, fiber length, and fiber/matrix formulation.
2. Cellulose fiber had better interface matrix adhesion than lignocellulose fiber.
3. Between two TPS matrices, less plasticizer resulted in higher mechanical strength. Glycerol produced a poor quality of adhesion at the filler and matrix interface.
4. Natural fiber addition enhanced the thermal resistance.
5. Lignocellulose fiber composite showed a higher degradation temperature than cellulose fiber.
6. The addition of biodegradable polyesters resulted in no significant improvement in the post-processing stability.

flexural modulus than permanganate treated composite.
4. Fiber with permanganate treatment showed better biodegradability (6.15% wt. loss in 15 days) than alkaline and untreated fiber composite.

(Averous & Boquillon, 2004)

(Continued)

TABLE 10.4 (Continued)
Works on Starch-Based Biopolymer

Polymer Matrix	Reinforced Natural Fiber and Plasticizer	Fiber Treatment and Preparation Method	Conclusion	Reference
Cornstarch thermoplastic	Eucalyptus urograndis pulp (16 wt %), 1 mm fiber length with 60 aspect ratio.Glycerin (30 wt%) as a plasticizer	No chemical treatmentPrepared in batch mixer at 170°C and hot pressed.	1. Investigated the improvement in mechanical properties of the biocomposite with fiber reinforcement. 2. The composite showed a 100% and 50% increase in tensile strength and modulus properties, respectively, compared to neat TPS. 3. SEM images showed good adhesion and interaction between the matrix and the fiber. 4. Fiber and glycerin had better interaction resulting in lower glass transition temperature and water absorption capacity of the matrix than the single TPS. 5. Improper dispersion due to self-agglomeration is a problem in composite synthesis.	(Curvelo et al., 2001)

LCA: life cycle analysis; TPS: thermoplastic starch; SEM: scanning electron microscopy; PVA: poly vinyl alcohol; PPF: paper pulp fibers

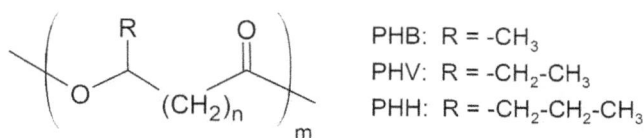

PHB: R = -CH$_3$
PHV: R = -CH$_2$-CH$_3$
PHH: R = -CH$_2$-CH$_2$-CH$_3$

FIGURE 10.6   PHA repeating units (Polymer Properties Database, 2021).

PHAs are produced from 100% bio-based material, and they are biocompatible and biodegradable. Some grades of PHAs exhibit good moisture resistance and similar mechanical properties of polypropylene (PP). It could be an alternative for PP and polyethylene (PE). The main drawback is the relatively higher cost of the polymer than the petroleum-based polymers (Verhoogt et al., 1994) and other biopolymers. However, due to their excellent biodegradation property, they are a suitable replacement for single-use plastics (Meereboer et al., 2020). The current market share of PHA products is relatively small, but it is expected to grow faster due to environmental requirements.

PHAs find applications in tableware articles such as cutlery, food containers, plates, trays, drinking cups; agricultural usages in soil protection sheet, waste bags, films; packaging material; and PHA fibers in surgical sutures, drug delivery, and medical devices with biodegradable behavior. Many PHA-based natural fiber composites have been prepared and studied so far, which showed enhanced mechanical, impact strength, and modulus than the pure polymer material. The addition of natural fibers as filler enhances the biodegradation rate of the biocomposites by increasing water absorption capacity. The cost of the composite is also reduced (Badia et al., 2014). Wood flour (Wu, 2006), flax (Shanks et al., 2004), abaca (Shibata et al., 2002), wheat straw (Avella et al., 2000), sisal (Moliner et al., 2018), hemp (Hermida & Mega, 2007), and kenaf (Joyyi et al., 2017) fibers have been incorporated in a PHA matrix-based composite material. Some of the recent studies on PHA-based biocomposites are presented in Table 10.5.

10.2.3   POLY (LACTIC ACID) (PLA)–BASED COMPOSITES

One of the most widely used, promising, and the oldest biodegradable polymer is poly-lactic acid-(PLA). It is a thermoplastic aliphatic polyester primarily sourced from agro-based products like sugarcane, corn starch, tapioca roots, and wheat. PLA polymer ((_CH(CH3)_CO_O_)n) (Figure 10.7) can be synthesized mainly by direct polymerization and ring-opening polymerization (ROP) reaction of lactide monomer (Siakeng et al., 2019). Enzymatic bioproduction route is also available. Processes are investigated for using non-toxic catalysts like calcium (Ca), zinc (Zn), magnesium (Mg), etc., in direct polymerization and ROP routes as a substitute for heavy metal catalysts (Inkinen et al., 2011). The innovation in synthesis techniques has dramatically reduced PLA pricing (Yu et al., 2006). PLA exhibits excellent biodegradability, compostability, processability, and recyclability properties. It finds many industrial applications (textile, packaging, agriculture, biomedical, electronics, and transportation) as a substitute for petroleum-based polymers

TABLE 10.5
Works on PHA-Based Biopolymer

Polymer Matrix	Reinforced Natural Fiber and Plasticizer	Fiber Treatment and Preparation Method	Conclusion	Reference
PHA	Wood flour (WF)50 wt%	No chemical treatment of wood flourTwin-screw extrusion method	1. Studied mechanical stability of the biocomposite under in-service conditions like indoor, outdoor, and soil environments. 2. The indoor condition had a smaller effect on the stability of biocomposite than the outdoor condition. 3. Neat PHA composite is more stable than wood flour loaded composite in the outdoor environment. However, the deterioration was polymer matrix type independent. 4. Soil environments had the most significant mechanical stability loss than other environments due to the additional actions of bacteria, enzymes, and fungi. PHA/WF-based composite had suffered more impact than PE and PLA-based composite in the soil environment. 5. Mechanical stability is directly influenced by water adsorption (hygroscopic behavior of WF) and matrix degradation rate by creating voids and cracks in the composite.	(Chan et al., 2020)

(Continued)

6. PHA/WF composite formulation could be used in applications where only time-bound mechanical performance is needed.

				(Wang et al., 2020)
Modified polyhydroxyalkanoate (MPHA) by obtained by interfacial compatibility modification	PLF and waste OSP.OSP acts as a natural antimicrobial agent.	OSP and PLF were thermally treated.PLF was further treated with 3% sodium hypochloriteHot press method	1. Investigated the cytocompatibility, biodegradability, and antibacterial properties of the synthesized composite. 2. The cell adhesion test confirmed non-cytotoxic behavior, and MTT assay OSP improved the antimicrobial character of the MPHA/PLF/OSP composites. 3. MPHA/PLF/OSP has more negligible water adsorption than PHA/PLF/OSP composites. 4. After being buried in soil compost, the weight loss of composites indicated that both were biodegradable, especially at high OSP/PLF substitution levels. 5. Composites with a modified PHA matrix showed improved tensile properties than the unmodified matrix due to the better distribution of PLF/OSP in the matrix. Water adsorption was less for MPHA/PLF/OSP composite. 6. Biodegradation in soil incubated environment was faster for MPHA/PLF/OSP than neat PHA but lower than PHA/PLF composite. An increase in fiber or OSP loading resulted in a higher biodegradation rate.	

TABLE 10.5 (Continued)
Works on PHA-Based Biopolymer

Polymer Matrix	Reinforced Natural Fiber and Plasticizer	Fiber Treatment and Preparation Method	Conclusion	Reference
			7. Composite could be used in 3D printing filament, catering objects, antibacterial films, stationery, and packaging applications.	
PHA (grafted with maleic anhydride)	Cotton fiber (CF)	No chemical treatmentDip-coating method	1. Studied the influence of grafting degree on the PHA/CF composite hydrophobicity character.	(Zhao et al., 2017)
			2. Hydrophobicity of the PHA was increased due to MA graft polymerization by reducing the surface energy.	
			3. At 0.05% grafting degree and 15% fiber loading, the composite film showed maximum contact angle (130°) and rested the aging with time without hydrophobicity reduction.	
			4. Grafting and fiber reinforcement increased the tensile property of the composite film by H$_2$ bond formation.	
			5. The composite film is suitable for packaging (paper-based) materials.	
Aliphatic polyesters PCL, PHBV, PBS, and PLA	Abaca fibers (10 wt%)	Treated with acetic anhydride (AA) and pyridine (molar ratio 1:1)Mixing fiber and polymer was done with a twin rotary	1. Investigated the biodegradability of the composite by the soil-burial test.	(Teramoto et al., 2004)

(Continued)

mixer. Desktop injection molding and hot press methods were used to prepare the composite sample.

2. Among the pure polymer matrices, biodegradability was in the order of was PCL>PHBV>PBS>PLA.
3. The biodegradation was independent of fiber treatment for PCL-based composite due to the high biodegradability of PCL polymer.
4. Fiber addition (particularly untreated fiber) enhanced the biodegradability for PBS and PHBV-based composite than pure PBS and PHBS.
5. PLA/AA-abaca composite showed no weight loss, but the PLA matrix-based untreated fiber composite exhibited 10% weight loss caused by the fiber's degradation.
6. Cracks were the primary reason for degradation in composites with untreated fiber, but with treated fiber composites, cracks were not developed, and therefore degradation was slow.

PHB

Flax fiber withFlax:PHB ratio is 1:1 (volume)

The fiber was dried and treated with GTA, TBC, and PEG as a plasticizer. Heat pressed method

1. Studied the morphological and mechanical behavior of the composite.
2. PHB and fiber exhibited better interfacial interaction. But it was decreased with the addition of plasticizer.
3. Plasticizers caused changes in the crystallinity and glass transition of the PHB.

(Wong et al., 2002)

TABLE 10.5 (Continued)
Works on PHA-Based Biopolymer

Polymer Matrix	Reinforced Natural Fiber and Plasticizer	Fiber Treatment and Preparation Method	Conclusion	Reference
			4. The addition of PEG and GTA improved the loss modulus (G'') and fiber/matrix interaction than the pure PHB composite.	
			5. PEG and GTA helped to get a higher distribution of fibers in the PHB matrix.	
			6. The composite displayed transcrystallinity with the addition of the plasticizer.	
			7. If absorption of water issue is controlled, then it will create a material with favorable properties.	

PHA: polyhydroxyalkanoate; PLF: pineapple leaf fiber; OSP: oyster shell powder; PCL: poly (ε-caprolactone); PHBV: poly (3-hydroxybutyrate-co-3-hydroxyvalerate); PBS: poly (butylene succinate); PLA: poly (lactic acid); PHB: polyhydroxybutyrate; GTA: glycerol triacetate (GTA); TBC: tributyl citrate (TBC); PEG: poly (ethylene glycol) (PEG)

$$H \left[O - CH - \underset{\underset{O}{\|}}{\overset{\overset{CH_3}{|}}{C}} \right]_n OH$$

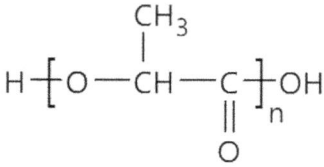

FIGURE 10.7 PLA repeating units (Polymer Properties Database, 2021).

(Saba et al., 2017; Siakeng et al., 2019). The complete biodegradation ability of PLA is very attractive in packaging, drug delivery, tissue engineering, and agricultural applications. Hydrolysis and enzyme action (composting) are effective processes for PLA degradation. However, poor toughness, brittleness, low break elongation, poor hydrophilicity, and slow degradation rate are significant limitations of PLA, which need extensive modifications (Inkinen et al., 2011). Polymer surface modification through plasticizer addition and copolymerization improves the mechanical properties and increases the thermal resistance and degradation rate (Jamshidian et al., 2010).

Green composites made of PLA and other natural fibers are investigated to a large extent (Kobashi et al., 2008; Yao et al., 2011). The addition of natural fibers improved the biocomposite's mechanical, thermal, and biodegradation properties than the pure PLA matrix (Dicker et al., 2014; Yao et al., 2011). Reinforcement with natural fibers reduced the post-application waste disposal cost of the bio-composite. Easy and compatible processing by compression, injection molding, and extrusion technique is another advantage of PLA-natural fiber composite synthesis (Hu et al., 2010). Life cycle assessment (LCA) of PLA reveals its suitability and appropriate application as interior components in automobiles since the transportation requirement is less for PLA than synthetic polymers (Bajpai et al., 2013; Tabone et al., 2010). Many fibers such as ramie, hemp, wood, kenaf, jute, sisal, flax, bamboo, and rice husk have been used as reinforcement with PLA polymer matrix to improve the mechanical, thermal, physical, biodegradation, and tribological behavior of the final composite material (Saba et al., 2017). Some of the recent studies on PLA-based biocomposites are presented in Table 10.6.

10.2.4 POLYBUTYLENE SUCCINATE (PBS)–BASED BIOCOMPOSITE

PBS is a known petroleum-derived biodegradable polymer with broad applications in agriculture, food packaging, tissue engineering, and biomedicine (Mochane et al., 2021). PBS can be considered renewable source-based if the share of the biosource is more than 50% (Satyanarayana et al., 2009). PBS is also known as poly-tetramethylene succinate and is an aliphatic, thermoplastic polyester resin containing $C_8H_{12}O_4$ as a repeating unit (Figure 10.8). There are mainly two routes for synthesizing PBS (i) trans-esterification with succinate diesters and (ii) direct esterification with diacid. The most common synthesis method is the direct esterification/polycondensation of 1,4-butanediol (BDO) and dimethyl succinate (succinic acid) (Jacquel et al., 2011). The monomers (BDO, succinic acid) can be produced either from petroleum-based or natural sources.

PBS exhibits comparable properties with polyethylene and polypropylene, having the additional advantage of biodegradability behavior (Nurul Fazita et al.,

TABLE 10.6

Works on PLA-Based Biopolymer

Polymer Matrix	Reinforced Natural Fiber and Plasticizer	Fiber Treatment and Preparation Method	Conclusion	Reference
PLA	Kraft softwood pulp in 10–35 wt%	Bleached treated.First mixing in gelimat kinetic mixer and then injection molded.	1. Investigated the impact strength and water absorption on the composite. 2. It reported lower dispersion with a higher cellulose content of the fiber. The problem was sorted with the addition of a dispersion agent. 3. Found impact resilience of the biocomposite comparable to PP-glass fiber. However, higher fiber content lowered the impact resilience linearly in the case of the un-notched composite due to the lowering of matrix material and the presence of fiber ends. 4. Biocomposite showed a smaller reduction in impact resilience than PP-glass fiber composite. 5. The resilience of the notched fiber composite was more than the neat PLA. 6. Pure PLA showed a 0.94% water intake, but with increased fiber loading, water absorption increased (max. 7.03% with 35 wt% of fiber) due to the hydrophilic nature of the natural fiber.	(Oliver-Ortega et al., 2020)

(Continued)

| PLA | Pine fiber (30 wt%, 90–180 μm mesh size) | Solvent-borne epoxy treatment (0.5–10 wt %) using impregnation method, hot pressing, and compression molding | 7. Pure PLA did not show degradation, but fiber composite showed <1% degradation.
1. Studied the tensile and Young's modulus of the epoxy fiber–treated biocomposite.
2. Compared to pure PLA, Young's modulus and tensile strength of the biocomposite increased by 82% and 20%, respectively, due to improved matrix–fiber adhesion by epoxy treatment.
3. The optimum epoxy concentration for treatment was (1.0 wt%), as it produced lesser voids than the untreated biocomposite and resulted in maximum value for tensile strength (71 MPa) and Young's modulus (5.4 GPa).
4. Epoxy treatment improved the fiber/matrix chemical compatibility, interface adhesion, fiber distribution, and fiber wettability. | (Zhao et al., 2020) |
| PLA | Kenaf (40 mesh size), 30 wt% fiber loading. | Acetylation treatmentTwin screw extruder | 1. Studied the fiber properties for time variation of acetylation treatment and investigated the thermal, water uptake, and mechanical properties of the treated fiber composite.
2. Fiber treatment made it more hydrophobic.
3. With increased acetylation time, the tensile and flexural strengths of the | (Chung et al., 2018) |

TABLE 10.6 (Continued)
Works on PLA-Based Biopolymer

Polymer Matrix	Reinforced Natural Fiber and Plasticizer	Fiber Treatment and Preparation Method	Conclusion	Reference
			composite were increased (acetylation level covering 25 wt% of fiber) after showing an initial dip (lower than untreated composite) during the small-time duration (0.5 hour).	
			4. Thermal stability was improved with acetylation treatment time. The optimum acetylation time was 2 hours.	
PLA sheets	Hemp and harakeke fiber in mat form but aligned discontinuously (1.5–40 wt%). The fiber mat was prepared by the DSF method.	Alkali treated (5 wt%) Hot press and compression mold method	1. Applied the Bowyer-Bader model to calculate fiber orientation factor (Kθ) and compared it with other discontinuous fiber composites' Kθ values. The evaluated Kθ will give more accurate stiffness and strength obtained by theoretical modeling rather than assuming a value for Kθ.	
			2. Composites prepared with DSF made fiber mat showed improved fiber alignment than injection molded and discontinuous fiber mat.	
			3. Evaluated (Kθ) values for both hemp and harakeke composites were higher than the reported value in literature (Injection molded and hot pressed random fiber mat	

(Sawpan et al., 2012)

PLA and unsaturated synthetic polyester

Hemp fiber in two forms short/random and long/aligned (0–50 wt% fiber loading)

Alkali (5 wt%) and silane (0.5 wt%) treated fiber.Composites were prepared by twin screw extruder and then Injection molding; and compression molding methods.

composites); however, it was marginally lower than the highest Kθ value for oriented non-woven fiber composite.

4. With the increase in fiber loading, the Kθ value decreased due to the fiber misalignment caused by the aggregation of fiber and higher applied processing pressure.

1. Investigated the flexural modulus, strength, and fiber defects of the PLA/hemp and UPE/hemp biocomposites.

2. Increased fiber loading for both the composites resulted in decreased flexural strength and increased flexural modulus.

3. Kinks (i.e., fiber defects) caused the lowering of flexural strength by creating stress points during testing. Due to the defects, debonding, crack, and failure occurs much before the complete load transfer from matrix to the fiber.

4. Maximum 30 wt% fiber loading is suitable for PLA-based composite using the conventional injection molding method.

5. Silane and alkali fiber treatment enhanced the matrix/fiber adhesion and resulted in higher flexural modulus and strength than the untreated biocomposite.

(Continued)

TABLE 10.6 (Continued)
Works on PLA-Based Biopolymer

Polymer Matrix	Reinforced Natural Fiber and Plasticizer	Fiber Treatment and Preparation Method	Conclusion	Reference
			6. Ordered fiber reinforcement resulted in better flexural properties than the random and short fiber reinforcement.	
			7. UPE/hemp fiber showed a similar trend in flexural property variation as PLA/hemp composite, but UPE/hemp had better performance with treated fiber.	

PP: polypropylene; DSF: dynamically sheet formed; UPE: unsaturated thermoset polyester

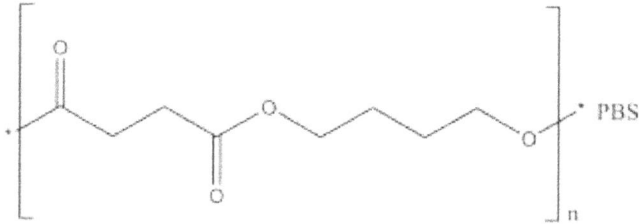

FIGURE 10.8 Repeating units of PBS (Siracusa et al., 2015).

2017; Xu & Guo, 2010). The copolymerization process can customize the mechanical and physical properties with different types and amounts of co-monomers like terephthalic acid, adipic acid, methyl succinic acid, 1,3-propanediol, ethylene glycol, and so on (Mochane et al., 2021). However, lower viscosity at melt conditions, lower gas resistance, high price, and poor tensile properties are some drawbacks limiting the PBS applications (Frollini et al., 2015; Sinha Ray et al., 2003). Incorporating fillers like graphene, carbon nanotubes, clay, etc., are commonly introduced into the matrix to improve the PBS properties (Siracusa et al., 2015). PBS polymer is suitable for processing with injection molding, extrusion, film blowing, and thermoforming methods. The application areas of PBS include packaging, compost bags, agriculture mulch films, and tissue engineering (Shaiju et al., 2020).

Natural fiber biocomposite with PBS as a biodegradable polymer matrix is gaining importance in several applications due to its favorable mechanical properties. Reinforcement with natural fibers improves the physical and mechanical behavior of the neat PBS polymer. However, the final composite's properties are highly affected by the fiber type, its surface modification, aspect ratio and placing orientation, fiber loading, the addition of fillers, fabrication method, and modification to the polymer matrix (Hong et al., 2021). Many natural-based fibers like cotton, jute, curaua, sisal, coir, sugarcane bagasse, etc., are used as reinforcement in PBS matrix polymer to get composites possessing better properties than the single polymer. Calabia et al. (2013) reported 15%–78% tensile strength improvement of untreated cotton fiber (10–40 wt%)/PBS composite than the pure PBS (Calabia et al., 2013). The tensile strength further increased by (25%–118%) with silane-treated fiber. The biodegradability was also increased with fiber addition. However, the thermal stability was decreased with an increase in fiber loading than the pure PBS matrix. The recent investigations on PBS/natural fiber composites are presented in Table 10.7.

10.2.5 POLY (ε-CAPROLACTONE) (PCL)–BASED BIOCOMPOSITE

PCL is a petroleum-derived semicrystalline aliphatic polyester having a linear structure. It can undergo biodegradation by the action of microorganisms (Arbelaiz et al., 2006). The polymeric chain contains C6H10O2 as a repeat unit (Elzein et al., 2004), as shown in Figure 10.9. The ring-opening of ε-caprolactone and subsequent polymerization of the monomer can produce PCL. The polymerization can proceed through a cationic coordination mechanism (Malikmammadov et al., 2018). The

TABLE 10.7

Works on PBS-Based Biopolymer

Polymer Matrix	Reinforced Natural Fiber and Plasticizer	Fiber Treatment and Preparation Method	Conclusion	Reference
PBS	Bamboo fiber with 40–60 mesh size and length of fiber less than 380 μm	Treated with PDA and 3-aminopropyl triethoxysilanecomposite was prepared by mixing, extrusion, and then hot pressing methods	1. Studied the mechanical and water intake properties of the treated fiber composite. 2. Chemical treatment and cross-linking with silane had enhanced the covalent adhesion between the matrix and fiber interface. 3. Tensile strength, tensile modulus, flexural strength, flexural modulus, and impact strength were improved by 70%, 25%, 37%, 24%, and 63%, respectively. 4. The achieved mechanical properties were superior to most of the natural fiber biocomposites. 5. Higher resistance to water and lower hygroscopic behavior was observed for the treated fiber composite. 6. The authors reported the combination to be suitable for the synthesis of high-performance biocomposite.	(Hong et al., 2021)
PBS	The bamboo fiber in powder form in 80 mesh size (i.e., 180 microns). Fiber loading: 10, 20, 30 wt%.	Alkaline (5M) treatedinjection molding method	1. Investigated the morphological and mechanical properties of the composite with treated and untreated fiber.	(Pivsa-Art & Pivsa-Art, 2021)

(*Continued*)

Material	Fiber	Method	Findings	References
			2. At any fiber content, treated fiber composite exhibited higher tensile strength than untreated fiber composite.	
			3. With an increase in fiber loading, Young's modulus was increased, but break elongation was decreased.	
			4. Pure PBS and fiber-reinforced composite had similar Izod impact strength.	
			5. Water intake was increased with an increase in fiber loading. Alkali-treated fiber composite showed higher water absorption than untreated fiber composite due to the removal of hemicellulose and lignin content, which are hydrophobic.	
			6. Prepared biocomposite was suitable for indoor furniture and packaging applications.	
PBS	Grounded alfa fiber. Fiber loading: 2, 5, and 10 wt%	Fibers were extracted using 2N NaOH solutionHot press compression molding method	1. Studied the rheological, morphological, and thermal properties of the biocomposites based on fiber loading.	(Arabeche et al., 2020)
			2. Crystallization of the PBS matrix was improved by fiber addition, which acts as a nucleating agent.	
			3. Fiber addition improved the dynamic behavior, increased the complex viscosity and modulus, and induced the percolation of the biocomposites.	

TABLE 10.7 (Continued)

Works on PBS-Based Biopolymer

Polymer Matrix	Reinforced Natural Fiber and Plasticizer	Fiber Treatment and Preparation Method	Conclusion	Reference
PBS	Curaua fiberCase 1: fixed-length 3 cm; fiber loading 10, 20 and 30 wt %.Case 2: Fixed fiber content 20 wt%; varying lengths of 1, 2, 3, and 4 cm.	No chemical treatmentCompression molding	4. Change in the PBS microstructure was observed due to the change and shift of slope in the Cole-Cole curve. 5. Homogeneous dispersion and good interfacial adhesion were the major factors for improvement in the mechanical properties of the composite. 1. Studied the composite's mechanical, morphology, and water absorption properties based on fiber content and length. 2. For case 1 samples, an increase in fiber content improved the flexural and impact strengths. The optimum fiber loading was 30 wt%, giving the maximum impact and flexural strength. 3. For case 2 samples, impact strength was higher for short fiber than long fiber due to the better homogeneous microstructure and load transferability. The optimum fiber length was found to be between 1–2 cm. However, length did not influence the flexural property majorly.	(Frollini et al., 2015)

(Continued)

PBS

Ramie fiber (0–30 wt%)

The fiber was treated with silane (0.5 wt%), alkali (5 wt %), acetic anhydride, and maleic anhydride (5 wt %).Co-rotating twin screw extruder and then injection molding

4. Fiber content rather than length was the chief factor influencing the water absorption behavior. Water intake was more for higher fiber content.

5. Fiber biocomposites had better mechanical properties than pure PBS and suitable for interior car parts or rigid packaging applications.

1. Investigated the suitability of a single fiber fragmentation test for selecting the effective surface treatment method. The mechanical and interfacial behavior was also studied for the biocomposite samples.

2. The single fiber fragmentation test measured IFSS to understand the fiber-matrix interfacial interaction.

3. Alkali treatment resulted in the maximum IFSS value, and it was selected as surface modification treatment for biocomposite synthesis.

4. SEM indicated the better dispersion and adhesion of alkali-treated fiber in the matrix than the untreated fiber.

5. Higher mechanical properties (tensile strength and modulus) were observed for treated fiber composite than untreated fiber composite.

(Zhou et al., 2013)

TABLE 10.7 (Continued)
Works on PBS-Based Biopolymer

Polymer Matrix	Reinforced Natural Fiber and Plasticizer	Fiber Treatment and Preparation Method	Conclusion	Reference
			6. This study established the suitability of IFSS measurement as a simple, cost and time-effective method to choose the proper surface treatment. 7. PBS/Ramie biocomposite was suitable for biodegradable and high-performance composites.	

PDA: polydopamine; IFSS: interfacial shear strength; SEM: scanning electron microscope

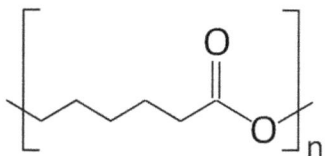

FIGURE 10.9 Repeating unit of PCL.

route of PCL synthesis is the free radical assisted ring-opening polymerization reaction of 2-methylene-1–3-dioxepane (Woodruff & Hutmacher, 2010).

Among the available biodegradable plastics, PCL is one of the most hydrophobic polymers. It has excellent mechanical properties, non-toxic, resistant to chlorine, oil, water, and biocompatibility with many other polymeric materials, and therefore finds wide-scale applications in drug delivery, medical devices, bone graft substitute, composite materials, and tissue engineering (Elzein et al., 2004; Malikmammadov et al., 2018). However, the major drawback is the slower degradation (lower than polylactides) of the PCL chain, which may take 2 to 3 years for complete biological degradation (Malikmammadov et al., 2018). Recently, interest and application for PCL-based natural fiber composites have increased due to environmental concerns of non-degradable synthetic polymer plastics. Biocomposites have been prepared with PCL as the polymeric matrix material, including natural fibers like bamboo root, jute, flax, abaca, etc., as fillers. Table 10.8 presents some recent studied on PCL biocomposites.

10.3 BIODEGRADATION OF COMPOSITES

Biological degradation is the most attractive property of biodegradable polymer-based composite with natural fiber as filler. Microorganisms' activity decomposes the polymer's complex molecules into simpler molecules like CH_4, CO_2, water, inorganic compounds, and biomass. Abiotic factors such as soil burial, water, compost, lipase solution, activated sludge, etc., are responsible for weakening the polymeric structure and starting the degradation process. Other physical factors like shear force, light, compression, and thermal actions also initiate the fragmentation of the composite/polymer material, thereby reducing the mechanical strength (Karthika et al., 2019). Thus, the biodegradation behavior of the material can be understood in terms of loss in molecular mass due to the formation of CO_2, CH_4, and H_2O in anaerobic and aerobic conditions. The rate conversion of carbon to CO_2 could be a suitable measure of biodegradation rate. The three phases of biodegradation behavior (lag, biodegradation, and plateau phase) are represented in Figure 10.10.

Biodegradation of a polymeric material is believed to consist of a two-step process. In the first step, the complex molecular chain is broken down into shorter molecular chains by the action of enzymes/microorganisms (biotic) and other abiotic factors. The second step is the bio-assimilation of oligomers (chemicals produced in the first step) by microbial action, called the demineralization process, producing water, salts, biomass, minerals, and gases. The whole degradation process can be categorized into aerobic or anaerobic processes depending on the oxygen availability (Karthika et al., 2019). CO_2 is produced in aerobic, and CH_4 is produced in the anaerobic degradation

TABLE 10.8
Works on PCL-Based Biopolymer

Polymer Matrix	Reinforced Natural Fiber and Plasticizer	Fiber Treatment and Preparation Method	Conclusion	Reference
PCL	Bamboo-root flour (40 μm mesh size). Fiber loading 5–30 wt%	NaOH treated (5 wt %).Co-rotating twin screw extruder and injection molding	1. Studied the mechanical properties of the biocomposite for suitability in cryo-packaging applications. 2. The UTS and %EB of the biocomposite were improved with the fiber addition than the neat PCL polymer material due to the higher fiber/matrix interaction resulted from the formation of trans-crystallization, nucleation, and nano fibrillar morphology. 3. With increased fiber content (up to 5 wt%.), tensile strength, toughness, and elongation were increased. 4. The mechanical properties were higher than LLDPE up to 10 wt% but comparable to LLDPE at 30 wt%. 5. The cost of PCL is higher than LLDPE, but the addition of natural fibers will improve the biocomposite's mechanical properties and economic value. 6. The glass transition temperature (Tg) was retained with 30 wt% fiber. It was suitable for applications in cryo-packaging.	(Bhagabati et al., 2021)

PCL	Jute (30–70 wt%)	No fiber chemical treatmentCompression molding method	1. Studied the mechanical and degradation properties of the ⁶⁰Co gamma-irradiated and non-irradiated biocomposite samples. 2. Gamma-irradiated 50 wt% fiber-loaded composites showed improved mechanical and physical properties. 3. A lower degradation rate was found for irradiated composites.	(Islam et al., 2009)
PCL-g-MA	Short flax fiber (0 to 60 wt%).The compatibilizer effect was studied with 30 wt% flax fiber load.	No fiber chemical treatmentInjection molding method	1. Mechanical properties were studied for the effect of fiber content and g-MA coupling agent. Thermal properties were investigated for flax/PCL-g-MA and flax/PCL composites. 2. Flexural, tensile strength, and crystallinity were highest for PCL-g-MA matrix-based biocomposite due to increased interfacial adhesion. 3. Thermal stability was slightly lowered with matrix modification and fiber loading. 4. The crystallinity and mechanical properties were decreased at a smaller coupling agent; however, both properties were improved at a higher content.	(Arbelaiz et al., 2006)
PCL	Short abaca fiber (5 mm length).Fiber loading: 0, 5, 10, 15, 20 wt%	Surface esterification of abaca fiber (acetic anhydride or butyric anhydride/pyridine)Melt mixing and injection molding	1. Studied the tensile properties of the PCL/abaca fiber and PCL/GF composites. Effects of natural fiber loading and its surface modification were investigated.	(Shibata et al., 2003)

(Continued)

TABLE 10.8 (Continued)
Works on PCL-Based Biopolymer

Polymer Matrix	Reinforced Natural Fiber and Plasticizer	Fiber Treatment and Preparation Method	Conclusion	Reference
			2. In the case of natural fiber composite, the tensile strength and modulus were increased with fiber content.	
			3. At same fiber wt%, PCL/abaca composites showed better tensile strength than PCL/GF composite.	
			4. Fiber treatment improved the tensile strength of the treated fiber composite than the untreated fiber composite due to the formation of better interfacial adhesion.	
PCL and MA as Compatibilizer	Wood flour + lignin	No chemical treatmentReactive extrusion and injection molding	1. Mechanical, thermal properties were investigated for PCL-g-MA and PCL-based biocomposite.	(Nitz et al., 2001)
			2. Grafting with MA enhanced the mechanical properties of the fiber composite.	
			3. At 40 wt% WF and 2.5 wt% PCL-g-MA, the tensile strength and Young's modulus was increased by 115% and 450%, respectively, than the neat PCL. Those properties were almost four times higher than the PP/WF composites.	

4. WF-based composites had higher mechanical properties than lignin-based composites.

5. Lignin-filled composites showed better biostability than the WF composite. Therefore, lignin could be a better candidate as a filler to increase the longevity of the mater in outdoor applications.

PCL: poly (ε-caprolactone); LLDPE: linear low-density polyethylene; UTS: ultimate tensile strength; EB: elongation at the break; g-MA: g-Maleic anhydride; PP: polypropylene

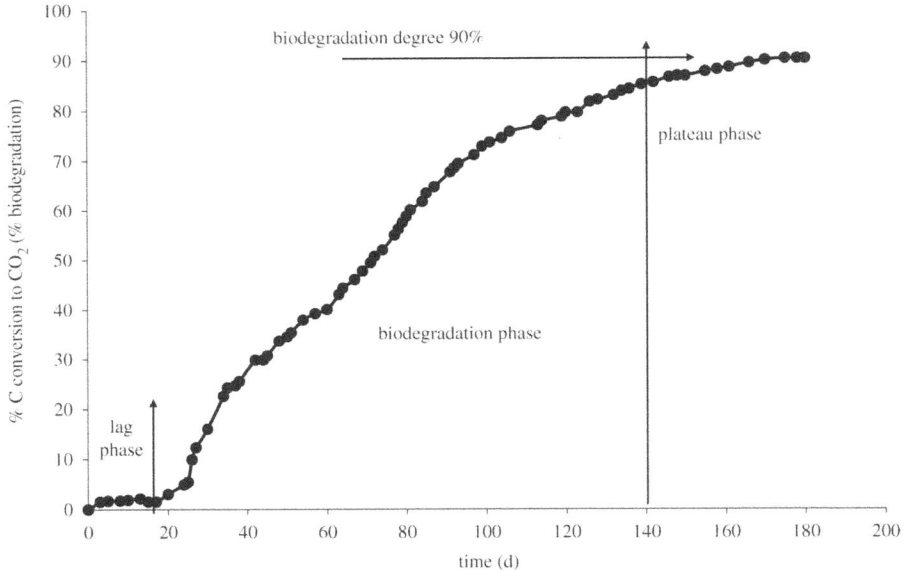

FIGURE 10.10   Three stages of the biodegradation process. Reprinted with permission (Song et al., 2009).

process (Equations 10.1, 10.2). The changes in the morphology are visible in SEM images. As the degradation advances with time, microbial action creates holes all over the sample surface and results in size reduction.

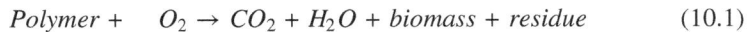

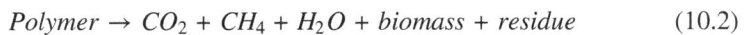

$$Polymer + \quad O_2 \rightarrow CO_2 + H_2O + biomass + residue \tag{10.1}$$

$$Polymer \rightarrow CO_2 + CH_4 + H_2O + biomass + residue \tag{10.2}$$

A composite material loses its strength by propagating fracture and delamination during the disintegration process, leading to the ultimate failure of the material. The composite biodegradation process is influenced by many factors such as type of polymer matrix, chemical structure and modification, type of natural fiber, fiber surface treatment, fiber dispersion, interfacial interaction, and composite fabrication techniques. Soil burial test for weight loss measurement is the commonly used method of investigating the degradation of polymer composite.

There are three mechanisms of composite biodegradation (Karthika et al., 2019):

- Enzymes or acids attacking the resin directly
- Crack propagation by gas generation
- Destabilization of the polymer through the sulfides and chlorides action

Polymers like PLA, starch, PHA, PCL, PBS, etc., are degradable under the specified condition, differing from the actual usage situation. The degradation behavior and extent are different for different polymer matrices. Biodegradability of PBS is observed in soil burial, lipase solution, compost, activated sludge, and water environment (Xu & Guo, 2010). PBS was tested for disintegration in the lab (BOD test) and field (seawater) environment and found negligible weight loss in seawater (Nakayama et al., 2019). Sekiguchi et al., 2011 also reported slow degradation of PBS in the deep marine environment (Sekiguchi et al., 2011). However, the improvement in biodegradability with the addition of natural fibers like sugarcane rind fiber, jute was reported in the literature (Mochane et al., 2021). However, fiber surface modification reduces the biodegradation of the composite due to better adhesiveness. PLA is degradable in the actual condition of usage. It can degrade in compost and soil by natural process. Fiber addition had improved the biodegradation of PLA-based composites, as reported from the flax/PLA biocomposite investigation (Bayerl et al., 2014). The higher content of fiber showed higher biodegradability, while neat PLA had only minimal degradation behavior. Nieddu et al. (2009) reported higher PLA/nano clay composite degradability in compost because of the better hydrophilic nature of clay material (Nieddu et al., 2009). In the case of the PCL matrix, Nitz et al., (2001) found the enhanced biodegradation property for PCL/WF (wood flour) composite than the PCL/lignin composite due to the better water-absorbing capacity WF (Nitz et al., 2001). Lignin will be a suitable long durable composite material. The biodegradation of PCL/ clay nanocomposite was better than pure PCL (Jimenez et al., 1997). Teramoto et al. (2004) studied the biodegradability of treated and untreated abaca fiber–reinforced composite with aliphatic polyesters like poly(butylene succinate), poly(3-hydroxybutyrate-co-3-hydroxyvalerate), (poly(ε-caprolactone), and poly(lactic acid) ((Teramoto et al., 2004). The effect of biodegradation was not significant for PCL/abaca composite (both treated and untreated) as PCL is itself highly degradable than the natural fiber. However, accelerated degradation was observed in PBS and PHBV-based untreated abaca fiber composite than their respective pure polymer material. In the case of PLA matrix, while there was no evidence of degradation for pure PLA and treated abaca/PLA composite, the PLA/untreated abaca fiber composite showed about 10% weight loss in 60 days.

10.4 APPLICATION OF BIODEGRADABLE COMPOSITES

It is observed that despite having the bio-disintegration ability, the number of products developed by using natural fiber–reinforced biodegradable polymer is less compared to the other biocomposites. However, those materials are applied in packaging, medical, housing construction, furniture, textile, electronics, automobile, municipal solid waste management, terrestrial, and aquatic environment applications. In automobiles, different parts like door panels, parcel shelves, armrest, seat shells, headrests, instrument panels, etc., can be prepared from biodegradable composite material. Applications in indoor products like furniture are also growing at a good pace.

PLA-based biocomposites are successfully used in the automobile, pharma, and food packaging industries. It is more suitable in those short-term applications, which are not required to be durable for a longer time. Researchers have tried to create biofuel cells (BFC) based on PLA nanocomposites (Saba et al., 2017). PLA-based keratin and chitosan composites were explored for biomedical applications (Tanase & Spiridon, 2014). Reinhardt et al., (2013) reported structural application of PLA/cellulose-based viscose (rayon) fiber biocomposite (Reinhardt et al., 2013). Biocomposites of PLA and natural fiber have been studied for applications in 3D printing, electrochemical biocell, and biosensors products (Saba et al., 2017).

Nanocrystal scaffolds of cellulose and PBS have been applied in tissue engineering (Mochane et al., 2021); date palm fiber and PBS biocomposite were studied for application in sustainable products (Chaari et al., 2020); composite of empty fruit brunch/PBS-modified tapioca starch was used to apply in the preparation of agriculture much film. Apart from these uses, literature has mentioned PBS-based biocomposite application in furniture, food packaging, biomedical, and building materials (Mochane et al., 2021).

Starch-based fiber composite can be used in agriculture and packaging applications. Xie et al., (2018) prepared a starch-based sisal fiber biocomposite for using as packing material, which can replace the expandable polystyrene (EPS) (Xie et al., 2018). Zhang et al. (2018) synthesized a biocomposite of starch/ureal plant fiber (OUF) with suitable mechanical and physical properties for transportation and packaging applications (Zhang et al., 2018). The fibers from cellulose are usable in packaging applications with biodegradable behavior. Nanofibrillated cellulose (NFC) is an excellent filler material for nano-composites suitable for adhesive and foam applications (Čolnik et al., 2020). Biocomposites of hydroxyethyl cellulose (HEC) and cellulose fibers (CF) were prepared by which could be applied as green packaging films (Sirviö et al., 2018).

The biocomposite prepared from palm fiber and maleic anhydride-grafted polyhydroxyalkanoate (PHA-g-MA) polymer matrix has been studied for applying as a 3D printing filament (Wu et al., 2017). The poly-hydroxybutyrate (PHB)/coconut fiber was analyzed and found appropriate for the application in the packaging industry also (da Silva Moura et al., 2019). The mechanical property and biodegradation behavior of PHB/hemp fiber composite was studied for accelerated weathering. Its applicability in temporary construction applications was observed where durability is not the principal criteria (Michel & Billington, 2012). Multilayer cotton fabric biocomposite with PHB and PLA blend is suitable for automotive interior parts, furniture, and building applications (Michel & Billington, 2012).

PCL-based biocomposites can be used in packaging, orthopedic, biomedicine, tissue engineering, sensor, and structural applications (Rocha & Rosa, 2017). PLA fiber-based PCL composite was synthesized for packaging application (Ju et al., 2015). PCL biocomposites of hemp, sisal, and coconut fibers were studied for their mechanical and biodegradability properties with an attempt to use in orthopedic applications (Sarasini et al., 2015).

CONCLUSION

Natural fiber–reinforced biodegradable composites have many attractive characteristics such as recyclability, bio-decomposition ability, light weight, and good mechanical properties. Still, there are a few aspects like cost of raw material, biodegradation process, difficulty in reproducing the composite's properties, and durability issues in case of high strength applications. Composite biodegradability behavior depends on the chemical structure of polymeric material rather than the source of the polymer. Starch-based, cellulose-based polymers, PBS, PLA, PHA, PCL, etc., are widely used polymer matrices in biodegradable composite preparation. Packaging and biomedical industries are the major potential areas where biodegradable composites find extensive applications. The combination of polymers, fibers, and coupling materials needs to be developed to get the desired and satisfactory performance level.

REFERENCES

Arabeche, K., Abdelmalek, F., Delbreilh, L., Zair, L., & Berrayah, A. (2020). Physical and rheological properties of biodegradable poly (butylene succinate)/Alfa fiber composites. *Journal of Thermoplastic Composite Materials, February*. 10.1177/0892705720904098

Arbelaiz, A., Fernández, B., Valea, A., & Mondragon, I. (2006). Mechanical properties of short flax fibre bundle/poly(ε-caprolactone) composites: Influence of matrix modification and fibre content. *Carbohydrate Polymers, 64*(2), 224–232. 10.1016/j.carbpol.2005.11.030

Arikan, E. B., & Ozsoy, H. D. (2015). A review: Investigation of bioplastics. *Journal of Civil Engineering and Architecture, 9*, 188–192.

Asyraf, M. R. M., Rafidah, M., Azrina, A., & Razman, M. R. (2021). Dynamic mechanical behaviour of kenaf cellulosic fibre biocomposites: A comprehensive review on chemical treatments. *Cellulose, 28*(5), 2675–2695. 10.1007/s10570-021-03710-3

Avella, M., Rota, G. L., Martuscelli, E., Raimo, M., Sadocco, P., Elegir, G., & Riva, R. (2000). Poly (3-hydroxybutyrate-co-3-hydroxyvalerate) and wheat straw fibre composites: Thermal, mechanical properties and biodegradation behaviour. *Journal of Materials Science, 35*(4), 829–836. 10.1023/A:1004773603516

Averous, L., & Boquillon, N. (2004). Biocomposites based on plasticized starch: Thermal and mechanical behaviours. *Carbohydrate Polymers, 56*(2), 111–122. 10.1016/j.carbpol.2003.11.015

Badia, J. D., Kittikorn, T., Strömberg, E., Santonja Blasco, L., Martínez Felipe, A., Ribes Greus, A., Ek, M., & Karlsson, S. (2014). Water absorption and hydrothermal performance of PHBV/sisal biocomposites. *Polymer Degradation and Stability, 108*, 166–174. 10.1016/j.polymdegradstab.2014.04.012

Bajpai, P. K., Singh, I., & Madaan, J. (2013). Tribological behavior of natural fiber reinforced PLA composites. *Wear, 297*(1), 829–840. 10.1016/j.wear.2012.10.019

Bayerl, T., Geith, M., Somashekar, A. A., & Bhattacharyya, D. (2014). Influence of fibre architecture on the biodegradability of FLAX/PLA composites. *International Biodeterioration & Biodegradation, 96*, 18–25. 10.1016/j.ibiod.2014.08.005

Ben Rebah, F., Prévost, D., Tyagi, R. D., & Belbahri, L. (2009). Poly-beta-hydroxybutyrate production by fast-growing rhizobia cultivated in sludge and in industrial wastewater. *Applied Biochemistry and Biotechnology, 158*(1), 155–163. 10.1007/s12010-008-8358-1

Bhagabati, P., Das, D., & Katiyar, V. (2021). Bamboo-flour-filled cost-effective poly(ε-caprolactone) biocomposites: A potential contender for flexible cryo-packaging applications. *Materials Advances, 2*(1), 280–291. 10.1039/D0MA00517G

Calabia, B. P., Ninomiya, F., Yagi, H., Oishi, A., Taguchi, K., Kunioka, M., & Funabashi, M. (2013). Biodegradable poly (butylene succinate) composites reinforced by cotton fiber with silane coupling agent. *Polymers, 5*(1), 128–141. 10.3390/polym5010128

Chaari, R., Khlif, M., Mallek, H., Bradai, C., Lacoste, C., Belguith, H., Tounsi, H., & Dony, P. (2020). Enzymatic treatments effect on the poly (butylene succinate)/date palm fibers properties for bio-composite applications. *Industrial Crops and Products*, *148*, 112270. 10.1016/j.indcrop.2020.112270

Chan, C., Vandi, L., Pratt, S., Halley, P., Richardson, D., Werker, A., & Laycock, B. (2020). Mechanical stability of polyhydroxyalkanoate (PHA)-based wood plastic composites (WPCs). *Journal of Polymers and the Environment*, *28*(5), 1571–1577. 10.1007/s1 0924-020-01697-9

Chen, G. Q., & Patel, M. K. (2012). Plastics derived from biological sources: Present and future: A technical and environmental review. *Chemical Reviews*, *112*(4), 2082–2099. 10.1021/cr200162d

Chi, H., Xu, K., Wu, X., Chen, Q., Xue, D., Song, C., Zhang, W., & Wang, P. (2008). Effect of acetylation on the properties of corn starch. *Food Chemistry*, *106*(3), 923–928. 10.1 016/j.foodchem.2007.07.002

Choi, J., & Lee, S. Y. (1999). Efficient and economical recovery of poly (3-hydroxybutyrate) from recombinant Escherichia coli by simple digestion with chemicals. *Biotechnology and Bioengineering*, *62*(5), 546–553. 10.1002/(sici)1097-0290(19990305)62

Chung, T. J., Park, J. W., Lee, H. J., Kwon, H. J., Kim, H. J., Lee, Y. K., & Yin Tze, W. T. (2018). The improvement of mechanical properties, thermal stability, and water absorption resistance of an eco-friendly PLA/kenaf biocomposite using acetylation. *Applied Sciences*, *8*(3), Article 376. 10.3390/app8030376

Čolnik, M., Knez-Hrnčič, M., Škerget, M., & Knez, Ž. (2020). Biodegradable polymers, current trends of research and their applications, a review. *Chemical Industry & Chemical Engineering Quarterly*, 26(4), 401–418. 10.2298/CICEQ191210018C

Curvelo, A. A. S., de Carvalho, A. J. F., & Agnelli, J. A. M. (2001). Thermoplastic starch–cellulosic fibers composites: Preliminary results. *Carbohydrate Polymers*, *45*(2), 183–188. 10.1016/S0144-8617(00)00314-3

da Silva Moura, A., Demori, R., Leão, R. M., Frankenberg, C. L. C., & Santana, R. M. C. (2019). The influence of the coconut fiber treated as reinforcement in PHB (poly-hydroxybutyrate) composites. *Materials Today Communications*, *18*, 191–198. 10.101 6/j.mtcomm.2018.12.006

Dicker, M. P. M., Duckworth, P. F., Baker, A. B., Francois, G., Hazzard, M. K., & Weaver, P. M. (2014). Green composites: A review of material attributes and complementary applications. *Composites Part A: Applied Science and Manufacturing*, *56*, 280–289. 10.1016/j.compositesa.2013.10.014

Elzein, T., Nasser, M. E., Delaite, C., Bistac, S., & Dumas, P. (2004). FTIR study of polycaprolactone chain organization at interfaces. *Journal of Colloid and Interface Science*, *273*(2), 381–387. 10.1016/j.jcis.2004.02.001

Espinach, F. X., Julián, F., Alcalà, M., Tresserras, J., & Mutjé, P. (2014). High stiffness performance alpha-grass pulp fiber reinforced thermoplastic starch-based fully biodegradable composites. *BioResources*, *9*(1), 738–755.

European bioplastics. (2021). *Global Production Capacities of Bioplastics 2020 & 2025*. https://www.european-bioplastics.org/wp-content/uploads/2020/11/Global_ Production_Capacity_Total_2020vs2025.jpg

Faruk, O., Bledzki, A. K., Fink, H.-P., & Sain, M. (2012). Biocomposites reinforced with natural fibers: 2000–2010. *Progress in Polymer Science*, *37*, 1552–1596. 10.1016/ j.progpolymsci.2012.04.003.

Frollini, E., Bartolucci, N., Sisti, L., & Celli, A. (2015). Biocomposites based on poly (butylene succinate) and curaua: Mechanical and morphological properties. *Polymer Testing*, *45*, 168–173. 10.1016/j.polymertesting.2015.06.009

Gholampour, A., & Ozbakkaloglu, T. (2020). A review of natural fiber composites: Properties, modification and processing techniques, characterization, applications. *Journal of Materials Science*, *55*(3), 829–892. 10.1007/s10853-019-03990-y

Gross, R. A., & Kalra, B. (2002). Biodegradable polymers for the environment. *Science*, *297*(5582), 803. 10.1126/science.297.5582.803

Gurunathan, T., Mohanty, S., & Nayak, S. K. (2015). A review of the recent developments in biocomposites based on natural fibres and their application perspectives. *Composites Part A: Applied Science and Manufacturing*, *77*, 1–25. 10.1016/j.compositesa.2015.06.007

Hermida, É. B., & Mega, V. I. (2007). Transcrystallization kinetics at the poly (3-hydroxybutyrate-co-3-hydroxyvalerate)/hemp fibre interface. *Composites Part A: Applied Science and Manufacturing*, *38*(5), 1387–1394. 10.1016/j.compositesa.2006.10.006

Hocking, P. J. (1992). The classification, preparation, and utility of degradable polymers. *Journal of Macromolecular Science, Part C*, *32*(1), 35–54. 10.1080/15321799208018378

Hong, G., Cheng, H., Zhang, S., & Rojas, O. J. (2021). Mussel-inspired reinforcement of a biodegradable aliphatic polyester with bamboo fibers. *Journal of Cleaner Production*, *296*, 126587. 10.1016/j.jclepro.2021.126587

Hu, J., Jahid, M. A., Harish Kumar, N., & Harun, V. (2020). Fundamentals of the fibrous materials. In J. Hu, B. Kumar, & J. Lu (Eds.), *Handbook of Fibrous Materials* (pp. 1–36). Wiley Online Books. 10.1002/9783527342587.ch1

Hu, R. H., Sun, M. Y., & Lim, J. K. (2010). Moisture absorption, tensile strength and microstructure evolution of short jute fiber/polylactide composite in hygrothermal environment. *Materials & Design*, *31*(7), 3167–3173. 10.1016/j.matdes.2010.02.030

Huyhua, S. (2010). Recycling plastics: New recycling technology and biodegradable polymer development. https://illumin.usc.edu/recycling-plastics-new-recycling-technology-and-biodegradable-polymer-development/

Iman, M., & Maji, T. K. (2015). Bionanocomposites: A greener alternative for future generation. In V. K. Thakur & M. R. Kessler (Eds.), *Green Biorenewable Biocomposites: From Knowledge to Industrial Applications* (pp. 516). Apple Academic Press.

Inkinen, S., Hakkarainen, M., Albertsson, A. C., & Södergård, A. (2011). From lactic acid to poly (lactic acid) (PLA): Characterization and analysis of PLA and its precursors. *Biomacromolecules*, *12*(3), 523–532. 10.1021/bm101302t

Islam, T., Khan, R. A., Khan, M. A., Rahman, M. A., Fernandez-Lahore, M., Huque, Q. M. I., & Islam, R. (2009). Physico-mechanical and degradation properties of gamma-irradiated biocomposites of jute fabric-reinforced poly (caprolactone). *Polymer-Plastics Technology and Engineering*, *48*(11), 1198–1205. 10.1080/03602550903149169

Jacquel, N., Freyermouth, F., Fenouillot, F., Rousseau, A., Pascault, J. P., Fuertes, P., & Saint-Loup, R. (2011). Synthesis and properties of poly (butylene succinate): Efficiency of different transesterification catalysts. *Journal of Polymer Science Part A: Polymer Chemistry*, *49*(24), 5301–5312. 10.1002/pola.25009

Jamshidian, M., Tehrany, E. A., Imran, M., Jacquot, M., & Desobry, S. (2010). Poly-lactic acid: Production, applications, nanocomposites, and release studies. *Comprehensive Reviews in Food Science and Food Safety*, *9*(5), 552–571. 10.1111/j.1541-4337.2010.00126.x

Jawaid, M., Sapuan, S. M., & Alothman, O. Y. (2016). *Green Biocomposites: Manufacturing and Properties*. Springer International Publishing. https://books.google.co.in/books?id=H0R4DQAAQBAJ

Jiang, T., Duan, Q., Zhu, J., Liu, H., & Yu, L. (2020). Starch-based biodegradable materials: Challenges and opportunities. *Advanced Industrial and Engineering Polymer Research*, *3*(1), 8–18. 10.1016/j.aiepr.2019.11.003

Jimenez, G., Ogata, N., Kawai, H., & Ogihara, T. (1997). Structure and thermal/mechanical properties of poly (ε-caprolactone)-clay blend. *Journal of Applied Polymer Science*, *64*(11), 2211–2220. 10.1002/(SICI)1097-4628(19970613)64:11<2211::AID-APP17>3.0.CO;2-6

John, M. J., & Thomas, S. (2008). Biofibres and biocomposites. *Carbohydrate Polymers*, *71*(3), 343–364. 10.1016/j.carbpol.2007.05.040

Joyyi, L., Ahmad Thirmizir, M. Z., Salim, M. S., Han, L., Murugan, P., Kasuya, K., Maurer, F. H. J., Zainal Arifin, M. I., & Sudesh, K. (2017). Composite properties and biodegradation of biologically recovered P(3HB-co-3HHx) reinforced with short kenaf fibers. *Polymer Degradation and Stability*, *137*, 100–108. 10.1016/j.polymdegradstab.2017.01.004

Ju, D., Han, L., Guo, Z., Bian, J., Li, F., Chen, S., & Dong, L. (2015). Effect of diameter of poly (lactic acid) fiber on the physical properties of poly (ε-caprolactone). *International Journal of Biological Macromolecules*, *76*, 49–57. 10.1016/j.ijbiomac.2015.01.059

Karthika, M., Shaji, N., Johnson, A., Neelakandan, M. S., Gopakumar, A. D., & Thomas, S. (2019). Biodegradation of green polymeric composites materials. In P. M. Visakh, O. Bayraktar, & G. Menon (Eds.), *Bio Monomers for Green Polymeric Composite Materials* (pp. 141–159). Wiley Online Books. 10.1002/9781119301714.ch7

Khalid, M. Y., Imran, R., Arif, Z. U., Akram, N., Arshad, H., Al Rashid, A., & García Márquez, F. P. (2021). Developments in chemical treatments, manufacturing techniques and potential applications of natural-fibers-based biodegradable composites. *Coatings*, *11*(3), 1–18. 10.3390/coatings11030293

Kobashi, K., Villmow, T., Andres, T., & Pötschke, P. (2008). Liquid sensing of melt-processed poly (lactic acid)/multi-walled carbon nanotube composite films. *Sensors and Actuators, B: Chemical*, *134*(2), 787–795. 10.1016/j.snb.2008.06.035

Kumar, S. R., Shaiju, P., O'Connor, K. E., & Babu P, R. (2020). Bio-based and biodegradable polymers - State-of-the-art, challenges and emerging trends. *Current Opinion in Green and Sustainable Chemistry*, *21*, 75–81. 10.1016/j.cogsc.2019.12.005

Liu, W., Thayer, K., Misra, M., Drzal, L. T., & Mohanty, A. K. (2007). Processing and physical properties of native grass-reinforced biocomposites. *Polymer Engineering & Science*, *47*(7), 969–976. 10.1002/pen.20611

Lu, Y., Tighzert, L., Dole, P., & Erre, D. (2005). Preparation and properties of starch thermoplastics modified with waterborne polyurethane from renewable resources. *Polymer*, *46*(23), 9863–9870. 10.1016/j.polymer.2005.08.026

Mahmud, S., Hasan, K. M. F., Jahid, M. E., Mohiuddin, K., Zhang, R., & Zhu, J. (2021). Comprehensive review on plant fiber-reinforced polymeric biocomposites. *Journal of Materials Science*, *56*(12), 7231–7264. 10.1007/s10853-021-05774-9

Malikmammadov, E., Tanir, T. E., Kiziltay, A., Hasirci, V., & Hasirci, N. (2018). PCL and PCL-based materials in biomedical applications. *Journal of Biomaterials Science, Polymer Edition*, *29*(7–9), 863–893. 10.1080/09205063.2017.1394711

Mathijsen, D. (2021). Fully bio-based fiber reinforced thermoplastics can now challenge polypropylene composites. *Reinforced Plastics*, *65*(2), 96–100. 10.1016/j.repl.2021.02.008

Meereboer, K. W., Misra, M., & Mohanty, A. K. (2020). Review of recent advances in the biodegradability of polyhydroxyalkanoate (PHA) bioplastics and their composites. *Green Chemistry*, *22*(17), 5519–5558. 10.1039/D0GC01647K

Michel, A. T., & Billington, S. L. (2012). Characterization of poly-hydroxybutyrate films and hemp fiber reinforced composites exposed to accelerated weathering. *Polymer Degradation and Stability*, *97*(6), 870–878. 10.1016/j.polymdegradstab.2012.03.040

Mochane, M. J., Magagula, S. I., Sefadi, J. S., & Mokhena, T. C. (2021). A review on green composites based on natural fiber-reinforced polybutylene succinate (PBS). *Polymers*, *13*(8), 1200. https://www.mdpi.com/2073-4360/13/8/1200

Mohanty, A. K., Wibowo, A., Misra, M., & Drzal, L. T. (2004). Effect of process engineering on the performance of natural fiber reinforced cellulose acetate biocomposites. *Composites Part A: Applied Science and Manufacturing*, *35*(3), 363–370. 10.1016/j.compositesa.2003.09.015

Mohnty, A. K., Misra, M., Drzal, L. T., Selke, S. E., Harte, B. R., & Hinrichsen, G. (2005). Natural fibers, biopolymers, and biocomposites: An introduction. In A. K. Mohanty, &

M. Misra (Eds.), *Natural Fibers, Biopolymers and Biocomposites* (pp. 1–35). Taylor & Francis.

Moliner, C., Badia, J. D., Bosio, B., Arato, E., Kittikorn, T., Strömberg, E., Teruel-Juanes, R., Ek, M., Karlsson, S., & Ribes-Greus, A. (2018). Thermal and thermo-oxidative stability and kinetics of decomposition of PHBV/sisal composites. *Chemical Engineering Communications*, *205*(2), 226–237. 10.1080/00986445.2017.1384921

Muthu, S. S. (2016). *Textiles and Clothing Sustainability: Implications in Textiles and Fashion*. Springer, Singapore. https://books.google.co.in/books?id=vrrhDAAAQBAJ

Nakayama, A., Yamano, N., & Kawasaki, N. (2019). Biodegradation in seawater of aliphatic polyesters. *Polymer Degradation and Stability*, *166*, 290–299. 10.1016/j.polymdegradstab.2019.06.006

Netravali, A. N., & Chabba, S. (2003). Composites get greener. *Materials Today*, *6*(4), 22–29. 10.1016/S1369-7021(03)00427-9

Nieddu, E., Mazzucco, L., Gentile, P., Benko, T., Balbo, V., Mandrile, R., & Ciardelli, G. (2009). Preparation and biodegradation of clay composites of PLA. *Reactive and Functional Polymers*, *69*(6), 371–379. 10.1016/j.reactfunctpolym.2009.03.002

Nitz, H., Semke, H., Landers, R., & Mülhaupt, R. (2001). Reactive extrusion of poly-caprolactone compounds containing wood flour and lignin. *Journal of Applied Polymer Science*, *81*(8), 1972–1984. 10.1002/app.1628

Nurul Fazita, M. R., Nurnadia, M. J., Abdul Khalil, H. P. S., Mohamad Haafiz, M. K., Fizree, H. M., & Suraya, N. L. M. (2017). Woven natural fiber fabric reinforced biodegradable composite: Processing, properties and application. In M. Jawaid, S. M. Sapuan, & O. Y. Alothman (Eds.), *Green Biocomposites: Manufacturing and Properties* (pp. 199–224). Springer International Publishing. 10.1007/978-3-319-46610-1_9

Oliver-Ortega, H., Tarrés, Q., Mutjé, P., Delgado-Aguilar, M., Méndez, J. A., & Espinach, F. X. (2020). Impact strength and water uptake behavior of bleached kraft softwood-reinforced PLA composites as alternative to PP-based materials. *Polymers*, *12*(9), Article 2144. 10.3390/POLYM12092144

Pan, H., Wang, X., Jia, S., Lu, Z., Bian, J., Yang, H., Han, L., & Zhang, H. (2021). Fiber-induced crystallization in polymer composites: A comparative study on poly (lactic acid) composites filled with basalt fiber and fiber powder. *International journal of biological macromolecules*, *183*, 45–54. 10.1016/j.ijbiomac.2021.04.104

Pivsa-Art, S., & Pivsa-Art, W. (2021). Eco-friendly bamboo fiber-reinforced poly (butylene succinate) biocomposites. *Polymer Composites*, *42*(4), 1752–1759. 10.1002/pc.25930

Plackett, D., Løgstrup Andersen, T., Batsberg Pedersen, W., & Nielsen, L. (2003). Biodegradable composites based on l-polylactide and jute fibres. *Composites Science and Technology*, *63*(9), 1287–1296. 10.1016/S0266-3538(03)00100-3

Polymer Properties Database. (2021). *Biodegradable Polyesters (Biobased) Aliphatic Polyesters*. https://polymerdatabase.com/polymer%20classes/Biodegradable%20Polyester%20type.html

Rai, P., Mehrotra, S., Priya, S., Gnansounou, E., & Sharma, S. K. (2021). Recent advances in the sustainable design and applications of biodegradable polymers. *Bioresource Technology*, *325*, 124739. 10.1016/j.biortech.2021.124739

Reddy, M. M., Vivekanandhan, S., Misra, M., Bhatia, S. K., & Mohanty, A. K. (2013). Biobased plastics and bionanocomposites: Current status and future opportunities. *Progress in Polymer Science*, *38*(10), 1653–1689. 10.1016/j.progpolymsci.2013.05.006

Reddy, S. R. T., Ratna Prasad, A. V., & Ramanaiah, K. (2021). Tensile and flexural properties of biodegradable jute fiber reinforced poly lactic acid composites. *Materials Today: Proceedings*, *44*, 917–921. 10.1016/j.matpr.2020.10.806

Reinhardt, M., Kaufmann, J., Kausch, M., & Kroll, L. (2013). PLA-viscose-composites with continuous fibre reinforcement for structural applications. *Procedia Materials Science*, *2*, 137–143. 10.1016/j.mspro.2013.02.016

Rocha, D. B., & Rosa, D. (2017). Biodegradable composites: Properties and uses. In V. K. Thakur, M. K. Thakur, & M. R. Kessler (Eds.), Handbook of Composites from Renewable Materials, Biodegradable Materials (Vol. 5, First Edition) (pp. 36). Wiley. 10.1002/9781119441632.ch91

Rojas-Bringas, P. M., De-la-Torre, G. E., & Torres, F. G. (2021). Influence of the source of starch and plasticizers on the environmental burden of starch-Brazil nut fiber bio-composite production: A life cycle assessment approach. *Science of The Total Environment*, *769*, 144869. 10.1016/j.scitotenv.2020.144869

Ruhul Amin, M., Mahmud, M. A., & Anannya, F. R. (2019). Natural fiber reinforced starch based biocomposites. *Polymer Science, Series A*, *61*(5), 533–543. 10.1134/S0965545 X1905016X

Saba, N., Jawaid, M., & Al-Othman, O. (2017). An overview on polylactic acid, its cellulosic composites and applications [Review]. *Current Organic Synthesis*, *14*(2), 156–170. 10.2174/1570179413666160921115245

Sarasini, F., Tirillò, J., Puglia, D., Kenny, J. M., Dominici, F., Santulli, C., Tofani, M., & De Santis, R. (2015). Effect of different lignocellulosic fibres on poly (ε-caprolactone)-based composites for potential applications in orthotics. *RSC Advances*, *5*(30), 23798–23809. 10.1039/C5RA00832H

Satyanarayana, K. G., Arizaga, G. G. C., & Wypych, F. (2009). Biodegradable composites based on lignocellulosic fibers—An overview. *Progress in Polymer Science*, *34*(9), 982–1021. 10.1016/j.progpolymsci.2008.12.002

Sawpan, M. A., Pickering, K. L., & Fernyhough, A. (2012). Flexural properties of hemp fibre reinforced polylactide and unsaturated polyester composites. *Composites Part A: Applied Science and Manufacturing*, *43*(3), 519–526. 10.1016/j.compositesa.2011. 11.021

Sekiguchi, T., Saika, A., Nomura, K., Watanabe, T., Watanabe, T., Fujimoto, Y., Enoki, M., Sato, T., Kato, C., & Kanehiro, H. (2011). Biodegradation of aliphatic polyesters soaked in deep seawaters and isolation of poly (ε-caprolactone)-degrading bacteria. *Polymer Degradation and Stability*, *96*(7), 1397–1403. 10.1016/j.polymdegradstab.2 011.03.004

Shaiju, P., Dorian, B. B., Senthamaraikannan, R., & Padamati, R. B. (2020). Biodegradation of poly (butylene succinate) (PBS)/stearate modified magnesium-aluminium layered double hydroxide composites under marine conditions prepared via melt compounding. *Molecules*, *25*(23), 5766. 10.3390/molecules25235766

Shanks, R. A., Hodzic, A., & Wong, S. (2004). Thermoplastic biopolyester natural fiber composites. *Journal of Applied Polymer Science*, *91*(4), 2114–2121. 10.1002/ app.13289

Shibata, M., Takachiyo, K., Ozawa, K., Yosomiya, R., & Takeishi, H. (2002). Biodegradable polyester composites reinforced with short abaca fiber. *Journal of Applied Polymer Science*, *85*(1), 129–138. 10.1002/app.10665

Shibata, M., Yosomiya, R., Ohta, N., Sakamoto, A., & Takeishi, H. (2003). Poly (ε-caprolactone) composites reinforced with short abaca fibres. *Polymers and Polymer Composites*, *11*(5), 359–367. 10.1177/096739110301100502

Siakeng, R., Jawaid, M., Ariffin, H., Sapuan, S. M., Asim, M., & Saba, N. (2019). Natural fiber reinforced polylactic acid composites: A review. *Polymer Composites*, *40*(2), 446–463. 10.1002/pc.24747

Singh, A. A., Genovese, M. E., Mancini, G., Marini, L., & Athanassiou, A. (2020). Green processing route for polylactic acid–cellulose fiber biocomposites. *ACS Sustainable Chemistry & Engineering*, *8*(10), 4128–4136. 10.1021/acssuschemeng.9b06760

Sinha Ray, S., Okamoto, K., & Okamoto, M. (2003). Structure–Property relationship in bio-degradable poly (butylene succinate)/layered silicate nanocomposites. *Macromolecules*, *36*(7), 2355–2367. 10.1021/ma021728y

Siracusa, V., Lotti, N., Munari, A., & Dalla Rosa, M. (2015). Poly (butylene succinate) and poly (butylene succinate-co-adipate) for food packaging applications: Gas barrier properties after stressed treatments. *Polymer Degradation and Stability*, *119*, 35–45. 10.1016/j.polymdegradstab.2015.04.026

Sirviö, J. A., Visanko, M., Ukkola, J., & Liimatainen, H. (2018). Effect of plasticizers on the mechanical and thermomechanical properties of cellulose-based biocomposite films. *Industrial Crops and Products*, *122*, 513–521. 10.1016/j.indcrop.2018.06.039

Song, J. H., Murphy, R. J., Narayan, R., & Davies, G. B. H. (2009). Biodegradable and compostable alternatives to conventional plastics. *Philosophical Transactions of the Royal Society B: Biological Sciences*, *364*(1526), 2127–2139. 10.1098/rstb.2008.0289

Soon, C. Y., Rahman, N. A., Tee, Y. B., Talib, R. A., Tan, C. H., Abdan, K., & Chan, E. W. C. (2019). Electrospun biocomposite: Nanocellulose and chitosan entrapped within a poly (hydroxyalkanoate) matrix for Congo red removal. *Journal of Materials Research and Technology*, *8*(6), 5091–5102. 10.1016/j.jmrt.2019.08.030

Tabone, M. D., Cregg, J. J., Beckman, E. J., & Landis, A. E. (2010). Sustainability metrics: Life cycle assessment and green design in polymers. *Environmental Science & Technology*, *44*(21), 8264–8269. 10.1021/es101640n

Tanase, C. E., & Spiridon, I. (2014). PLA/chitosan/keratin composites for biomedical applications. *Materials Science & Engineering C-Materials for Biological Applications*, *40*, 242–247. 10.1016/j.msec.2014.03.054

Teramoto, N., Urata, K., Ozawa, K., & Shibata, M. (2004). Biodegradation of aliphatic polyester composites reinforced by abaca fiber. *Polymer Degradation and Stability*, *86*(3), 401–409. 10.1016/j.polymdegradstab.2004.04.026

Toriz, G., Gatenholm, P., Seiler, B. D., & Tindall, D. (2006). Cellulose fiber reinforced cellulose esters: Biocomposite for the future. In*Natural Fibers, Biopolymers, and Biocomposites* (1st ed.). CRC Press. https://doi.org/10.1201/9780203508206

Van de Velde, K., & Kiekens, P. (2002). Biopolymers: Overview of several properties and consequences on their applications. *Polymer Testing*, *21*(4), 433–442. 10.1016/S0142-9418(01)00107-6

Verhoogt, H., Ramsay, B. A., & Favis, B. D. (1994). Polymer blends containing poly (3-hydroxyalkanoate)s. *Polymer*, *35*(24), 5155–5169. 10.1016/0032-3861(94)90465-0

Visakh, P. M., Bayraktar, O., & Menon, G. (2019). *Bio Monomers for Green Polymeric Composite Materials*. Springer.

Vroman, I., & Tighzert, L. (2009). Biodegradable polymers. *Materials*, *2*(2), 307–344. https://www.mdpi.com/1996-1944/2/2/307

Wang, C., Dung-Yi, W., & Shan, S. (2021). Preparation, characterization, and functionality of bio-based polyhydroxyalkanoate and renewable natural fiber with waste oyster shell composites. *Polymer Bulletin*, *78*, 4817–4834. 10.1007/s00289-020-03341-x

Wilkinson, S. L. (2001). Nature's pantry is open for business: Coffee grounds, molasses, and cellulose yield plastics, composites, and high-end paint. *Chemical and Engineering News(USA)*, *79*(4), 61–62.

Wojnowska-Baryła, I., Kulikowska, D., & Bernat, K. (2020). Effect of bio-based products on waste management. *Sustainability*, *12*(5), 2088. https://www.mdpi.com/2071-1050/12/5/2088

Wollerdorfer, M., & Bader, H. (1998). Influence of natural fibres on the mechanical properties of biodegradable polymers. *Industrial Crops and Products*, *8*(2), 105–112. 10.1016/S0926-6690(97)10015-2

Wong, S., Shanks, R., & Hodzic, A. (2002). Properties of poly (3-hydroxybutyric acid) composites with flax fibres modified by plasticiser absorption. *Macromolecular Materials and Engineering*, *287*(10), 647–655. 10.1002/1439-2054(200210)287:10<647::AID-MAME647>3.0.CO;2-7

Woodruff, M. A., & Hutmacher, D. W. (2010). The return of a forgotten polymer—Polycaprolactone in the 21st century. *Progress in Polymer Science*, *35*(10), 1217–1256. 10.1016/j.progpolymsci.2010.04.002

Wu, C. S. (2006). Assessing biodegradability and mechanical, thermal, and morphological properties of an acrylic acid-modified poly (3-hydroxybutyric acid)/wood flours bio-composite. *Journal of Applied Polymer Science*, *102*(4), 3565–3574. 10.1002/app.24817

Wu, C. S., Liao, H. T., & Cai, Y. X. (2017). Characterisation, biodegradability and application of palm fibre-reinforced polyhydroxyalkanoate composites. *Polymer Degradation and Stability*, *140*, 55–63. 10.1016/j.polymdegradstab.2017.04.016

Xie, Q., Li, F., Li, J., Wang, L., Li, Y., Zhang, C., Xu, J., & Chen, S. (2018). A new biodegradable sisal fiber–starch packing composite with nest structure. *Carbohydrate Polymers*, *189*, 56–64. 10.1016/j.carbpol.2018.01.063

Xu, J., & Guo, B. H. (2010). Poly (butylene succinate) and its copolymers: Research, development and industrialization. *Biotechnology Journal*, *5*(11), 1149–1163. 10.1002/biot.201000136

Yao, M., Deng, H., Mai, F., Wang, K., Zhang, Q., Chen, F., & Fu, Q. (2011). Modification of poly (lactic acid)/poly (propylene carbonate) blends through melt compounding with maleic anhydride. *Express Polymer Letters*, *5*(11), 937–949. 10.3144/expresspolymlett.2011.92

Yu, L., Dean, K., & Li, L. (2006). Polymer blends and composites from renewable resources. *Progress in Polymer Science*, *31*(6), 576–602. 10.1016/j.progpolymsci.2006.03.002

Yusof, F. M., Wahab, N. A., Abdul Rahman, N. L., Kalam, A., Jumahat, A., & Mat Taib, C. F. (2019). Properties of treated bamboo fiber reinforced tapioca starch biodegradable composite. *Materials Today: Proceedings*, *16*, 2367–2373. 10.1016/j.matpr.2019.06.140

Zaini, E. S., Azaman, M. D., Jamali, M. S., & Ismail, K. A. (2018). Synthesis and characterization of natural fiber reinforced polymer composites as core for honeycomb core structure: A review. *Journal of Sandwich Structures & Materials*, *22*(3), 525–550. 10.1177/1099636218758589

Zhang, C. W., Li, F. Y., Li, J. F., Li, Y. L., Xu, J., Xie, Q., Chen, S., & Guo, A. F. (2018). Novel treatments for compatibility of plant fiber and starch by forming new hydrogen bonds. *Journal of Cleaner Production*, *185*, 357–365. 10.1016/j.jclepro.2018.03.001

Zhao, C. Z., Junrong, L., Beihai, H., & Lihong, Z. (2017). Fabrication of hydrophobic biocomposite by combining cellulosic fibers with polyhydroxyalkanoate [Original Paper]. *Cellulose*, *24*(5), 2265–2274. 10.1007/s10570-017-1235-8

Zhao, X., Li, K., Wang, Y., Tekinalp, H., Larsen, G., Rasmussen, D., Ginder, R. S., Wang, L., Gardner, D. J., Tajvidi, M., Webb, E., & Ozcan, S. (2020). High-strength polylactic acid (PLA) biocomposites reinforced by epoxy-modified pine fibers. *ACS Sustainable Chemistry & Engineering*, *8*(35), 13236–13247. 10.1021/acssuschemeng.0c03463

Zhou, M., Yan, J., Li, Y., Geng, C., He, C., Wang, K., & Fu, Q. (2013). Interfacial strength and mechanical properties of biocomposites based on ramie fibers and poly (butylene succinate). *RSC Advances*, *3*(48), 26418–26426. 10.1039/C3RA43713B

Zumstein, M. T., Narayan, R., Kohler, H. P. E., McNeill, K., & Sander, M. (2019). Dos and do nots when assessing the biodegradation of plastics. *Environmental Science & Technology*, *53*(17), 9967–9969. 10.1021/acs.est.9b04513

11 Applications of Natural Fibers–Reinforced Composites (I)

Priyanka Soni and Shishir Sinha
Department of Chemical Engineering, Indian Institute of
Technology Roorkee, India

11.1 PLANT-BASED FIBERS

The natural fibers comprise of a wide scale of plant and mineral fibers. The primary structural component of plant fibers is cellulose, while animal fibers primarily comprise of protein. Even though mineral fibers consisted of an asbestos group of minerals, these are now not used as reinforcement in composites because of related health issues (Pickering et al., 2016). The different types of plant-based fibers that are utilized as reinforcement in natural fiber polymer composites follow.

Figure 11.1. shows the different kinds of plant-based fibers. Some of them are described below.

11.1.1 BAST FIBERS

Bast fibers are generally used as reinforcement in polymer composite preparation because the availability of these fibers is higher than other fibers. These fibers are taken from the external layer of the stem of the plant by different fiber extraction methods.

11.1.1.1 Flax Fibers

These fibers are pulled out from the linseed/flax plant. The flax plant is well known because we get these fibers from the stem of the plant and linseed oil (utilized in many industries). The tensile strength of flax fiber is much better in comparison to glass fiber. It has high tensile strength, very low density, and is stiff in nature.

11.1.1.2 Hemp Fibers

The fibers of hemp are pulled out from the stem of the hemp plant. Hemp fibers are strong and stiff, and less elastic than any other natural fiber. Hemp fibers have good flexural strength, tensile strength, and fire retardancy, making them suitable as substitute glass fibers in polymer composites (Syath Abuthakeer et al., 2016).

DOI: 10.1201/9781003201724-11

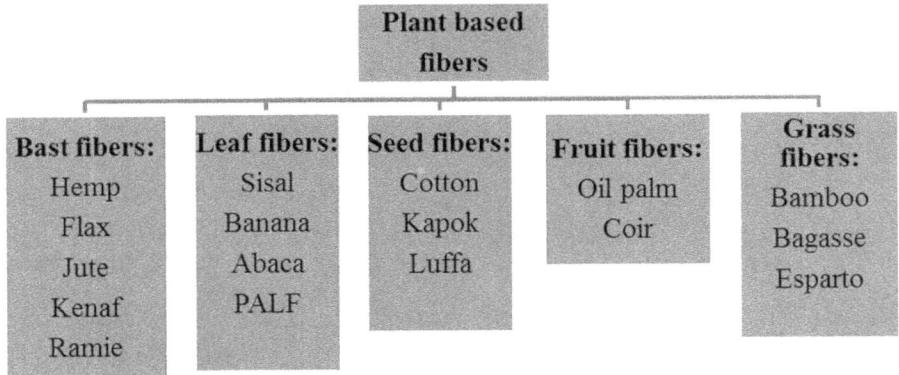

FIGURE 11.1   Classification of plant fibers.

11.1.2   LEAF FIBERS

These fibers are extracted from the plant's leaf. These fibers are categorized as hard fibers because they are usually long and stiff.

11.1.2.1   Sisal Fibers

Sisal fiber is pulled out from leaves by different parting processes such as retting and mechanical extraction. The cultivation time for sisal production is significantly less (Saxena et al., 2011). The tensile strength, corrosion resistance, resistance to salt water, and acid and alkali resistance are high for sisal fiber (Peças et al., 2018).

11.1.2.2   Palm Fibers

Palm fiber is extracted from the leaves of the dwarf palm (which is a member of the palm family). These fibers are pulled out from the palm tree leaves by different parting processes like retting, mowing, hackling, combing, and spinning. Date palm fibers are having the highest cellulose than other fibers. It is having high tensile strength, low density, and easy availability (Ghori et al., 2018).

11.1.3   SEED FIBERS

These fibers are extracted from seeds and seed cases of plants.

11.1.3.1   Cotton Fibers

These fibers are mainly consumed in the textile industry. The collection of cotton fiber is done by picking, which is commonly accomplished by hand (Saxena et al., 2011).

Cotton fiber has outstanding penetrability. Cotton signifies 46% of the world's creation of fibers (Peças et al., 2018).

11.1.3.2   Luffa Fibers

These fibers are not a single strand like other fibers but a package of cellulose fibrils making the fibrous circulatory system. The primary source of *Luffa cylindrica* or luffa fiber is ripe and dried fruit (Saw et al., 2013). The luffa fibers absorb water quickly.

11.1.4 FRUIT FIBERS

Fruit fibers are parted from the dried fruits of plants.

11.1.4.1 Coir Fiber

The coconut fibers (coir fibers) are extracted from the shell of the coconut palm fruit.

These fibers are coated with lignin which makes them more substantial than other natural fibers. Coir fibers are resistant to salt water and microbial degradation. It is readily available in the market, and its durability is greater than other natural fibers (Syath Abuthakeer et al., 2016).

11.1.5 GRASS FIBERS

The fibers of grass are pulled out from the plant's stem and leaves.

11.1.5.1 Bamboo Fibers

This fiber is a renewed plant fiber extracted from the bamboo tree. Bamboo fiber has a high growing frequency, tensile strength, and fixing carbon dioxide (CO_2), making it more suitable for different applications as reinforcement in composites (Syath Abuthakeer et al., 2016).

11.2 NATURAL FIBER–REINFORCED POLYMER COMPOSITES

Natural fiber–reinforced polymer composites combine different kinds of polymers (matrix) and outstanding strength plant fibers (reinforcement) such as hemp, jute, PALF, sisal, coir, and kenaf, etc.

The polymers that are used as matrices in composites are categorized into two groups: thermoplastics and thermosets. At high temperatures, the thermoplastic polymer tends to be softer and reattain its properties after cooling down, while the thermoset has a very high cross-linked structure (Asim et al., 2018) (Figure 11.2).

Flax, kenaf, jute, and hemp fibers are mainly used as reinforcement for composite applications because they are easily available along with their excellent mechanical properties and high strength in comparison to other bast fibers (Lotfi et al., 2021).

Natural fibers are capable material that substitutes the synthetic things and its associated goods for the reduced mass and energy conservancy applications. The utilization of natural fiber–reinforced polymer composites for the replacement of present synthetic or glass fiber–reinforced polymer composites is vast (Sanjay et al., 2016).

Natural fibers are having many exclusive properties such as sustainability, non-toxic nature, excellent performance, adaptability, low processing cost, non-irritable

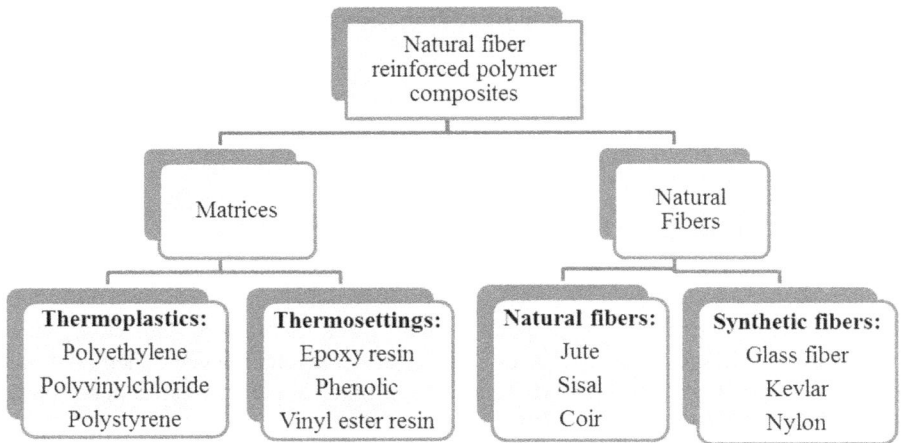

FIGURE 11.2   Natural fiber–reinforced composites.

to the respiratory system, skin, and resistance to corrosion. Because of these properties, natural fiber polymer composites have extra consideration compared to synthetic fiber polymer composite materials. The energy utilization during the fabrication of synthetic fiber composites is more compared to natural fiber composites. The application of natural fiber–reinforced polymer composites (NFRCs) increased because of their eco-friendly manufacturing and recycling process (Kumar et al., 2019).

The several types of plant fibers like coconut, jute, hemp, oil palm, sisal, and bamboo used in the fabrication of natural fiber–reinforced polymer composites has gained significant importance in different sectors. These are the various fields in which plant fiber–reinforced composites are used, i.e., automotive industries, structural component production, packing industries, and in the construction sector. Natural fiber–reinforced polymer composites have applications in numerous other areas, such as in electric industries, aerospace industries, regeneration equipment, watercraft, machinery, workplace goods, and so on (Mohammed et al., 2015) (Table 11.1).

11.3   LEADING ISSUES IN USING NATURAL FIBERS

Although there are many advantages to utilizing plant fiber reinforced composites, there are likewise some drawbacks. There are various challenges in utilizing natural fibers as reinforcement in composites (Figure 11.3) such that:

   a. There is a weak interfacial bonding between natural fiber and polymer matrix.
   b. Plant fibers are not long-lasting (not durable).
   c. The mechanical strength of natural fibers is less.
   d. Natural fibers are hydrophilic (water-absorbing tendency) in nature.

TABLE 11.1

Type of Natural Fibers, Producers, and Applications (Lotfi et al., 2021; Mohammed et al., 2015; Peças et al., 2018; Singh Dhaliwal, 2020)

Fibers	Producers	Applications
Hemp fiber	China, Canada, USA	Construction industries, textile products, paper and packaging industries, home furnishing, electrical equipment, manufacture of pipes, insulation materials
Jute fiber	India, Bangladesh, China	Building instrumental panels, roof interior coating, door supports, the door closes, transport industries, packing industries, geotextiles, chipboards, office panels, furnishings for houses, structural supports for agricultural slips
Flax fiber	China, Italy, Tunisia	Window framework, dashboards, flooring, railing products, fencing, badminton racquet, cycle frame, fork, chair post, snowboarding, laptop covers, green wall panel, textile industries, non-woven materials, insulation carpets, and specialist broadside
Coir fiber	India, Indonesia, SriLanka	Pulleys, mats, brushes, cushion, geotextiles, horticultural goods, containers, boxes, trays, building panels, flush gate shutters, roofing slips, storing tank, head covering materials and post containers, projector cover, voltage stabilizer cover, brushes, and brooms
Cotton fiber	India, China, USA	Home furnishing industry, textile industries, clothing, cordage, textile fabric, apparel, upholstery, non-wovens, specialty paper, health, and sanitary supplies
Kenaf fiber	India, China, Malaysia	Packaging materials, mobile phone covers, bags, insulations, clothing, soilless potting mixes, animal bedding, and material that grips oil and fluids
Ramie fiber	China, Brazil, India	Industrial stitching thread, stuffing materials, fishing trap, filter materials, materials for household furnishings, clothing, paper industries, textile fabrics
Sisal fiber	Brazil, Tanzania, Kenya	In the construction industries such as boards, gates, shutting panels, rooftops, home furnishing, packaging, fabrication of paper and pulp products

When natural fibers are exposed to the external environment, their degradation starts very fast. Several factors are responsible for the deterioration of fiber (Figure 11.4) such that:

a. Biological deterioration
b. Water deterioration
c. Mechanical deterioration
d. Climate deterioration
e. Deterioration due to fire

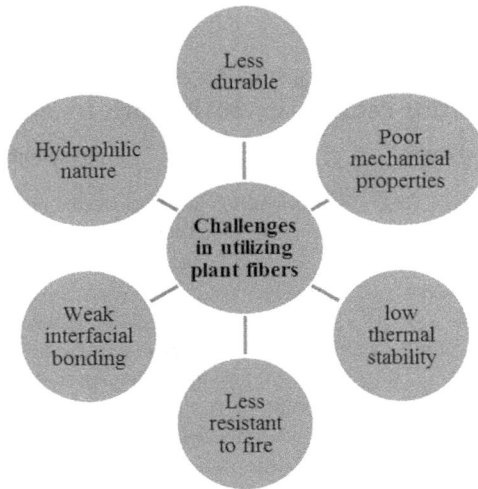

FIGURE 11.3   Main challenges in utilizing natural fibers in composites.

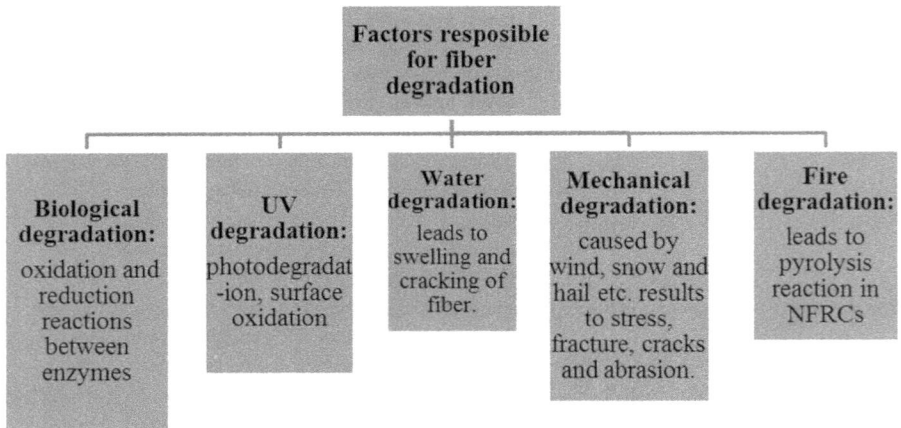

FIGURE 11.4   Various factors responsible for natural fiber degradation (Azwa et al., 2013; Kumar et al., 2019).

11.3.1   BIOLOGICAL DEGRADATION

Biodegradation of fibers occurred because of microorganisms present in the environment (bacteria and fungi), by enzymatic reactions (Tripathi & Yadav, 2017). Microorganisms or enzymes react with hemicelluloses present in natural fiber and convert them into consumable units. Because of the enzymatic reaction, the interfacial bonding between fiber-matrix weakens, resulting in reduced strength of polymer composites. In the biological degradation process, oxidation and reduction of fibers also occurred (Kumar et al., 2019).

11.3.2   WATER DEGRADATION

Water degradation is also a serious issue associated with natural fiber polymer composites. Natural fiber–reinforced polymer composites (NFPCs) absorb moisture when they are exposed to the environment. The primary sources for moisture absorption are ice, dew, and the ocean. Hemicellulose is the main component of fiber that absorbs moisture. Because of water absorption, fiber starts to swell and contract when it dries, resulting in the cracking of natural fiber polymer composites (NFPCs).

The mechanical strength of natural fiber polymer composites is decreased because of absorbed moisture bonds with the hydroxyl group of fiber (Kumar et al., 2019).

11.3.3   UV DEGRADATION

When natural fiber–reinforced polymer composites are exposed to the outdoor environment, fiber starts to degrade because of ultraviolet radiation (UV). The lignin present in fiber is accountable for UV degradation (Figure 11.5).

11.3.4   MECHANICAL DEGRADATION

The key factors responsible for the mechanical degradation of plant fiber polymer composites are wind, dirt, ice, hail, etc., resulting in stress, fracture, cracks, and abrasion of natural fiber composites.

11.3.5   FIRE DEGRADATION

The poor fire resistance is also a severe problem regarding natural fiber–reinforced polymer composites. The pyrolysis process (oxidation, hydrolysis) takes place during the fire degradation of fiber. The thermal degradation of natural fibers occurs in a two-stage method. The first process occurred in the range of 220°C–280°C temperature, while the second one occurred at 280°C–300°C (Kumar et al., 2019).

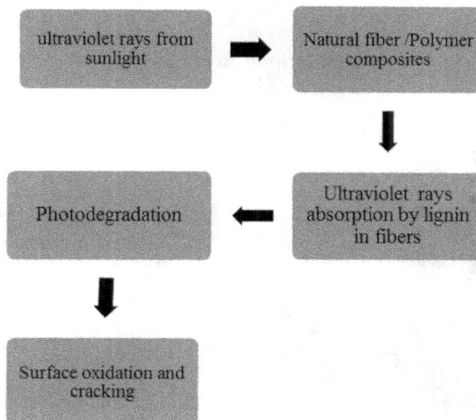

FIGURE 11.5   Ultraviolet radiation (UV) degradation of plant fibers (Omrani et al., 2016).

11.4 SURFACE MODIFICATION OF PLANT FIBERS

Fiber surface modification can improve the different fiber issues when utilized as reinforcement in polymer composites such as fiber and matrix interfacial bonding, the irregularity of fiber surface, wettability, and decrease water absorbing nature of the natural fiber (Ahmad et al., 2019).

Several kinds of surface treatments are used to improve the properties of the fiber (Figure 11.6).

Every surface modification has its advantages and disadvantages. Several advantages and disadvantages of physical treatments of the natural fibers are given below (Table 11.2).

11.4.1 CHEMICAL MODIFICATIONS OF NATURAL FIBERS

Several chemical treatments are used for the modification of fiber surfaces. Various chemicals are used to treat natural fibers such as alkali, silane coupling agent, peroxides, permanganates, etc. These chemical treatments can improve the mechanical properties of fibers along with removing the weak components of fibers such as hemicelluloses and lignin (Cruz & Fangueiro, 2016).

Several advantages and disadvantages of chemical treatments of plant fibers are given (Table 11.3).

11.5 VARIOUS APPLICATIONS OF NATURAL FIBER–REINFORCED POLYMER COMPOSITES

The utilization of plant fibers rapidly obtained the attention from different industrial fields. Properties of plant fibers are affected by the the rising natural environment, age of the plants, types, temperatures, moisture, and top soil properties (Sanjay et al., 2016).

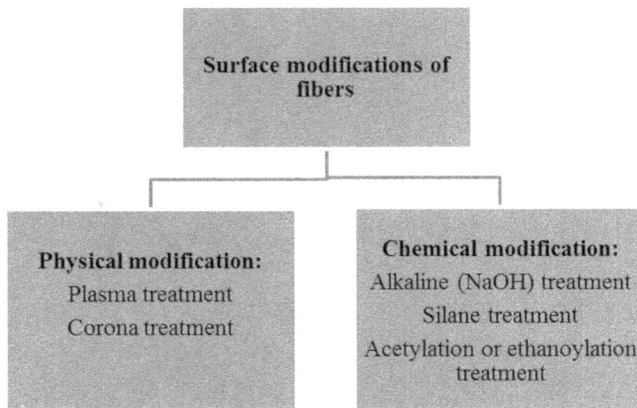

FIGURE 11.6 Different types of surface treatments used for natural fibers.

TABLE 11.2

Several Advantages and Disadvantages of Physical Treatments of Plant Fibers (Ahmad et al., 2019; Cruz & Fangueiro, 2016; Mukhopadhyay & Fangueiro, 2009)

Treatment	Advantages and Disadvantages
Stretching	**Advantages:** Even load distribution of fiber in composite. Reduction of fiber density. Heat treatment during stretching will improve the fiber strength. **Disadvantages:** The high heating rate throughout stretching can produce contraction on the surface of fiber.
Calendering	**Advantages:** Increment in the surface accessible for fiber and polymer matrix interaction. **Disadvantages:** A lot of damage occurs to the fiber surface during the calendaring process.
Solvent extraction	**Advantages:** By solvent extraction process, huge fiber cellulose content is obtained from plants. **Disadvantages:** By-products from the solvent extraction process are dangerous to ecosystems.
Corona treatment	**Advantages:** Fiber wettability improved. The compatibility between hydrophilic fibers and the hydrophobic matrix can be enhanced by corona treatment. **Disadvantages:** Surface etching can decrease the fiber sustainability.
Plasma treatment	**Advantages:** Plasma treatment can change the surface of natural fiber without changing the properties of the fiber.Enhance the fiber-matrix interfacial bonding. **Disadvantages:** An etching mechanism that occurs during plasma treatment creates pits on the fiber surface, utilizing the plasma properties.

Natural fiber–reinforced polymer composites are used in numerous fields; some of them are described below (Figures 11.7 and 11.8).

11.5.1 Automoible Applications of Natural Fiber Polymer Composites

The industrial sectors want to substitute heavyweight products with lightweight materials without negotiating strength. Natural fiber polymer composite materials are used for making lightweight materials. Natural fiber composites are chosen because of their excellent strength, light weight, non-corrosive nature, and excellent dimensional stability. The rising attention to more eco-friendly composite materials

TABLE 11.3

Advantages and Disadvantages of Chemical Modification of Natural Fibers (Ahmad et al., 2019; Gupta et al., 2015)

Treatment	Advantages and Disadvantages
Alkaline treatment	**Advantages:** Alkaline treatment rises the unevenness of the fiber surface. The maximum amount of lignin, wax, and oils appear in fiber is eliminated by alkaline treatment. **Disadvantages:** Increased alkaline concentration damages the fiber surface and decreases the tensile strength.
Silane treatment	**Advantages:** Fiber hydroxyl groups present in the fiber-matrix surface are reduced by a silane coupling agent. Silane treatment makes fiber surface tougher and decreases the water adsorption capacity of the fibers. **Disadvantages:** Tensile strength of fibers decreased after silane treatment.
Acetylation treatment	**Advantages:** The dimensional stability of natural fiber composites is increased. Increases crystallinity. Reduces water absorption capacity of fibers. **Disadvantages:** Because of this treatment, the mechanical properties of fibers decreased because of degradation of fiber and cracking of fiber surfaces.
Peroxide treatment	**Advantages:** Increases crystallinity and mechanical properties of the fiber surface. Improves interfacial bonding between fiber-matrix and also reduces the moisture absorption capacity of the fiber.Improves thermal stability. **Disadvantages:** The chemical bonding at the fiber surface depends on the surface morphology and chemical structure of fiber.

highlights the utilization of bio-based composite (natural fiber reinforced composites) (Syath Abuthakeer et al., 2016).

Natural fiber–reinforced polymer composites are employed in automobile industries for making lightweight components. Lightweight automobiles improve fuel efficiency and decrease greenhouse gas emissions (Asim et al., 2018).

The utilization of plant fibers in the automobile industry began in the 1930s when Henry Ford recommended the utilization of plant fibers. Different kinds of plant fibers such as flax, coir, hemp, sisal, and kenaf have been utilized by vehicle constructors to reinforce the car's plastic body parts and interior parts (Lotfi et al., 2021).

In the 1950s, natural fibers were utilized in the construction of the East German Trabant. Natural fiber polymer composites embrace significant possibilities for the

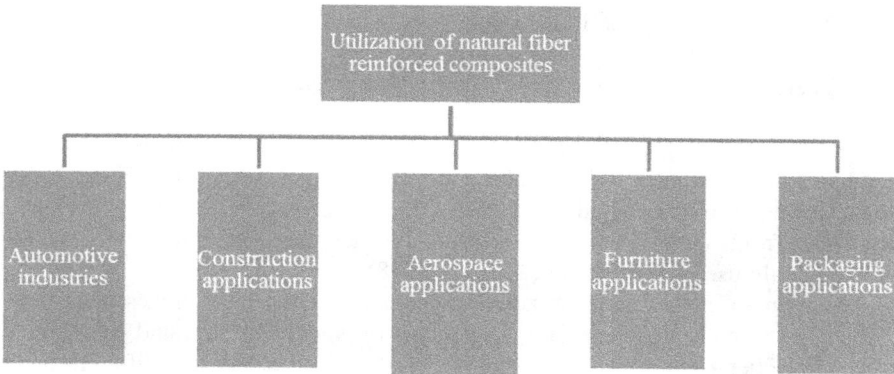

FIGURE 11.7   Various uses of natural fiber–reinforced polymer composites in distinct sectors.

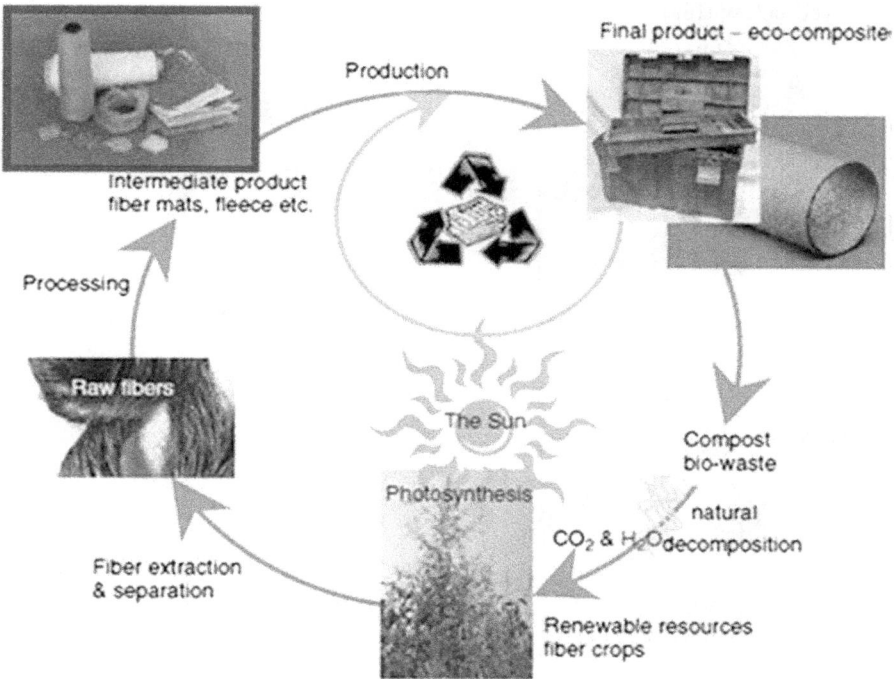

FIGURE 11.8   The life span of natural fiber–reinforced composites (Asim et al., 2018) (with permission).

automobile industry because of the greater demand for lightweight and environment friendly items. According to different studies, the manufacturing cost of automotive parts can reduce by 20%, and the overall weight reduction of automotive parts is 30%. The use of lightweight automotive parts have many advantages:

a. The fuel consumption is reduced.

b. The recycling of these parts is possible.

c. There is a decline in the trash disposal and greenhouse gas releases.

In different regions of the world, different varieties of plant fibers are used. The flax and hemp fibers are mainly used by the European automotive industry. The jute and kenaf fibers are from Bangladesh and India, banana fiber from the Philippines, and sisal fiber from South Africa and the United States. The German automotive industry mainly used flax fiber (Peças et al., 2018).

A maximum of the car manufacturing companies nowadays utilizes plant fiber composites to make different parts such as interiors, door linings, and paneling.

Natural fiber–reinforced composites are used for construction of different parts of a vehicle, which are discussed below (Figures 11.9 and 11.10).

Different kinds of plant fibers (primarily hemp and flax) are used for automotive applications.

Various uses of natural fiber polymer composites in automotive industries are displayed below (Figure 11.11).

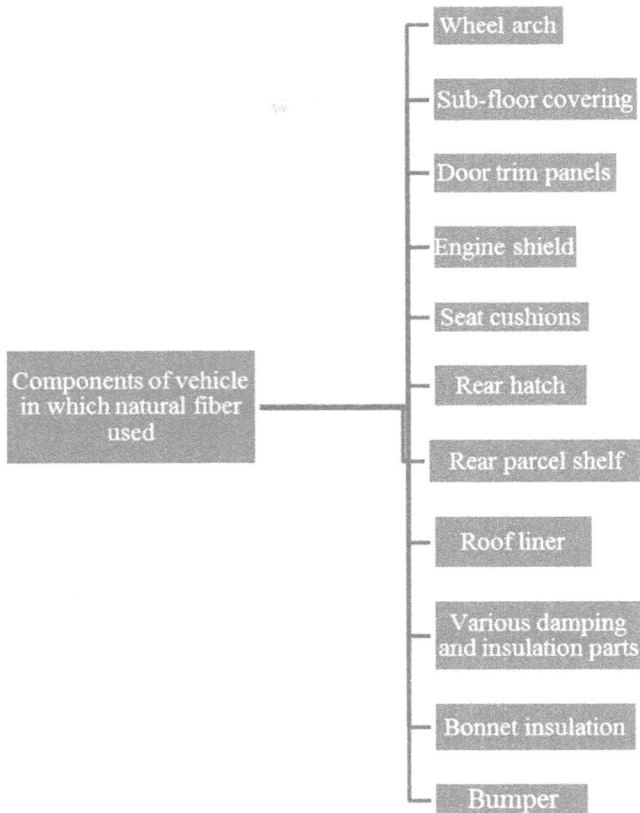

FIGURE 11.9 Different parts of a vehicle in which plant fibers are used (Sonar et al., n.d.).

FIGURE 11.10 Various parts of a vehicle made from natural fiber polymer composites (Asim et al., 2018).

(a)

(b)

(c)

(d)

FIGURE 11.11 Various automobile applications of natural fiber polymer composites: (a) the biomobile; (b) the electric scooter; (c) helmet of bicycle; and (d) racing car (Pil et al., 2016).

Most car manufacturing companies use natural fibers to produce interior parts such as inner gate panels, seat wraps, and insulation, and the commonly utilized fibers for these applications are flax, hemp, sisal, and kenaf. The polymers utilized as a matrix in arrangement with these plant fibers are mainly polypropylene and PLA (Peças et al., 2018) (Figure 11.12).

The main processes used by automotive industries for processing natural fiber composites are compression molding, injection molding, and structural reaction injection molding.

11.5.1.1 Thermoplastic and Thermoset Polymers

A thermoset or thermoplastic is utilized as a matrix in the manufacturing of plant fiber–reinforced polymer composites. The most commonly utilized thermoplastic in automobile application is polypropylene, particularly for manufacturing nonstructural components.

The most frequent use of polypropylene is due to its outstanding properties; for example, low density, excellent processability, good mechanical and electrical properties, and excellent dimensional constancy and impact strength. Various thermoplastic polymers are used in composites, with polyethylene (PE), polystyrene, and polyamide.

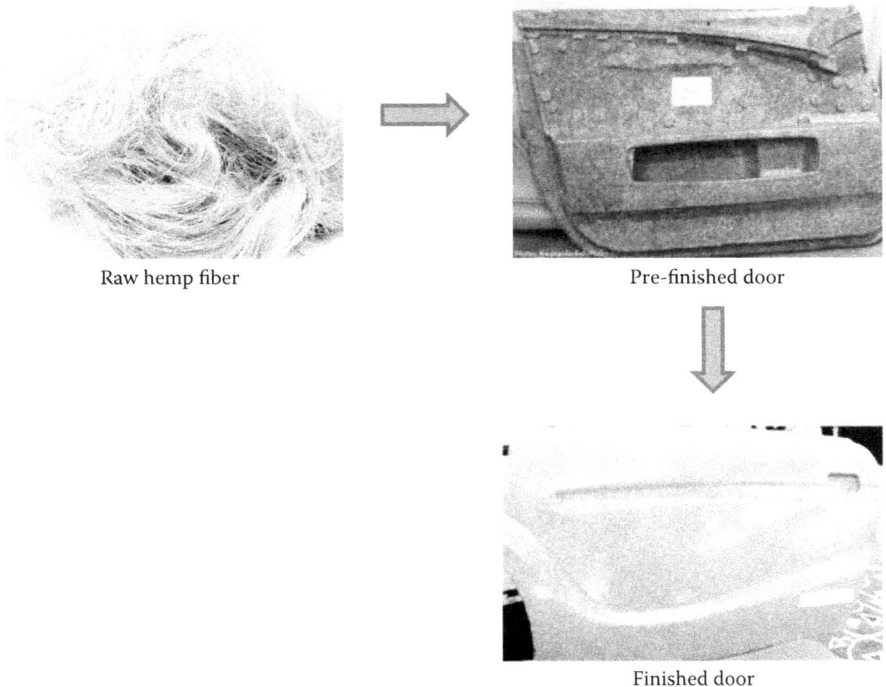

Raw hemp fiber   Pre-finished door

Finished door

FIGURE 11.12   Door panel of a car made from hemp fiber polymer composite.

The main limitation in fabrication of natural fiber thermoplastic composites is processing temperature because, at high temperatures, fibers start to degrade. The thermoset resins that are used in natural fiber–reinforced composites in automobile industries are polyester, vinyl ester, and epoxy resins. Epoxy resins are used frequently in automobile trades because of their superior strength and easy degradation to the environment. The applications of epoxy resins are limited because their curing time is long and have a high monomer cost.

Vinylester resins are newly added resins in the family unit of thermosetting resins. Vinylester resins have exceptional chemical resistance, great thermal and mechanical properties, and comparative smooth handling. They are more resistant to moisture than epoxy resins when cured at room temperature (Holbery & Houston, 2006).

Different kinds of fibers are used in the manufacturing of various components of a vehicle. Some of them are discussed below (Figure 11.13).

11.5.1.2 Advantages and Disadvantages of Natural Fiber Polymer Composites in Automotive Applications

There are a number of advantages and disadvantages of using plant fiber composites in automotive applications. Some of them are discussed below (Table 11.4).

11.5.1.3 Fabrication of Composites in Automotive Industries

There are various production processes utilized for natural fiber–centered automotive components. Some of them are described below (Figure 11.14).

In automobile industries, natural fiber polymer composites are utilized in the manufacturing of interior and external components.

11.5.1.4 Interior Applications

Natural fibers are used to manufacture interior components of a vehicle, such as door panels, seatback inside layer, and package drops (the area at the rear of the seats of cars), etc. Coir fiber is utilized in the development of different parts of

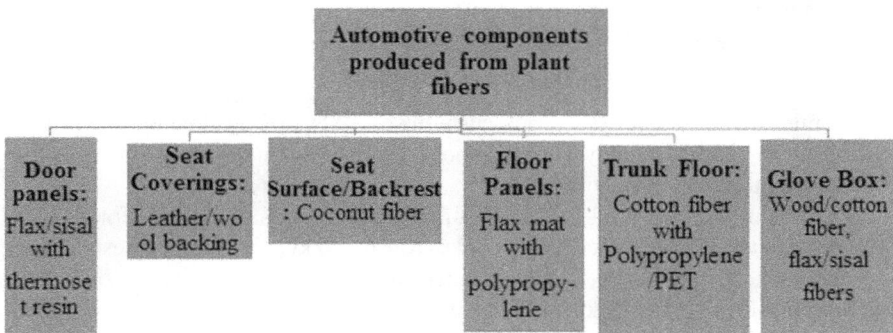

FIGURE 11.13 Different automotive parts manufactured from plant fibers (Holbery & Houston, 2006).

TABLE 11.4

Advantages and Disadvantages of Using Natural Fiber Polymer Composites in Automotive Applications (Fogorasi & Barbu, 2017)

Advantages	Disadvantages
There is an overall weight reduction (up to 10%–20%).	The thermal stability of natural fiber composites is minimal.
The fuel consumption is reduced.	They absorbed moisture when exposed to the environment, which results in swelling and cracking of natural fiber polymer composites.
There are several health benefits as compared to the synthetic fiber during the manufacturing process of composites because of no emission of toxic compounds.	The strength of plant fiber polymer composites is less, particularly the impact strength.
The greenhouse gas emission is reduced.	The durability of natural fiber reinforced polymer composites is low compared to synthetic fiber polymer composites.
The manufacturing cost is decreased.	Because of the breaking of fibers during the manufacturing process, the performance of composites decreased.
There are many socio-economic benefits, i.e., utilization of natural fiber polymer composites in automotive applications will create new opportunities in agriculture.	The disintegration of properties of natural fibers is also a limitation of composites.
In case of collisions and burning, natural fiber composites show good accident performance and high stability.	The inhomogeneous mixing of fiber and polymer matrix is also a limitation of natural fiber composites.
There are many technical benefits, i.e., lesser energy loss, good wear safeguard, and an enlarged life span of tools.	
The manufacturing and degradation are eco-friendly because natural fiber composites are renewable resources.	

automobiles, such as seat bases, back cushions of seats, and head belts. Abaca fiber is utilized in underfoot body panels. About 5–10 kg of plant fibers are used per vehicle in automotive industries.

The weight of fibers used in the manufacturing of automotive parts follws. The natural fibers used to manufacture seatbacks are 1.6–2.0 kg, while in headrests 2.5 kg, head door-liners 1.2–1.8 kg, rear gate-liners 0.8–1.5 kg, parcel shelves less than 2.0 kg, boot-liners 1.5–2.5 kg, sunroof inner shields less than 0.4 kg, and in the vibration and harshness materials less than 0.5 kg.

All leading automobile companies now utilize natural fiber–reinforced composites in various applications. A leading automobile manufacturer of Germany,

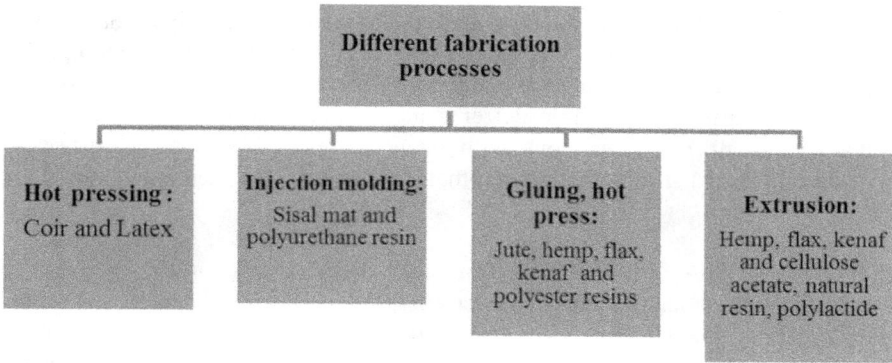

FIGURE 11.14   Different techniques used in the production of natural fiber–centered automotive components (Huda et al., 2008a).

Daimler-Benz, uses various natural fibers such as sisal, coconut, hemp, and flax to reinforce polypropylene components to substitute glass fibers. Different automobile parts are manufactured by Daimler-Benz using natural fiber composites such as dashes and center armrest cabinets, with seat shells and lining on seatbacks. Another automobile manufacturer, Daimler Chrysler, enlarged natural fibers' use in several vehicles equal to 98%. In Mercedes S-class vehicles, almost 27 parts are manufactured from natural fiber–reinforced composites.

In 2004, almost 10,000 tons of plant fiber were used by BMW Groups in its vehicles. Nearly 24 kg of fiber is used in every BMW 7-series car, along with flax and sisal plant fiber in the interior gate coatings and sheets, the cotton fiber in the noise proofing, wool fabric in the padding, and wood fiber in the seat support pillows.

Toyota produced a recyclable plastic that is developed from starch taken from sweet potatoes and other vegetables. The plastic was utilized as a matrix with plant fibers for utilizing the fuel-effective ES3 idea car's support enhancer and additional interior components. Kenaf fiber has been utilized in the frame structure of Toyota's i-foot and i-unit model automobiles. Kenaf fibers are used by Ford (Germany) in the Mondeo model. The door plates of the Mondeo have been developed from kenaf fiber polypropylene (PP) composites.

VolcaLite is a natural fiber–reinforced composite manufactured with long-cut basalt fiber and polypropylene (PP). VolcaLite has many advantages, such as higher process temperatures, greater modulus, and good chemical resistance compared to glass fiber. It is utilized as a replacement for equally glass and carbon fiber in polymer composites. It has been used for headliners, offering ultra-skinny profiles up to 3 mm.

Automotive component dealer Cambridge Industries manufacture flax fiber–reinforced polypropylene composites for Freightliner Century COE C-2 heavy-duty automobiles. Plant fibers are also being utilized as reinforcement in polyurethane composite materials. The plant fiber polyurethane composites are employed in the inner gate panel for S-class Mercedes-Benz, manufactured in Germany.

Becker Group Europe produced a natural fiber–reinforced polyurethane composite called Fibropur. According to Becker reports, this natural fiber composite is elastic and does not pull apart, equal in extremely contoured parts such as gate pull cups and arm rests. Chemie Pelzer manufactures panels by utilizing natural and synthetic fiber combinations with phenolic resin. Mercedes-Benz uses a mixture of flax and sisal with a polyurethane polymer matrix to manufacture door inner trim sheets.

11.5.1.5 Exterior Applications

The application of natural fiber polymer composites is presently restricted to the interior parts of the automobiles. Recent innovations are directed to the utilization of natural fiber–reinforced polymer composites to manufacture the outer parts. Daimler Chrysler uses abaca fiber in exterior underfloor paneling on the Mercedes A-class. The composites of abaca fiber have great elasticity and remarkable tensile strength. Other exterior parts of vehicle (forward-facing bumper, under floor trim of automobiles, etc.) manufactured from natural fiber–reinforced polymer composites should be accessible soon. The flax fibers are used by Mercedes-Benz to manufacture exterior parts for Travego and other models.

Various exterior parts of vehicles, for example the bumper, steering wheel, glass housing, lenses, and body structures, are manufactured using natural fiber–reinforced composites.

11.5.1.6 Bumpers

Bumpers are used to protect a vehicle during any collision. They are fixed on the back and forward-facing end of cars. They guard the car parts such as headlights and taillights, hood, fender, exhaust, and chilling system.

11.5.1.7 Body-in-White (BIW)

BIW is defined as a group of components that tolerate stagnant and dynamic loads and also import torsion toughness to the motor vehicle. It contained several marked steel components, which are connected together to form BIW. Currently, steel is substituted by natural fiber–reinforced polymer composite in the development of BIW. Ford Motors will develop BIW utilizing natural fiber–reinforced composite in its Composite Concentrated Vehicle project (Agarwal et al., 2020).

Different kinds of natural fiber–reinforced polymer composites are utilized in the buildup of automotive parts. Some of them are described below (Table 11.5).

Many automobile manufacturers, for instance Mercedes-Benz, Audi, and Daimler, substitute several glass fiber–built composites with natural fiber–reinforced composites (Table 11.6).

11.5.2 Aircraft Applications

At the foundation of aerospace companies, aircraft configuration were manufactured with wood composite (natural fiber composite), wire, and cloth. Recently, aircraft parts are manufactured by utilizing plant fiber, an alternate to synthetic fibers; for example, carbon, glass, and Kevlar (Asim et al., 2018).

TABLE 11.5

Utilization of Natural Fiber Polymer Composites in Manufacturing of Different Automotive Parts (Singh Dhaliwal, 2020)

Natural Fiber (Reinforcement)	Matrix	Automotive Applications
Abaca fiber	Epoxy and polypropylene resin	Under-deck panel and body panels
Coconut fiber	Epoxy, polypropylene, and polyester resin	Seat bases, rear cushions, head restraints, internal trim, seatback cushioning, seat surfaces/back rests
Bast fibers (flax, hemp, kenaf, sisal, etc.)	Polypropylene (PP) and polyester resin	Carrier for enclosed door sheets, shielded components for instrument boards, covered inserts, carrier for tough and soft arm rests, seat support panels, door panels, gate bolsters, headliners, side and rear walls, seatbacks, rear deck salvers, pillars, center consoles, load floors, box trim
Banana fiber	LDPE (low-density polyethylene)	Wrapping paper
Coir fiber	Polyester resin	Car seat shields, mattresses, door mats, carpets
Cotton fiber	Polypropylene	Noise proofing, trunk board, insulation
Fibrowood reprocessed	Polypropylene granules	Plastic deposit for seatback board
Flax fiber	Polypropylene, epoxy resin	Seatbacks, covers, back parcel shelves, other internal trim, floor trays, pillar panels, and principal consoles, floor board
Flax and hemp fiber	Epoxy resin	Carrier for enclosed door boards
Flax and sisal fiber	Thermoset resins	In the interior gate linings and boards, door sheets
Kenaf fiber	Polypropylene	Door inside board
Kenaf and flax fiber	Thermoset resins	Package plates and door board inserts
Kenaf and hemp fiber	Polypropylene, polyester resin	Door sheets, back parcel shelves, other inside trim, Lexus package shelves, door boards
Wood fiber	Acrylic resin and synthetic fiber	Carrier for enclosed door boards, covered or frothed instrument sheets, covered inserts and parts, covered seatback sheets, the fiber in the seatback cushions, inserts, spare tire, shields
Wood Flour	PP or polyolefin (POE)	Carrier for enclosed gate panels, carrier for the arm rest, transportation for shielded inserts
Wool fibers	Leather as matrix	Upholstery, seat casings

TABLE 11.6

Applications of Natural Fiber–Reinforced Composites (NFRCs) in the Manufacturing of Different Automotive Parts (Mohammed et al., 2015; Peças et al., 2018; Syduzzaman et al., 2020)

Automobile Manufacturers (Vehicle Models)	Utilizations
Rover 2000 and others	Rear depot rock shelf and insulations
Opel (Vectra, Astra, and Zafira)	Entrance interior panels, pillar covering panel, headliner board, and instrumental sheet
AUDI (series A2, A3, A4 Avant, A6, A8)	Boot-lining, extra tire-lining, side and backward door panel, seatbacks, and cap rack
BMW (series 3, 5, and 7, and other Pilot)	Seatback, headliner section, boot-lining, gate panels, noise insulation sheets, and molded footwell coatings
Volkswagen (Passat Variant, A4, Bora)	Seatback, door board, boot-lid finish sheet, and boot-lining
Daimler Chrysler (A, C, and S grades, EvoBus) (in outer components)	Support shield panel, gate interior panels, car wind shield, and business desks
VAUXHALL (Astra, Corsa, and Zafira)	Headliner board, interior door sheets, pillar coverup panel, and instrument Board
Lotus Eco Elise	Body sheets, spoiler, seat covers, and interior mats
Citroen C5	Interior gate lining
Mercedes Benz (A, S, E, and C grades)	gate panels (sisal and wood fibers combined with epoxy resin matrix), glove compartment (cotton fibers/wood molded), instrument board bear, insulations (cotton fiber), trunk board (cotton with Polypropylene/PET resins), and seat exterior/backrest (coir fiber/natural rubber)
Mercedes Benz (dumper truck)	Interior engine covering, engine insulation, sun shield, inside insulation, buffer, and ceiling supports
Peugeot (405, 406, and 407)	Head and back gate panels, seatbacks, and package shelf
Fiat (Punto, Brava,Marea, Alfa Romeo)	Door sheets
General Motors (Cadillac De Ville and Chevrolet TrailBlazer)	Seat supports, cargo area floormat
Mitsubishi (Fiat SpA)	Cargo area decks, door sheets, and instrumental board
Toyota (Raum, Brevis, Celsior)	Floor mats, spare exhaust cover, door sheets, and seat supports
Renault (Clio, Twingo)	Rear portion layer
Saab (9S)	Door panels
Ford (Mondeo CD162)	Deck trays, gate inserts, door shets, B-pillar, and boot-lining
Volvo (XC70)	Stool cushions, natural froths, and cargo deck tray
Saturn (L300)	Bundle trays and door sheets
Toyota	Support garnish and other interior componets

Natural fiber polymer composites have a few advantages; for example, high strength, light weight, excellent fatigue, and resistance to corrosion. Natural fiber composites are used in several aircraft industries to manufacture different Airbus models such as the Boeing 757, 767, and 777, and in Europe, the Airbus versions A310, A320, A330, and A340 aircraft. The aircraft manufacturer in Dallas-Fort Worth (USA) used natural fiber composites to manufacture airplane blades in their aircraft models (412, 407, 427, 214, 609, OH58D, V22).

There are many applications of natural fiber–reinforced composites materials in Airbus airliner models (Figure 11.15) (Table 11.7).

FIGURE 11.15 Application of plant fiber composites in aircraft components (interior) (Asim et al., 2018).

TABLE 11.7

Different Applications of Natural Polymer Fiber–Reinforced Composites Materials in Airbus Airplane Models (Mansor et al., 2019)

Aircraft Models	Natural Fiber Composite Applications
A300	Fairings, radar dome
A310	Rudder, spoilers, window brakes, winches, VTP check box
A320	Flaps, dried up HTP check box, LG gate, engine hoods
A330 and A340	Moist HTP check box, ailerons
A340–600	Rear divider, keel shaft, J-nose
A380	Center ground wing box, wing beams, rear unpress fuselage, crossover beams
A400	Higher wing
A350	Guest wing box, fuselage

11.5.3 NAVAL APPLICATIONS

Natural fiber–reinforced polymer composites are used as an alternate solution for boat manufacturers. Natural fiber composites to substitute synthetic fibers goods finds countless considerations inreducing non-renewable resources. The plant fiber–reinforced polymer composites have low density, good sound absorption ability, excellent impact strength, and good stiffness. Sugar palm fiber–reinforced composites are used in many naval applications (Sapuan et al., 2019).

11.5.4 APPLICATIONS IN SPORTING GOODS

Natural fiber–reinforced polymer composites are utilized to manufacture sports gear because of their light weight, high strength, and biodegradable nature. Flax fiber–reinforced polymer composites are employed in the manufacturing of tennis rackets. Some applications of natural fiber composites are shown below (Figure 11.16).

11.5.5 APPLICATION IN SOIL SAFETY AND EROSION MANAGEMENT

Soil safety utilizing plant fibers and more biodegradable materials consists of leaves, straws, and vegetable remains. Currently, woven and non-woven textiles and covers manufactured from wheat straw, rice hay, wood shavings, coir, and jute are utilized as soil protection goods. These goods are divided into two classes. The first one is manufactured utilizing coir and jute fibers called erosion control meshes (ECM). The second one is manufactured from plant fibers or synthetic fibers linked by knits known as erosion control blankets (ECB). The straw or coir packages are kept together by meshes which can be utilized as sediment retention natural fiber rolls (Singh Dhaliwal, 2020).

11.5.6 APPLICATIONS OF NATURAL FIBER POLYMER COMPOSITES IN CONSTRUCTION

Numerous natural fibers are utilized as reinforcement in construction industries. Generally, the natural fiber polymer composites that are used in construction can be categorized into four main sections. These division are as follows:

- particle/small fiber reinforced,
- lengthy natural fiber reinforced,
- nano-fiber composites, and
- combined composites.

The fibers used for manufacturing long fiber composites are flax, sisal, and jute fibers. Many fabrication techniques are used to manufacture these composites, such as compression, vacuum grabbing, injection, wire winding, and pultrusion. Long natural fiber composites (LNFCs) have great mechanical properties for example tensile properties and biodegradable.

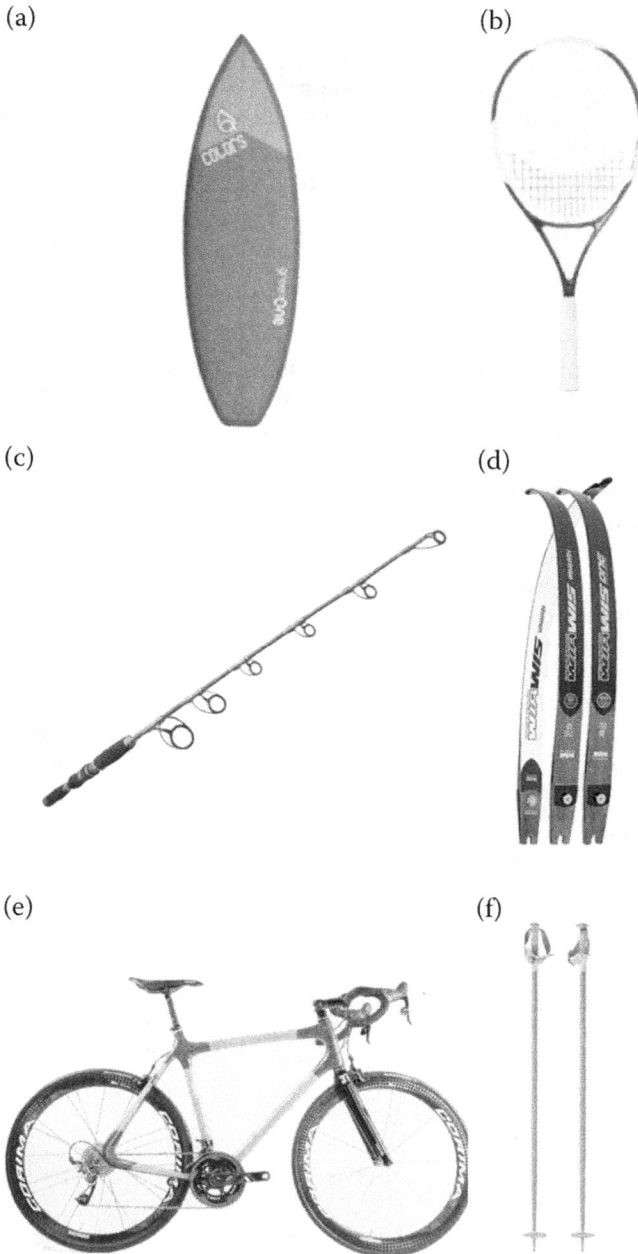

FIGURE 11.16   Different uses of flax fiber polymer composites in sporting products: (a) Notox surf board; (b) tennis racquet; (c) Caperlan fishing bar; (d) ArcWin Archery; (e) Le Ventoux bamboo-flax fiber bike; and (f) Kang flax ski rods (Pil et al., 2016) (with permission).

In the roofing structures of building construction, different kinds of natural fiber polymer composites are utilized; for example I-joists as ceiling beams, LVL (laminated veneer lumber), and wood-built composites as building blocks. The I-joists are employed for insulation and venting intention. Stressed skin sections are manufactured by wood-built polymer composites. They have been used for the off-site manufacturing of roofing parts, which improves the efficiency of building construction.

There are a variety of applications of natural fiber polymer composites in building construction.

11.5.6.1 Applications in Wall Building

Natural fiber polymer composites are used in wall construction. A combination of natural fiber–reinforced polymer composites and wood-reinforced composites is used in the development of a fence system in building construction.

11.5.6.2 Floorboards and Ceiling Structures

Fiber Composites used in flooring and ceiling structure applications are divided into two categories. The initial is for the load-carrying such as joists, and the next is for the semi-structural fiber composites for floor and ceiling coverings. The I-joists are utilized to prevent twisting and deflection as I-shape is effective in bending strength and refraction control.

11.5.6.3 Composites Beams and Pillars

Natural fiber polymer composites are commonly utilized for structure beams and pillars, continually substituting steel. Natural fibers in the polymer composite should be located equivalent to the primary stress direction to raise the strength of composites (Sanal & Verma, 2018) (Figure 11.17).

Different kinds of natural fibers are used in construction applications. Some of them are discussed below.

11.5.6.4 Flax Fibers

Flax fiber polymer composites are mainly used for structural applications, automotive applications, and customer applications. There are many advantages of using flax fiber composites, such as durability, further improvement in mechanical properties, and water resistance.

11.5.6.5 Jute Fibers

Jute fiber is the cheapest fiber among all natural fibers. The jute fiber is biodegradable and having high tensile strength. The jute fiber is of a very adaptive nature that has been utilized in natural fabrics for the building and agricultural sectors (Sanal & Verma, 2018).

11.5.6.6 Sisal Fibers

The industry mainly uses sisal fiber in three different categories. The lower-grade sisal fiber has a high cellulose and hemicellulose content, used primarily by the

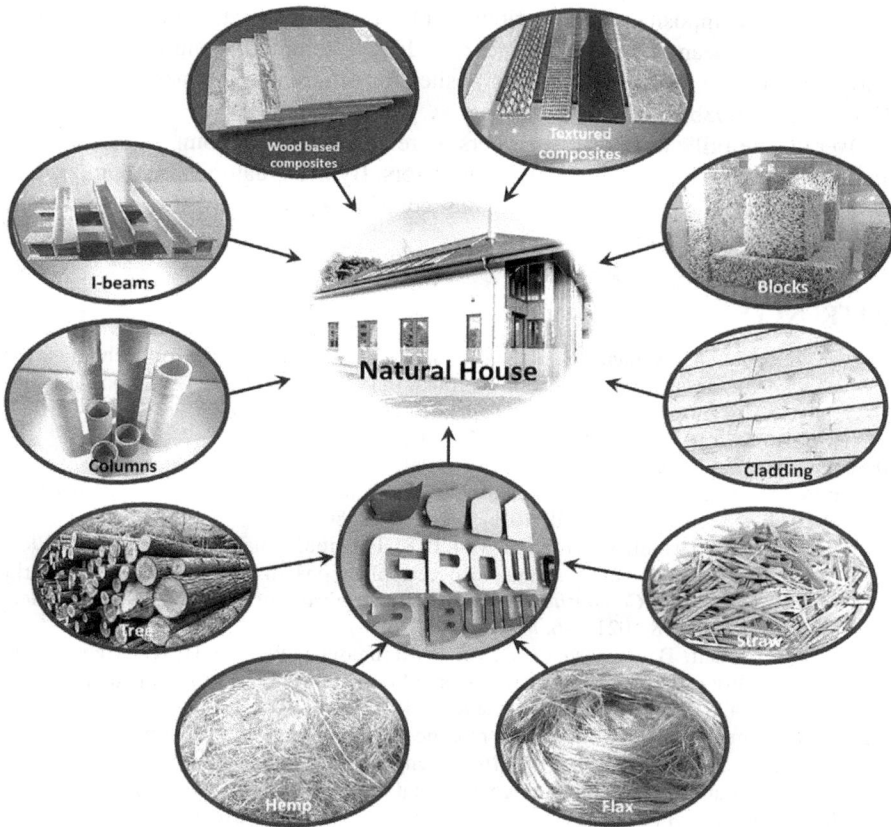

FIGURE 11.17 Different kinds of natural fiber composites used in building construction (Fan, 2017) (with permission).

paper trades. The moderate-grade sisal fibers are extensively employed by naval and farming sectors. The upper-grade sisal fiber is mainly used by the carpet industry.

11.5.6.7 Coconut Fibers

Coconut fiber is obtained from the inside husk of the coconut. These short and stiff fibers can be combined for many applications. The coconut fibers have the maximum lignin content and low cellulose content, making them stiff and strong (Sanal & Verma, 2018).

CONCLUSION

Natural fiber–reinforced composites with excellent properties have many applications in the current industrial sector. The good mechanical properties, low manufacturing cost, excellent thermal and acoustic properties, and biodegradability make them a potential alternative as reinforcement in the composites (Kumar et al., 2019).

Natural fiber composites have various applications in the automotive industry. Numerous European, German, Japanese, and U.S. automobile industries are presently making natural fiber–based composites for different components of vehicles (Huda et al., 2008b).

Overall, the utilization of plant fibers as reinforcement in composites can help create occupations in both rural and town sectors. By using natural fiber composites, waste production is also decreased, thus contributing to a healthy environment (Satyanarayana et al., 2009).

REFERENCES

Agarwal, J., Sahoo, S., Mohanty, S., & Nayak, S. K. (2020). Progress of novel techniques for lightweight automobile applications through innovative eco-friendly composite materials: A review. *Journal of Thermoplastic Composite Materials, 33*(7), 978–1013. SAGE Publications. 10.1177/0892705718815530

Ahmad, R., Hamid, R., & Osman, S. A. (2019). Physical and chemical modifications of plant fibres for reinforcement in cementitious composites. *Advances in Civil Engineering, 2019,* 1–18. Hindawi Limited. 10.1155/2019/5185806

Asim, M., Saba, N., Jawaid, M., & Nasir, M. (2018). Potential of natural fiber/biomass filler-reinforced polymer composites in aerospace applications. In M. Jawaid , & M. Thariq (Eds.), *Sustainable Composites for Aerospace Applications* (pp. 253–268). Elsevier. 10.1016/B978-0-08-102131-6.00012-8

Azwa, Z. N., Yousif, B. F., Manalo, A. C., & Karunasena, W. (2013). A review on the degradability of polymeric composites based on natural fibres. *Materials and Design, 47,* 424–442. Elsevier. 10.1016/j.matdes.2012.11.025

Cruz, J., & Fangueiro, R. (2016). Surface modification of natural fibers: A review. *Procedia Engineering, 155,* 285–288. 10.1016/j.proeng.2016.08.030

Fan, M. (2017). Future scope and intelligence of natural fibre based construction composites. In M. Fan, & F. Fu (Eds.), *Advanced High Strength Natural Fibre Composites in Construction* (pp. 545–556). Elsevier Inc. 10.1016/B978-0-08-100411-1.00022-4

Fogorasi, M. S., & Barbu, I. (2017). The potential of natural fibres for automotive sector – Review. *IOP Conference Series: Materials Science and Engineering, 252*(1). 10.1088/1757-899X/252/1/012044

Ghori, W., Saba, N., Jawaid, M., & Asim, M. (2018). A review on date palm (phoenix dactylifera) fibers and its polymer composites. *IOP Conference Series: Materials Science and Engineering, 368*(1). 10.1088/1757-899X/368/1/012009

Gupta, M. K., Srivastava, R. K., Bisaria, H., Gupta, M. K., & Srivastava, R. K. (2015). Potential of jute fibre reinforced polymer composites: A review. . *International Journal of Fiber and Textile Research, 5*(3), 30–38. http://www.urpjournals.com

Holbery, J., & Houston, D. (2006). Natural-fiber-reinforced polymer composites in automotive applications. *The Journal of The Minerals, 58,* 80–86. https://doi.org/10.1007/s11837-006-0234-2

Huda, M. S., Drzal, L. T., Ray, D., Mohanty, A. K., & Mishra, M. (2008a). Natural-fiber composites in the automotive sector. In K. L. Pickering (Ed.), *In Woodhead Publishing Series in Composites Science and Engineering, Properties and Performance of Natural-Fibre Composites* (pp. 221–268). Woodhead Publishing. 10.1533/9781845 694593.2.221

Kumar, R., Ul Haq, M. I., Raina, A., & Anand, A. (2019). Industrial applications of natural fibre-reinforced polymer composites – Challenges and opportunities. *International Journal of*

Sustainable Engineering, 12(3), 212–220. Taylor and Francis. 10.1080/19397038.201 8.1538267

Lotfi, A., Li, H., Dao, D. V., & Prusty, G. (2021). Natural fiber–reinforced composites: A review on material, manufacturing, and machinability. *Journal of Thermoplastic Composite Materials, 34*(2), 238–284. SAGE Publications. 10.1177/089270571 9844546

Mansor, M. R., Nurfaizey, A. H., Tamaldin, N., & Nordin, M. N. A. (2019). Natural fiber polymer composites: Utilization in aerospace engineering. In D. Verma, E. Fortunati, S. Jain, & X. Zhang (Eds.), *Biomass, Biopolymer-Based Materials, and Bioenergy: Construction, Biomedical, and other Industrial Applications* (pp. 203–224). Elsevier. 10.1016/B978-0-08-102426-3.00011-4

Mohammed, L., Ansari, M. N. M., Pua, G., Jawaid, M., & Islam, M. S. (2015). A review on natural fiber reinforced polymer composite and its applications. *International Journal of Polymer Science, 2015*. Hindawi Publishing. 10.1155/2015/243947

Mukhopadhyay, S., & Fangueiro, R. (2009). Physical modification of natural fibers and thermoplastic films for composites – A review. *Journal of Thermoplastic Composite Materials, 22*(2), 135–162. 10.1177/0892705708091860

Omrani, E., Menezes, P. L., & Rohatgi, P. K. (2016). State of the art on tribological behavior of polymer matrix composites reinforced with natural fibers in the green materials world. *International Journal of Engineering Science and Technology, 19*(2), 717–736. Elsevier B.V. 10.1016/j.jestch.2015.10.007

Peças, P., Carvalho, H., Salman, H., & Leite, M. (2018). Natural fibre composites and their applications: A review. *Journal of Composites Science, 2*(4), 66. 10.3390/jcs2040066

Pickering, K. L., Efendy, M. G. A., & Le, T. M. (2016). A review of recent developments in natural fibre composites and their mechanical performance. *Composites Part A: Applied Science and Manufacturing, 83*, 98–112. Elsevier. 10.1016/j.compositesa.2 015.08.038

Pil, L., Bensadoun, F., Pariset, J., & Verpoest, I. (2016). Why are designers fascinated by flax and hemp fibre composites? *Composites Part A: Applied Science and Manufacturing, 83*, 193–205. 10.1016/j.compositesa.2015.11.004

Sanal, I., & Verma, D. (2018). Construction materials reinforced with natural products. In L. M. Torres Martínez , O. Vasilievna Kharissova, & B. Ildusovich Kharisov (Eds.), *Handbook of Ecomaterials* (pp. 1–24). Springer International Publishing. 10.1007/ 978-3-319-48281-1_75-1

Sanjay, M. R., Arpitha, G. R., Naik, L. L., Gopalakrishna, K., & Yogesha, B. (2016). Applications of natural fibers and its composites: An overview. *Natural Resources, 7*(03), 108–114. 10.4236/nr.2016.73011

Sapuan, S. M., Nurazzi Norizan, M., & Mahamud, A. (2019). *Potential of natural fibre composites for transport industry: A review.* In Proceedings of the Prosiding Seminar Enau Kebangsaan 2019, Bahau, Malaysia, 2–11. https://www.researchgate. net/publication/332168594

Satyanarayana, K. G., Arizaga, G. G. C., & Wypych, F. (2009). Biodegradable composites based on lignocellulosic fibers – An overview. *Progress in Polymer Science (Oxford), 34*(9), 982–1021. 10.1016/j.progpolymsci.2008.12.002

Saw, S. K., Purwar, R., Nandy, S., Ghose, J., & Sarkhel, G. (2013). Fabrication, characterization, and evaluation of Luffa cylindrica fiber reinforced epoxy composites. *BioResources, 8*(4), 4805–4826.

Saxena, M., Pappu, A., Sharma, A., Haque, R., & Wankhede, S. (2011). Composite materials from natural resources: Recent trends and future potentials. In *Advances in Composite Materials –Analysis of Natural and Man-Made Materials*. InTech. 10.5772/18264

Singh Dhaliwal, J. (2020). Natural fibers: Applications. In *Generation, Development and Modifications of Natural Fibers* (pp. 1–23). IntechOpen. 10.5772/intechopen.86884

Sonar, T., Patil, S., Deshmukh, V., & Acharya, R. (2015). Natural fiber reinforced polymer composite material – A review. *IOSR Journal of Mechanical and Civil Engineering (IOSR-JMCE)*, *3*, 142–147. www.iosrjournals.org

Syath Abuthakeer, S., Vasudaa, R., & Nizamudeen, A. (2016). Application of natural fiber composites in engineering industries: A comparative study. *Applied Mechanics and Materials*, *854*, 59–64. 10.4028/www.scientific.net/amm.854.59

Syduzzaman, M., Faruque, M. A. Al, Bilisik, K., & Naebe, M. (2020). Plant-based natural fibre reinforced composites: A review on fabrication, properties and applications. *Coatings*, *10*(10), 1–34. MDPI AG. 10.3390/coatings10100973

Tripathi, P., & Yadav, K. (2017). Biodegradation of natural fiber & glass fiber polymer composite – A review. *International Research Journal of Engineering and Technology*, *4*(4), 1224–1228. www.irjet.net

12 Applications of Natural Fibers–Reinforced Composites (II)

Dharam Pal
Department of Chemical Engineering, National Institute of Technology Raipur, Raipur, Chhattisgarh, India

Manash Protim Mudoi
Department of Chemical Engineering, Indian Institute of Technology, Roorkee, Uttrakhand, India

Department of Chemical Engineering, University of Petroleum and Energy Studies, Dehradun, India

Santosh bahadur Singh
Department of Chemical Engineering, National Institute of Technology Raipur, Raipur, Chhattisgarh, India

Shishir Sinha
Department of Chemical Engineering, Indian Institute of Technology, Roorkee, Uttrakhand, India

INTRODUCTION

Packaging is an integral part of several industries, and they are used to perform diversified functions. The success of food security is related to efficient food packaging, where the quality and hygiene of the stored products are maintained at the desired level. Food can be degraded due to various factors like contaminates, moisture, microbes, light, temperature, air, and fragrances. The packaging should provide the resistance barrier to the food degradation factors (Sydow & Bieńczak, 2019). Proper and appropriate packaging should effectively separate the food products from the surrounding atmosphere to ensure food safety and quality. Temperature stabilization is another transport condition required to be maintained by the packaging material (Zwierzycki et al., 2011). Therefore, the suitable thermal, optical, and mechanical characterization of the packaging material is very much necessary (Rhim et al., 2013).

There are generally six categories of packaging: rigid plastic, glass, flexible packaging, paperboard, corrugated board, and others. The heavier glass bottles and metal cans are increasingly replaced with a cheaper and lighter polymer like

DOI: 10.1201/9781003201724-12

polyethylene terephthalate (PET). The utilization of flexible plastics is growing fast due to the increased demand in the food segment. Analysis of MarketsandMarkets is projecting the growth size of the market for global flexible plastic packaging from $160.8 billion (in 2020) to $200.5 billion (by 2025) (Teal, 2021). This represents a CAGR of 4.5% for the duration of the time period. The fast-growing segment under the flexible plastic category is the pouches (annual 6% increase through 2024), surpassing the other flexible and rigid packaging types. The market size for the barrier film is expected to reach $4.1 billion by 2025 (CAGR of 5.3% in 2020–2025 time duration). The market size growth for the packaging materials is due to the higher demand for a longer shelf-life requirement of food products, customer-friendly packaging, and the development of retail chains in developing countries.

Synthetic polymers are an attractive material for packaging due to their light weight, low cost of production, convenient use, suitable chemical and physical properties, and aesthetic appearance. The use of plastic in various sectors in 2015 is presented in Figure 12.1. The packaging sector had the maximum utilization of plastic (more than 35%) among all the competitive industries. However, the biggest drawback is the generation of waste (single-use plastic) is highest for the packaging industry.

Environment-friendly disposal is the biggest problem of synthetic polymer utilization. Food packaging polymers do not degrade easily, and therefore after disposal, they create an environmental hazard (Majeed et al., 2013). The stringent regulations and rising environmental concerns have motivated the scientific community to look for novel biodegradable polymers as an alternative to synthetic polymers. Polycaprolactone (PCL), poly (butylene succinate) (PBS), poly

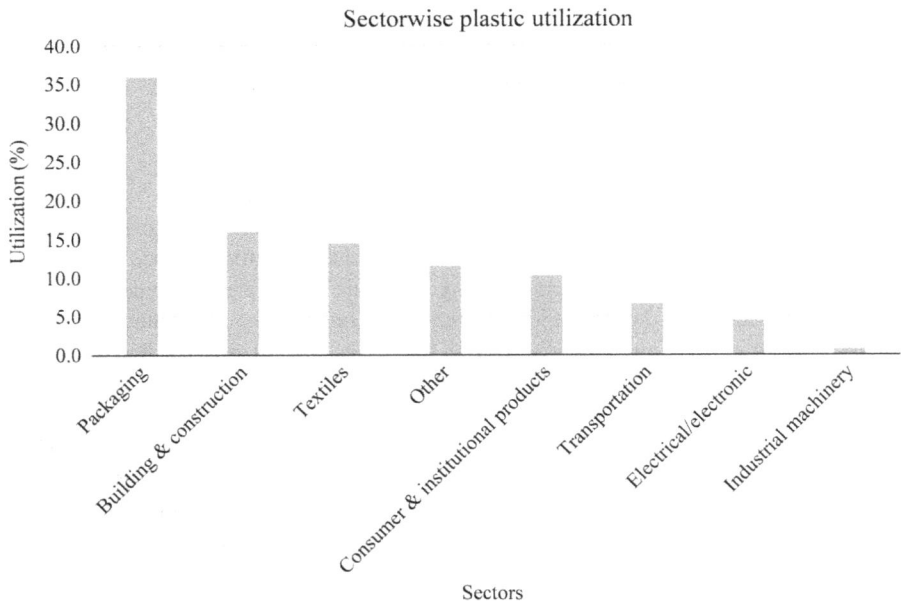

FIGURE 12.1 Plastic utilization in 2015 (Beckman, 2018).

(lactic acid) (PLA), poly (lactide-co-glycolide) (PLGA), etc., are some of the biodegradable polymers. These polymers can be decomposed by microbial actions. The higher cost of biodegradable polymers is an obvious obstacle in their wide-scale application in the packaging industry (Vroman & Tighzert, 2009). The issues related to the material cost, biodegradability, and required strength could be compensated by using appropriate fillers such as natural fiber along with the polymer matrix material. Lignocellulosic natural fibers like ramie, sisal, kenaf, hemp, nettle, etc., can be used as suitable fillers to synthesize polymeric composite for food packaging purposes. The literature for composite properties in this category is available in limited number, which are critical to maintain the stored food quality. This chapter discusses the present utilization of natural fiber composites in food packaging application and the state of the technical know-how of the application behavior.

12.1 APPLICATIONS IN FOOD PACKAGING

The different polymer materials can be used as matrix material to prepare the natural fiber composite in food packaging applications. The selection of fiber also depends on the desired properties of the end product. The polymer material can be of two types: thermosetting and thermoplastic. Air (oxygen) transport properties of the barrier material are essential for maintaining the quality of the stored food. Water sorption characteristics also affect the suitability of a polymeric material in food packaging applications.

Air or oxygen penetration is detrimental because it causes oxidation and spoiling of the food items stored inside the package. The barrier potential of the packaging material for gas transport is critical for safe food storage. Shields et al. (2008) used a design procedure to evaluate the permeability of oxygen in a new type of microfibril-reinforced poly(ethylene terephthalate) (PET) (30 wt%)/polyethylene (PE) (70 wt%) composite material (Shields et al., 2008). Those microfibril-reinforced polymer composites (MFC) were created with simple techniques and without using expensive coatings. The composites exhibited better oxygen barrier properties, tensile strength, and modulus compared to the neat PE. Thinner microfibril films of thickness 150–200 µm were produced through various fiber orientation, cooling, and manufacturing condition, which showed improved properties for oxygen barrier than PE films. The barrier potential was increased with the increase in the crystallinity of the film. The parameters producing a more significant influence on the oxygen permeability were cooling and manufacturing conditions than the orientation of fibrils. Composites of maleic anhydride grafted polyethylene (PE-MA)/HDPE (high-density polyethylene)/cellulose fiber were prepared by Fendler et al. (2007) to study the barrier properties for limonene and oxygen (Fendler et al., 2007). Between two sample preparation methods (slow cooled and quenched), higher resistance for oxygen permeability was observed for slowly cooled samples due to the more excellent crystallinity. The oxygen permeability was reduced significantly with a fiber content of more than 2 wt% in the composite. A highest reduction of 65% in oxygen permeability was observed for quenched samples at 50 wt% fiber content than the pure HDPE. It was due to the existence of cellulose crystals that are impermeable to gas transport (Sanchez-Garcia et al., 2008). A steep decrease in permeability was observed in 20–50 wt% since the crystallinity was not reduced dramatically (Fendler et al., 2007).

Permeability of water vapor is considered an important factor in designing the packaging material for respiring fresh foodstuff like strawberries. The durability of storing such food items increases considerably if the water vapor permeability increases. Berthet et al. (2015) investigated the water vapor permeability behavior of biodegradable PHBV (polyhydroxy-co-3-butyrate-co-3-valerate)/wheat straw fiber composite film as a replacement of polyolefins packets used for storing respiring food items (Berthet et al., 2015). With the incorporation of wheat straw fibers (max-content 30 wt% and max size 1.8 mm) to the polymer matrix (PHBV), the water vapor transmission rate (WVTR) was increased maximum up to 110.2 $g/m^2/day$, which is about a 1,010% increase than the pure PHBV (Berthet et al., 2015). The cost of the final packaging material could be reduced by 30% with the application of wheat straw fiber in PHBV as the cost for PHBV is very high (Berthet et al., 2015). The WVTR increases with the increase in fiber content, which can be explained by the enhanced hydrophilic behavior of the composite with the increase in fiber content. Water vapor permeability (WVP) was increased in cotton/PCL and hydrolyzed cellulose/PCL composite than in the neat PCL (Ludueña et al., 2012). The three primary mechanisms affecting the WVP in natural fiber composite are mentioned below (Sanchez-Garcia et al., 2008).

1. The matrix becomes more permeable to H_2O molecules as the degree of crystallinity decreases with the filler addition.
2. Tortuosity of the transport path increases with filler addition.
3. Voids are created in the interface of filler and polymer due to the fiber's agglomeration and weaker interfacial strength and allowing the easy transport of H_2O molecules.

Yang et al. (2006) observed a slight increase in water absorption and swelling in thickness for rice-husk flour/polyolefin bio-composite. Still, those values were very small compared to those of solid wood and other wood composites (Yang et al., 2006). The authors reported the suitability of the prepared biocomposite for food packaging applications. Both swelling of thickness and water absorption quantity varied with filler weight content and nature of matrix polymer used. LDPE (low-density polyethylene) showed higher water absorption and swelling in thickness than the PP (polypropylene) matrix due to poor PE chain/filler interfacial adhesion. In contrast, the increase in filler content increased the value for both the parameters due to the higher hydrophilicity of natural fillers. Higher hydrophilicity of fillers is undesirable as it reduces mechanical strength (Essabir et al., 2016).

Thermal stability is another critical parameter requiring due consideration in packaging applications. Higher thermal stability of the material is desirable for better performance. Sezgin et al. (2021) investigated the waste cotton fiber/waste PP and waste cotton fiber/waste PE (polyethylene) composites for thermal behavior and reported the improvement in the thermal resistance up to 0.11 m2oC/W (Sezgin 2021). Cotton, for reinforcement, was derived from the waste denim fabrics, while polyethylene and polypropylene granules for the matrix were prepared from the waste food, cleaning containers, and bottle caps. Thermal stability was analyzed for biocomposite of moringa oleifera fruit fiber (MOF)/polyethylene terephthalate (PET) thermoplastic polymer, and enhanced thermal stability was observed for composite with 20 wt% fiber loading (Nayak & Khuntia, 2019). The mechanical properties of the composite were also improved than the neat PET. Higher thermal stability was observed for PLA, and cellulose nanofiber reinforced (derived from areca catechu husk) composite than the neat PLA (Soman et al., 2017). In another study, a composite made from poly (3-hydroxybutyrate-co-3-hydroxyvalerate) (PHBV) and ceiba pentandra bark fibers was investigated the thermal stability for the packaging of fresh strawberries (Varghese et al., 2020). The thermal stability for the composite was found to be greater than 250°C, which was more than the neat PHBV and the composite was suitable for withstanding applications involving high temperature. Fiber loading of 10 wt% showed maintenance of fruit freshness than unpacked fruit.

UV resistance property of packaging material is important for the protection of the stored material. Safe packaging of colored items is dependent on the UV protection character. Natural fiber reinforced packaging material may provide the UV blocking behavior since natural fibers like hemp and flax possess excellent UV radiation absorption capacity due to lignins, pectins, waxes, and pigments. The porosity, structure, and density parameters of natural fibers influence the UV resistance properties. However, excessive chemical treatment during the processing of fibers to enhance compatibility with polymer matrix may degrade the UV resistance properties. Processing with nanolignin solution is an option for improving the UV blocking properties (Zimniewska & Batog, 2012).

The mechanical strength of the composite material is another property requiring thorough investigation. Some researchers found improved mechanical strength (Lopez-Gil et al., 2014; Nayak et al., 2021; Sahoo et al., 2021), while few researchers observed reducing mechanical properties of natural fiber–reinforced composite material (Berthet et al., 2015, 2017). Sahoo et al. (2021) investigated the rattan fiber–reinforced PET (polyethylene terephthalate) thermoplastic composite and found enhanced mechanical properties with fiber loading. Chemical-treated fiber composite had superior properties to the untreated fiber composite. The authors observed that at optimum fiber loading of 20 wt%, the composite exhibited excellent thermal and mechanical behavior, which were suitable for packaging applications (Sahoo et al., 2021). Biocomposites of LF (ladies finger) fiber with PET polymer matrix was studied by Nayak et al. (2021) at different fiber loading of 0, 10, 20, 30, and 40 wt% (Nayak et al., 2021). The treated fiber at 20 wt% loading had the highest mechanical properties than the neat PET and untreated fiber composites.

Another trend in packaging material is the development of active packages, which can actively alter the inside conditions of the package to keep the stored material's life and quality preserved. Food products are protected from bioterrorism and pathogens by using active packaging systems extending the shelf life (Han, 2014). This helps in the convenient distribution, consumption, retailing, and processing of food items. Many technologies related to active packaging are commercialized, and the food industry uses them for various applications. Antimicrobial, CO_2 absorbing, O_2 scavenging, and moisture scavenging systems are also incorporated in food packaging applications. Another class of active packaging has the ability to release substances, which can protect the food from microorganisms. Using natural fibers, antimicrobial agents (essential oils, lysozyme, and herbs) and antioxidant agents (thymol (2-isopropyl-5-methylphenol)) as fillers with polymeric matrix extend the shelf life and quality of the stored material (Tawakkal et al., 2017). Apart from antioxidant activity, thymol shows antimicrobial activity also against yeast, fungi, mold, and bacteria. Tawakkal et al. (2017) studied the antibacterial activity of thymol against E.coli in PLA/kenaf fiber composite films. The authors observed a reduced *E. coli* population on chicken slices with the addition of thymol in composite films. Some natural fibers (flax, hemp, bamboo, kapok) are inherently antimicrobial (Borsa, 2012).

12.1.1 FACTORS INFLUENCING BIOCOMPOSITES' BEHAVIOR

Deterioration of mechanical properties of natural fiber–reinforced composite is a major drawback for food packing applications. Some investigations have reported slight deterioration of parameters like tensile strength, impact resistance, and break elongation for biocomposite with natural filler addition (Yang et al., 2006). Low adhesion and wetting of natural fibers and polymer matrix is the primary reason for observing reduced mechanical properties. Opposing affinity for moisture, polymer matrices are hydrophobic, and natural fibers are hydrophilic, which leads to poor compatibility and interfacial adhesion between the fibers and matrices. Fiber surface modification with physical and chemical treatment enhances interfacial behavior and significantly affects mechanical properties (Devnani & Sinha, 2019). Treatment with plasma or corona (acts as a source for ionized gas), steam explosion, and laser treatment are the primary physical methods of fiber treatment. Natural fibers have hydrophilic properties that cause moisture absorption and consequently swelling of the fibers. However, in the context of the materials for respiring food packaging, this property is considered an advantage (Venkatachalam et al., 2016). Low-pressure plasma treatment did not alter the chemical composition of fibers vastly and enhanced the surface properties by dehydration and cleaning (Enciso et al., 2017). In the chemical treatment methods, the hydroxyl groups present in the natural fibers react with the chemicals. Acetylation, silane treatment, NaOH treatment, and benzoylation are the common methods of chemical treatment. In most of the methods, reactions sites are created due to the breaking of hydrogen bonds. In addition, waxes, lignin, and fatty acids are removed from the fiber surface, improving the fibers' compatibility with the polymer matrix. The composites' mechanical and physical properties with treated natural fibers are improved considerably (Devnani & Sinha, 2020; Jain et al., 2019). However, the chemical concentration and soaking time is very important parameters to optimize the achievable mechanical properties (Milan et al., 2018; Sepe et al., 2018). However, the chemical treatment is critical for food packaging applications as the food items will come into the direct contact of the chemicals; hence, the appropriate chemical is to be chosen for the fiber surface modification. Due to this limitation, the suggested and preferred treatment method is the physical treatment.

12.2 VARIOUS DIVERSE APPLICATIONS

12.2.1 NFRCs in Bioengineering and Environmental Engineering Applications

Bioengineering is the application of concepts and methods of the chemical science, physical science, and mathematics in biological system to solve problems like repair, reconstruction of lost and damaged or deceased tissues/organs/bones (heart valve, artificial heart, internal and external fracture fixators, artificial hip and knee joints, skin repair templates, dental implants, etc.) using engineering approaches. Composite materials are used for this purpose and are refers as biomaterials. Conventional composites (contain glass, carbon, or aramid fiber–reinforced polyurethanes, epoxy, unsaturated polyester, etc.) as implant/biomaterials have various disadvantages i.e., difficulties in removal/separation after end-of-life time, porosis (uneven growth of cells/ tissues underneath and surrounding the implant material/plates), and other surgical issues arises during removal of implant materials. Therefore, composite materials having less stiffness, more biodegradability and more bioresorbablity can solve the problem associated with conventional composite implant materials. In this regard, natural fiber–reinforced composites (Purnomo et al., 2019; Shesan et al., 2019; Tavares et al., 2020) offer significant opportunities for improved implant/biomaterials from renewable resources (natural fibers) with enhanced support for natural sustainability. Cheung et al. (2009) very nicely describe the application and future perspectives of natural fiber (animal- and plant-based fibers)–reinforced composites for various bioengineering as well as environmental engineering uses (Cheung et al., 2009).

12.2.2 NFRCs in Military Applications

NFRCs are the composite materials that have at least one fiber derived from natural renewable resources (plants and animals both). There is a lot of possibility to promote its high-level application in military applications such as bulletproof helmets, vests, and other various armor accessories. Nurazzi et al. (2021) very recently reviewed the various applications of natural fiber–reinforced composites (NFRCs) in bulletproof and ballistic applications Kevlar fabrics and aramid-reinforced composite show good performance in bulletproof and ballistic application but their disposal have great environmental concerns. They disrupt the ecosystem and pollute the environment. Use of natural fiber–reinforced composites in place of Kevlar fabric and aramid can reduce the environmental concern associated with Kevlar and aramid-reinforced composites. Easy availability, low cost, and easy of manufacturing the natural fibers have grasped the attention of scientific communities throughout world in order to explore the applications of NFRCs in heavy and durable armory equipment/products. Recently, many researchers worldwide explored the application of various eco-friendly natural fibers i.e., bamboo, bagasse, jute, rami, malva, kenaf, cocosnucifera sheath, etc.

as a green alternative of kevlar fabric in military equipment/products and reported that NFRCs have great reinforcement properties in the ballistic structures (Costa et al., 2019; Naveen et al., 2020; Pereira et al., 2020). Ballistic-resistant materials also developed by (Monteiro et al., 2016) using natural fiber–reinforced composites. Bagasse fiber–reinforced epoxy composites were fabricated in the form of plates with 30% loading of the fiber. The comparison of performance for ballistic applications was made with Kevlar TM plates, which can be utilized in a commercial multilayer armored system. Outcomes satisfied the standards and these materials were suggested for the application.

12.2.3 ELECTRICAL AND SENSOR DEVICES

The high thermal decomposition property can make natural fiber polymer composites for use in electrical and sensor devices. Coir dust (source of carbon) and sepiolite (binder) were mixed and composites were prepared (Bispo et al., 2011) and experimental findings recommend to use these material in electrodes to be used in sensors and energy storage devices.

12.2.4 PRINTED CIRCUIT BOARDS

These greener materials can be applied for the development of printed circuit boards. The green composites were prepared from soybean oil resins, feathers of chickens, and E glass and having a potential that can replace the costly and environmentally hazardous E glass fiber–based epoxy composite, which is commonly used in PCB. The flame retardant property of the printed circuit board was obtained by halogen-free melamine polyphosphate and diethylphosphinic salt. Experimental findings like mechanical characteristics, electrical properties, flame retardancy, and peel strength were found suitable for the application in PCB (Zhan & Wool, 2013).

12.2.5 INSULATION BOARDS

Waste newspaper were mixed with urea formaldehyde resins and composites were fabricated. Fire retardant–treated waste paper board exhibits have good incombustible properties that are appropriate for application as insulation boards or interior finishing material (Yang et al., 2002).

12.2.6 HOUSEHOLD FURNITURE APPLICATIONS

Sapuan & Maleque (2005) discussed the design concept, detail design, and development of woven banana fabric composite telephone. The developed stand was unique in presentation and was very aesthetically pleasing, golden brown in color, and can easily be matched with typical colors of furniture. The telephone stand was found suitable for both domestic purposes and for official use. Banana fiber was recommended reinforcement to fabricated reinforced composites as a novel and greener material for household furniture applications at low cost. It is always possible to replace conventional non-biodegradable metallic and nonmetallic materials and plastic by natural fiber–reinforced composite for a telephone stand.

12.2.7 CONSUMER PRODUCTS

With other well-known applications, NFRCs have growing applications in the field of consumer products. Wood and natural fiber–reinforced composites are frequently used by various companies in the manufacturing of consumer goods like toys, combs, trays, watchcases, etc. with the unique look and feel to attract customers. In comparison to pure plastic consumer products, NFRC-based consumer products have a more refined look, they have more natural look, have a nice feel, and need less plastic, which benefits the environmental footprint. Bamboo fiber–reinforced composites have frequent applications in various consumer products (Partanen & Carus, 2016).

12.2.8 SPORTS INDUSTRY

The sports industry is a fast-growing industry worldwide. Technological innovation is always needed in this industry for manufacturing and processing of materials to achieve durability, recyclability, reusability, efficiency, and cost effectiveness. Generally, glass/carbon fiber–reinforced composites are widely used in the sport industry but NFRCs have very good potential to replace glass/carbon fiber–reinforced composite materials from sport industry. Rashid et al. (2020) reported the application of banana fiber–reinforced composites in the sport industry. Based on various experimental findings and observations, they concluded that banana fiber–reinforced composites with comparable and even greater load-withstanding capabilities, may be a good alternative of pure glass/carbon fiber–reinforced composite (Al Rashid et al., 2020). Application of NFRCs for sports products would cut costs and lower the environmental impact. It will also help a domestic economy by using domestic/local resources.

12.2.9 AEROSPACE SECTOR

NFRCs due to their biodegradability, light weight, and lower cost open new prospects in the aerospace industry. Natural fiber–reinforced composites possess better properties i.e., heat and flame resistance, easy recyclability, easy and cheaper disposal, required for aircraft interior panels than conventional sandwich panels (Rajak et al., 2019). Balakrishnan et al. (2016) explored the application of NFRCs in construction of aircraft and spacecraft materials. Conventional aluminium alloy–based composites are most frequently used in the aerospace industry. The main reasons behind the use of composite materials in the aerospace industry are to increase specific stiffness and strength, to reduce the total weight, extent of fatigue life, and minimize the corrosion problem. NFRCs have numerous advantages over synthetic fiber–reinforced composites i.e., light weight, low cost, environmental benign, non-toxic/hazardous, renewability, and so on, which make NFRCs more promising materials in the sector of the aerospace industry (Balakrishnan et al., 2016).

12.3 FUTURE PERSPECTIVES

In the last decade, a tremendous amount of work has been reported by the scientific community on natural fiber extraction and their reinforced composites. Various extraction methods have been explored. Different surface treatment methods like physical, chemical, and biological have been attempted to improve the compatibility between fibers and polymer matrix. Innovations in fabrication and characterization methods were also exercised to get the superior quality products. Now the researchers are looking at multidimensional applications other that structured and semi-structured.

CONCLUSIONS

Natural fiber/polymer matrix composites are emerging as a good candidate for food packaging applications because of specific characters assisting in the longer life of food items. Additionally, biocomposites are environmentally friendly as the degradability character is introduced with the natural fibers. The tailored biocomposites with improved thermal stability, mechanical properties, and antibacterial/antimicrobial characters can replace the petroleum-based synthetic polymer packaging materials. For storing high respiring food materials, natural fibers' hydrophilic property (high water absorption) is desirable. The drawbacks of incompatibility between the fibers and matrix can be improved significantly with chemical and physical treatments. Hybrid composites of natural fibers/nano-clay fillers are advantageous to impart higher mechanical and physical strength. Cost reduction is another added advantage of using natural fibers in composite preparation for food packaging applications. Apart from food packaging, various industrial sectors are using these composites for diverse applications.

REFERENCES

Al Rashid, A., Khalid, M. Y., Imran, R., Ali, U., & Koc, M. (2020). Utilization of banana fiber-reinforced hybrid composites in the sports industry. *Materials*, *13*(14), 3167. 10.3390/ma13143167

Beckman, E. (2018). The world of plastics, in numbers. *The Conversation*. http://theconversation.com/the-world-of-plastics-in-numbers-100291

Balakrishnan, P., John, M. J., Pothen, L., Sreekala, M. S., & Thomas, S. (2016). Natural fibre and polymer matrix composites and their applications in aerospace engineering. In S. Rana, & R. Fangueiro (Eds.), *Woodhead Publishing Series in Composites Science and Engineering* (pp.365–383). Woodhead Publishing. 10.1016/B978-0-08-100037-3.00012-2

Berthet, M. A., Angellier-Coussy, H., Chea, V., Guillard, V., Gastaldi, E., & Gontard, N. (2015). Sustainable food packaging: Valorising wheat straw fibres for tuning PHBV-based composites properties. *Composites Part A: Applied Science and Manufacturing*, *72*, 139–147. 10.1016/j.compositesa.2015.02.006

Berthet, M. A., Mayer-Laigle, C., Rouau, X., Gontard, N., & Angellier-Coussy, H. (2017). Sorting natural fibres: A way to better understand the role of fibre size polydispersity on the mechanical properties of biocomposites. *Composites Part A: Applied Science and Manufacturing*, *95*, 12–21. 10.1016/j.compositesa.2017.01.011

Bispo, T. S., Barin, G. B., Gimenez, I. F., & Barreto, L. S. (2011). Semiconductor carbon composite from coir dust and sepiolite. *Materials Characterization, 62*(1), 143–147. 10.1016/j.matchar.2010.10.010

Borsa, J. (2012). Antimicrobial natural fibres. In R. M. Kozłowski (Ed.), *Handbook of Natural Fibres* (Vol. 2, pp. 428–466). Woodhead Publishing. 10.1533/978085709551 0.2.428

Cheung, H. yan, Ho, M. po, Lau, K. tak, Cardona, F., & Hui, D. (2009). Natural fibre-reinforced composites for bioengineering and environmental engineering applications. *Composites Part B: Engineering, 40*(7), 655–663. 10.1016/j.compositesb.2009.04.014

Costa, U. O., Nascimento, L. F. C., Garcia, J. M., Monteiro, S. N., Luz, F. S. da, Pinheiro, W. A., & Garcia Filho, F. da C. (2019). Effect of graphene oxide coating on natural fiber composite for multilayered ballistic armor. *Polymers, 11*(8), 1–18. 10.3390/polym11081356

Devnani, G. L., & Sinha, S. (2019). Epoxy-based composites reinforced with African teff straw (Eragrostis tef) for lightweight applications. *Polymers and Polymer Composites, 27*(4), 189–200. 10.1177/0967391118822269

Devnani, G. L., & Sinha, S. (2020). African teff straw as a potential reinforcement in polymer composites for light-weight applications: Mechanical, thermal, physical, and chemical characterization before and after alkali treatment. *Journal of Natural Fibers, 17*(7), 1011–1025. 10.1080/15440478.2018.1546640

Enciso, B., Abenojar, J., & Martínez, M. A. (2017). Influence of plasma treatment on the adhesion between a polymeric matrix and natural fibres. *Cellulose, 24*(4), 1791–1801. 10.1007/s10570-017-1209-x

Essabir, H., Raji, M., Bouhfid, R., & Qaiss, A. E. K. (2016). Nanoclay and natural fibers based hybrid composites: Mechanical, morphological, thermal and rheological properties. In M. Jawaid, A. E. K. Qaiss, & R. Bouhfid (Eds.), *Nanoclay Reinforced Polymer Composites: Natural Fibre/Nanoclay Hybrid Composites* (pp. 29–49). Springer, Singapore. 10.1007/978-981-10-0950-1_2

Fendler, A., Villanueva, M. P., Gimenez, E., & Lagarón, J. M. (2007). Characterization of the barrier properties of composites of HDPE and purified cellulose fibers. *Cellulose, 14*(5), 427–438. 10.1007/s10570-007-9136-x

Han, J. H. (2014). *Innovations in Food Packaging*. Elsevier Academic Press. https://books.google.co.in/books?id=wfdJAQAACAAJ

Jain, J., Sinha, S., & Jain, S. (2019). Compendious characterization of chemically treated natural fiber from pineapple leaves for reinforcement in polymer composites. *Journal of Natural Fibers, 18*(6), 845–856. 10.1080/15440478.2019.1658256

Lopez-Gil, A., Rodriguez-Perez, M. A., & De Saja, J. A. (2014). Strategies to improve the mechanical properties of starch-based materials: Plasticization and natural fibers reinforcement. *Polímeros, 24*(SPE), 36–42. 10.4322/polimeros.2014.054

Ludueña, L., Vázquez, A., & Alvarez, V. (2012). Effect of lignocellulosic filler type and content on the behavior of polycaprolactone based eco-composites for packaging applications. *Carbohydrate Polymers, 87*(1), 411–421. 10.1016/j.carbpol.2011.07.064

Majeed, K., Jawaid, M., Hassan, A., Bakar, A. A., Khalil, H. P. S. A., Salema, A. A., & Inuwa, I. (2013). Potential materials for food packaging from nanoclay/natural fibres filled hybrid composites. *Materials & Design, 46*, 391–410. 10.1016/j.matdes.2012.10.044

Milan, S., Christopher, T., & Jappes, J. T. W. (2018). Investigation on mechanical properties and chemical treatment of litterous fibre reinforced polymer composites. *International Journal of Computer Aided Engineering and Technology, 10*(1–2), 102–110.

Monteiro, S. N., Candido, V. S., Braga, F. O., Bolzan, L. T., Weber, R. P., & Drelich, J. W. (2016). Sugarcane bagasse waste in composites for multilayered armor. *European Polymer Journal, 78*, 173–185. 10.1016/j.eurpolymj.2016.03.031

Naveen, J., Jayakrishna, K., Hameed Sultan, M. T. Bin, & Amir, S. M. M. (2020). Ballistic performance of natural fiber based soft and hard body armour – A mini review. *Frontiers in Materials*, *7*(5), 440. Frontiers Media S.A. 10.3389/fmats.2020.608139

Nayak, S., Jena, P. K., Samal, P., Sahoo, S., Khuntia, S. K., & Behera, J. R. (2021). Improvement of mechanical and thermal properties of polyethylene terephthalate (PET) composite reinforced with chemically treated ladies finger natural fiber. *Journal of Natural Fibers*, 1–12. 10.1080/15440478.2021.1932680

Nayak, S., & Khuntia, S. K. (2019). Development and study of properties of Moringa oleifera fruit fibers/polyethylene terephthalate composites for packaging applications. *Composites Communications*, *15*, 113–119. 10.1016/j.coco.2019.07.008

Nurazzi, N. M., Asyraf, M. R. M., Khalina, A., Abdullah, N., Aisyah, H. A., Rafiqah, S. A., Sabaruddin, F. A., Kamarudin, S. H., Norrrahim, M. N. F., Ilyas, R. A., & Sapuan, S. M. (2021). A review on natural fiber reinforced polymer composite for bullet proof and ballistic applications. *Polymers*, *13*(4), 1–42. 10.3390/polym13040646

Partanen, A., & Carus, M. (2016). Wood and natural fiber composites current trend in consumer goods and automotive parts. *Reinforced Plastics*, *60*(3), 170–173. 10.1016/j.repl.2016.01.004

Pereira, A. C., Lima, A. M., Demosthenes, L. C. da C., Oliveira, M. S., Costa, U. O., Bezerra, W. B. A., Monteiro, S. N., Rodriguez, R. J. S., Deus, J. F. de, & Anacleto Pinheiro, W. (2020). Ballistic performance of ramie fabric reinforcing graphene oxide-incorporated epoxy matrix composite. *Polymers*, *12*(11), 1–17. 10.3390/polym12112711

Purnomo, Setyarini, P. H., & Cahyandari, D. (2019). Potential natural fiber-reinforced composite for biomedical application. *IOP Conference Series: Materials Science and Engineering*, *494*(1), 012018. 10.1088/1757-899X/494/1/012018

Rhim, J. W., Park, H. M., & Ha, C. S. (2013). Bio-nanocomposites for food packaging applications. *Progress in Polymer Science*, *38*(10), 1629–1652. 10.1016/j.progpolymsci.2013.05.008

Sahoo, S. K., Mohanty, J. R., Nanda, B. K., Nayak, S., Khuntia, S. K., Panda, K. R., Jena, P. K., & Sahu, R. (2021). Fabrication and characterization of acrylic acid treated rattan fiber reinforced polyethylene terephthalate composites for packaging industries. *Journal of Natural Fibers*. 10.1080/15440478.2021.1889439

Sanchez-Garcia, M. D., Gimenez, E., & Lagaron, J. M. (2008). Morphology and barrier properties of solvent cast composites of thermoplastic biopolymers and purified cellulose fibers. *Carbohydrate polymers*, *71*(2), 235–244. 10.1016/j.carbpol.2007.05.041

Sapuan, S. M., & Maleque, M. A. (2005). Design and fabrication of natural woven fabric reinforced epoxy composite for household telephone stand. *Materials and Design*, *26*(1), 65–71. 10.1016/j.matdes.2004.03.015

Sepe, R., Bollino, F., Boccarusso, L., & Caputo, F. (2018). Influence of chemical treatments on mechanical properties of hemp fiber reinforced composites. *Composites Part B: Engineering*, *133*, 210–217. 10.1016/j.compositesb.2017.09.030

Sezgin, H., Kucukali-Ozturk, M., Berkalp, O. B., & Yalcin-Enis, I. (2021). Design of composite insulation panels containing 100% recycled cotton fibers and polyethylene/polypropylene packaging wastes. *Journal of Cleaner Production*, *304*, 127132. 10.1016/j.jclepro.2021.127132

Shesan, O. J., Stephen, A. C., Chioma, A. G., Neerish, R., & Rotimi, S. E. (2019). Improving the mechanical properties of natural fiber composites for structural and biomedical applications. In (Eds.), *Renewable and Sustainable Composites* (pp. 1–27). IntechOpen. A. Pereira & F. Fernandes 10.5772/intechopen.85252

Shields, R. J., Bhattacharyya, D., & Fakirov, S. (2008). Oxygen permeability analysis of microfibril reinforced composites from PE/PET blends. *Composites Part A: Applied Science and Manufacturing*, *39*(6), 940–949. 10.1016/j.compositesa.2008.03.008

Soman, S., Chacko, A. S., & Prasad, V. S. (2017). Semi-interpenetrating network composites of poly(lactic acid) with cis-9-octadecenylamine modified cellulose-nanofibers from Areca catechu husk. *Composites Science and Technology, 141,* 65–73. 10.1016/j.compscitech.2017.01.007

Sydow, Z., & Bieńczak, K. (2019). The overview on the use of natural fibers reinforced composites for food packaging. *Journal of Natural Fibers, 16*(8), 1189–1200. 10.1080/15440478.2018.1455621

Tavares, T. D., Antunes, J. C., Ferreira, F., & Felgueiras, H. P. (2020). Biofunctionalization of natural fiber-reinforced biocomposites for biomedical applications. *Biomolecules, 10*(1). 10.3390/biom10010148

Tawakkal, I. S. M. A., Cran, M. J., & Bigger, S. W. (2017). Effect of poly(Lactic acid)/kenaf composites incorporated with thymol on the antimicrobial activity of processed meat. *Journal of Food Processing and Preservation, 41*(5), e13145. 10.1111/jfpp.13145

Teal, D. (2021). *Packaging Market Outlook.* https://www.packagingstrategies.com/articles/96035-packaging-market-outlook?v=preview

Varghese, S. A., Pulikkalparambil, H., Rangappa, S. M., Siengchin, S., & Parameswaranpillai, J. (2020). Novel biodegradable polymer films based on poly (3-hydroxybutyrate-co-3-hydroxyvalerate) and Ceiba pentandra natural fibers for packaging applications. *Food Packaging and Shelf Life, 25,* Article 100538. 10.1016/j.fpsl.2020.100538

Venkatachalam, N., Navaneethakrishnan, P., Rajsekar, R., & Shankar, S. (2016). Effect of pretreatment methods on properties of natural fiber composites: A review. *Polymers & Polymer Composites, 24*(7), 555–566. 10.1177/096739111602400715

Vroman, I., & Tighzert, L. (2009). Biodegradable polymers. *Materials, 2*(2), 307–344. 10.3390/ma2020307

Yang, H. S., Kim, D. J., & Kim, H. J. (2002). Combustion and mechanical properties of fire retardant treated waste paper board for interior finishing material. *Journal of Fire Sciences, 20*(6), 505–517. 10.1177/0734904102020006471

Yang, H. S., Kim, H. J., Park, H. J., Lee, B. J., & Hwang, T. S. (2006). Water absorption behavior and mechanical properties of lignocellulosic filler–polyolefin bio-composites. *Composite Structures, 72*(4), 429–437. 10.1016/j.compstruct.2005.01.013

Zhan, M., & Wool, R. P. (2013). Design and evaluation of bio-based composites for printed circuit board application. *Composites Part A: Applied Science and Manufacturing, 47*(1), 22–30. 10.1016/j.compositesa.2012.11.014

Zimniewska, M., & Batog, J. (2012). 4 - Ultraviolet-blocking properties of natural fibres. In R. M. Kozłowski (Ed.), *Handbook of Natural Fibres* (Vol. 2, pp. 141–167). Woodhead Publishing. 10.1533/9780857095510.1.141

Zwierzycki, W., Bieńczak, K., Bieńczak, M., Stachowiak, A., Tyczewski, P., & Rochatka, T. (2011). Thermal damage to the load in cold chain transport. *Procedia - Social and Behavioral Sciences, 20,* 761–766. 10.1016/j.sbspro.2011.08.084

Index

For Product Safety Concerns and Information please contact our EU
representative GPSR@taylorandfrancis.com
Taylor & Francis Verlag GmbH, Kaufingerstraße 24, 80331 München, Germany

9 781032 063218